NE, A.T.

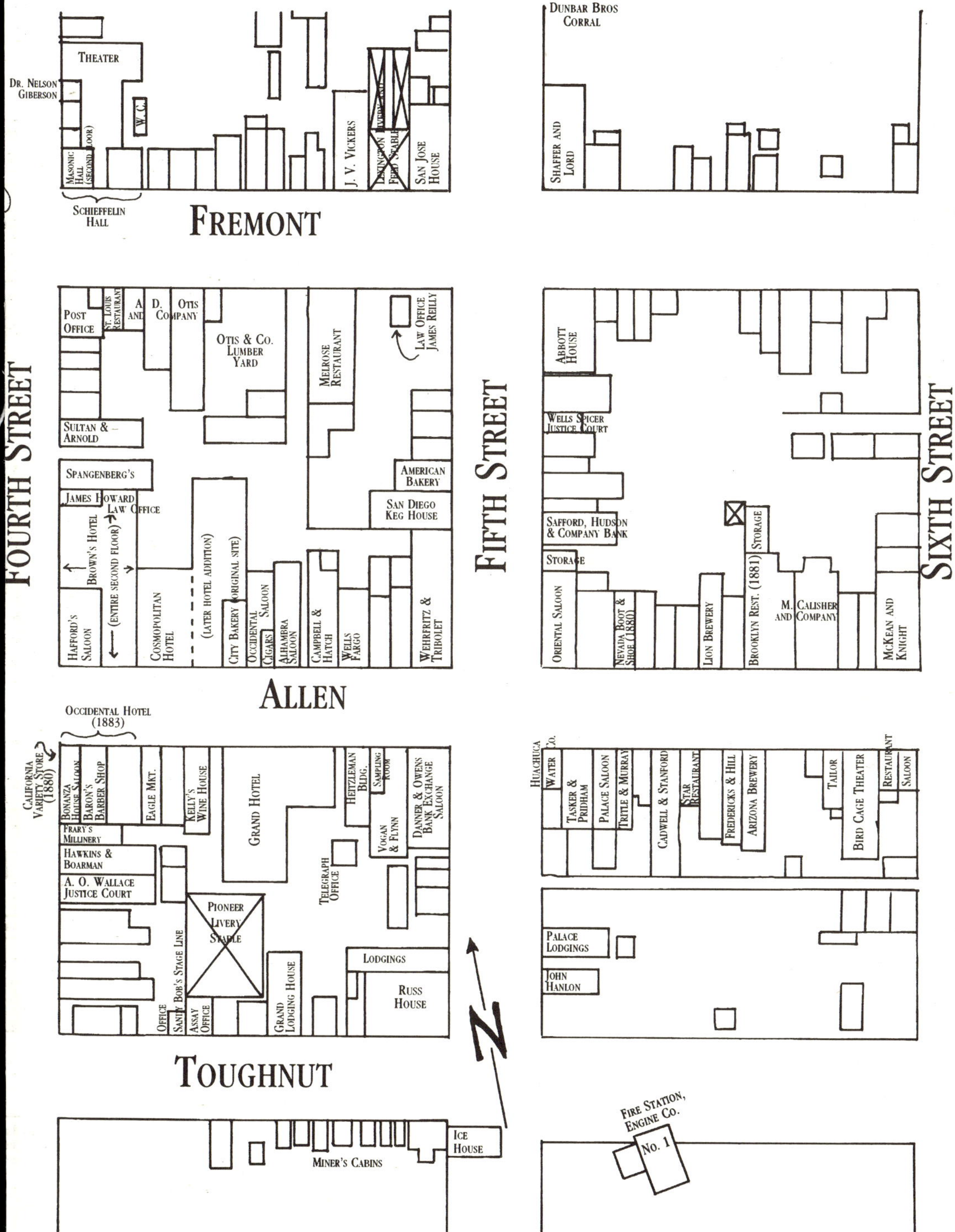
Dunbar Bros Corral
Theater
Dr. Nelson Giberson
W. C.
Masonic Hall (second floor)
Schieffelin Hall
J. V. Vickers
Lexington Livery and Feed Stable
San Jose House
Shaffer and Lord
Fremont
Post Office
St. Louis Restaurant
A. and D. Otis Company
Otis & Co. Lumber Yard
Melrose Restaurant
Law Office James Reilly
Sultan & Arnold
Spangenberg's
James Howard Law Office
American Bakery
San Diego Keg House
Brown's Hotel
(entire second floor)
Hafford's Saloon
Cosmopolitan Hotel
(later hotel addition)
City Bakery (original site)
Occidental Saloon
Cigars
Alhambra Saloon
Campbell & Hatch
Wells Fargo
Wehrfritz & Tribolet
Abbott House
Wells Spicer Justice Court
Safford, Hudson & Company Bank
Storage
Oriental Saloon
Nevada Boot & Shoe (1880)
Lion Brewery
Brooklyn Rest. (1881)
Storage
M. Calisher and Company
McKean and Knight
Fourth Street
Fifth Street
Sixth Street
Allen
Occidental Hotel (1883)
California Variety Store (1880)
Bonanza House Saloon
Baron's Barber Shop
Eagle Mkt.
Kelly's Wine House
Grand Hotel
Heitzleman Bldg.
Sampling Room
Danner & Owens Bank Exchange Saloon
Frary's Millinery
Hawkins & Boarman
A. O. Wallace Justice Court
Vogan & Flynn
Telegraph Office
Pioneer Livery Stable
Sandy Bob's Stage Line
Office
Assay Office
Grand Lodging House
Lodgings
Russ House
Huachuca Water Co.
Tasker & Pridham
Palace Saloon
Tritle & Murray
Cadwell & Stanford
Star Restaurant
Fredericks & Hill
Arizona Brewery
Tailor
Bird Cage Theater
Restaurant
Saloon
Palace Lodgings
John Hanlon
N
Toughnut
Ice House
Miner's Cabins
Fire Station, Engine Co. No. 1

Western Lands and Waters Series
XIX

TOMBSTONE
-IVERY AND
ED STABLE

TOMBSTONE, A.T.

A History of Early Mining, Milling, and Mayhem

by

WM. B. SHILLINGBERG

THE ARTHUR H. CLARK COMPANY
Spokane, Washington
1999

Arthur H. Clark Company
P.O. Box 14707
Spokane, WA 99214

LIBRARY OF CONGRESS CATALOG CARD NUMBER 98-23094
ISBN-0-87062-273-0

Shillingberg, Wm. B. (William B.)
Tombstone, A.T.: a history of early mining, milling and mayhem/
by Wm. B. Shillingberg.
400 p. cm. —(Western lands and waters series ; 19)
Includes bibliographical references and index.
ISBN 0-87062-273-0 (alk. paper)
1. Tombstone (Ariz.)—History—19th century. 2. Frontier and
pioneer life—Arizona—Tombstone. I. Title. II. Series.
F819.T6S45 1998
979.1'53—dc21 98-23094
CIP

In Memory of

Everett L. Brownsey

1912–1989

CONTENTS

ILLLUSTRATIONS

Chapter I

Beginnings

Once nearly forgotten, Tombstone, Arizona, is now trapped in myth and legend, a prisoner of that maddening twentieth-century entertainment spiral. Even walking along its quiet streets one finds it hard to separate truth from illusion and remember that this was a real town, not some Hollywood fantasy, its rough and rowdy exploits reported from San Francisco to New York.

Decades have come and gone since those same streets echoed the angry roar of gunfire or trembled beneath the monotonous rumble of ore wagons, once a seemingly endless procession of grinding wheels, cursing teamsters, cracking bullwhips, and the strain of leather in harness. Steam whistles from long-abandoned silver mines no longer mark shift changes or disasters. Only playful imagination can again crowd saloons with those boastful young men, puffing on cigars as they elbow their way toward gaming tables topped by whiskey tumblers and shuffled cards. From prospectors and peace officers to mining engineers and businessmen, all veterans of the wild days are gone. Little of what they knew and understood has survived.

Today Tombstone slumbers in the shadow of its faded glory, supported by clouded memory and the tourist dollar. Thoughtful visitors, whether staring down streets that seem to go nowhere or searching for the few original buildings left standing, are surprised by how small it all is. But then, the place is dominated by an even stranger atmosphere: the muted cadence of a carnival sideshow. History offers little solace; this town fell victim to hucksters and literary charlatans years ago. Dozens of fanciful stories fill the air in a constant chatter of deluded imagination, leaving one to walk away hopelessly confused about what really happened here.

Despite the shallow morality play usually dressed up and paraded as Tombstone's history—or perhaps hoping to identify with this carefully crafted farce—people are still drawn to the place. They have come too late, of course; but once it was something special. As one historian reflected as early as 1916: "There was nothing prosaic about the richness of Tombstone's mines. They were founded on romance, and romance and excitement dominated the days of their operation. Romance was in their location."[1]

As if crowning the late 1870s, the Tombstone Mining District rose like a colossus in the southeastern corner of the Arizona Territory, twenty-five miles north of the international border and sixty miles west of the New Mexico line. The town itself straddles an irregular plateau, originally called Goose Flats, at the deceivingly high elevation of 4,530 feet. From the base of the Tombstone Hills, south-southwest of the townsite, the area slopes gently westward toward the San Pedro River and its junction with the Babocomari.

Land nearest the town is evenly shaped and contoured but farther out shows more abrupt, steep-sided ridges. Everywhere the surface is cut by shallow gulches that flood during late summer storms. Vegetation is varied but characteristic. Rather high for most cacti, though several species are found, the area is generally covered by a variety of shrubs. Common are cat's claw, greasewood, and creosote. Mesquite and ocotillo are seen on surrounding slopes. Mesquite and palo verde also shade many arroyos. There is no timber on site sufficient for mining.

The district is surrounded by rugged mountain ranges. To the north-northeast are the Dragoons and to the southeast the Mules, home to the picturesque copper camp of Bisbee. Eastward, beyond these two ranges, lies the once-notorious Sulphur Springs Valley, followed by the Swisshelm and Chiricahua mountains, the San Simon Valley, and the New Mexico border. West of the district, beyond the San Pedro and some twenty or more miles from Tombstone, are the Whetstones, the smaller Mustangs, and the towering Huachucas.

For centuries various Indian tribes occupied this land, until pushed aside by invading Apache warriors. Yet even before these aggressive newcomers arrived, the vanguard of a more imposing threat trespassed the

[1]McClintock, *Arizona*, 2:410.

region. In 1539 Franciscan missionary Fray Marcos de Niza, a man as interested in finding wealth as in saving souls, led an expedition from Spanish settlements far to the south in search of the fabled Seven Cities of Cibola. There is some question whether Fray Marcos even reached Arizona, but his exaggerations excited Francisco Vásquez de Coronado. Marching through the San Pedro Valley the very next year, that young conquistador commanded a menagerie of men and animals on a historic journey into the interior of North America. Coronado found no golden cities. Yet, ironically, he did pass over land that in time yielded tremendous wealth.

Over the next two hundred years Spanish adventurers returned periodically to the isolated river valley. In 1775 fifty soldiers built a small fortification on the San Pedro less than a dozen miles from the Tombstone site. Within months Hugo Oconor, an Irish expatriate charged with reorganizing Spanish frontier defenses, ordered the post moved to an abandoned Sobaipuri Indian village on the west bank. It proved difficult trying to keep Apaches from raiding westward towards the Santa Cruz Valley and its fledgling village of Tucson. Soldiers lost two commanders and many comrades to ambush by the end of the decade. The place was then abandoned.

By the early 1800s several brave souls scattered along the river tried scratching out small farms and cattle ranches for themselves. Authorities eventually designated three land grants in the area: the San Ignacio del Babócomari—Arizona's largest—and two along the San Pedro, the San Rafael del Valle, and the San Juan de las Boquillas y Nogales. Renewed Apache raids during the 1830s and 1840s, however, forced yet another European retreat.

Arizona came under Mexican rule after 1821. Agriculture flourished on protected sites, but mining remained a comparatively small industry. "Of mining operations in Arizona," wrote one early historian, "records are barely sufficient to show that a few mines were worked, and that the country was believed to be rich in silver and gold."[2]

In 1846 the United States went to war with Mexico and life in Arizona abruptly changed. Heavily armed mountain men had operated out of New Mexico for years, trapping beaver along the Gila and San Pedro

[2]Bancroft, *History of Arizona and New Mexico*, 399.

rivers, but with war outsiders poured in as never before. American dragoons under General Stephen Watts Kearny traveled into history along the Gila River. Pursuing the same goal, Lt. Colonel Philip St. George Cooke's Mormon Battalion marched through portions of the San Pedro Valley. Cooke wrote that he and his men reached the river after covering twenty-seven waterless miles. Crossing over to the smoother side of the valley, they chose a campsite six miles north. The luckier ones gorged themselves on trout. Following the stream another fifteen miles the next morning, the soldiers began seeing evidence of deserted *ranchos* and noticed more and more wild cattle.

On the afternoon of December 11, 1846, near the later site of Charleston—a satellite milling town of Tombstone—the battalion was attacked by enraged bulls. The tall grass favored the animals. One caught a man in the thigh "and threw him clear over his body lengthwise; then charged on a team, ran its head under the first mule and tore out the entrails of the one beyond." The bulls killed another mule, badly bruised a sergeant, and hit one wagon hard enough to lift "the hind part of it out of the road." Nervous soldiers loaded their muskets for a defensive stand. One fearless trooper killed a charging bull, Cooke recalled, "almost at our feet."[3]

In the confusion of battle Lieutenant George Stoneman, the unit's quartermaster and Stonewall Jackson's West Point roommate, was wounded in the thumb by an errant musket ball, according to one account, and in another version from the accidental discharge of his own five-shot Paterson Colt. In the end the men greeted with relief orders to resume their march. Crossing the Babocomari, Cooke named the stream Bull Run—a designation later found on some military maps—in memory of that strange encounter. Thankful soldiers, having no thoughts of silver mines, quickly captured Tucson before continuing on to California, the prize of the Mexican War.

With peace in 1848 the United States acquired a vast western empire, including that portion of Arizona north of the Gila River. Soon small parties of easterners began stumbling through in their mad dash to reach California's newly discovered gold fields. To help satisfy demands

---

[3]Cooke, *Conquest of New Mexico and California*, 144-146; and "Cooke's Journal of the March of the Mormon Battalion," Bieber, ed., *Exploring Southwestern Trails*, 142-145.

for a southern railroad route to the West Coast, the United States Senate ratified the Gadsden Treaty in 1854. A $15 million provision expanded American holdings in the Southwest to their present limits.

Three years later Congress authorized a stagecoach route through this new territory. The Butterfield Overland made its first east-to-west run in the fall of 1858. Rolling westward from Dragoon Springs, young Waterman L. Ormsby, a special correspondent for the *New York Herald* and the line's only through passenger, jotted down impressions of the San Pedro Valley. He wrote of gullies, bluffs, dry creek beds, sloping hills, and the absence of "a respectably sized tree." Then, of particular interest to nineteenth-century travelers, he revealed: "However uninviting this valley may appear, it is said to be very fertile; but so long as it is left, as now, a prey to merciless Indians, no man can settle there in safety."[4] The need for military protection may have impressed Ormsby, but others chose to ignore the army.

Americans swarming into southern Arizona began a vigorous exploration for valuable minerals. The first serious mining activity in what became the Tombstone District centered on a location named in honor of Frederick Brunckow. Born in Berlin in 1820 to a German mother and Russian father, he received an excellent education and was fluent in German, Russian, French, and English. Brunckow studied classics at the University of Westphalia before graduating with honors from the mining school at Freiburg, Saxony. He decided to travel abroad following the upheaval of Europe's democratic revolutions of 1848. The young engineer had proven to be more than just a casual observer of those historic events.

Brunckow toiled at several mundane jobs after landing in New York in 1850. He worked as a deck hand on a Mississippi steamboat, then became a shingle-maker at the German community of New Braunfels, Texas. There, in 1856, he heard of a mining expedition being organized by the Sonora Exploring and Mining Company, a Cincinnati-based corporation.[5] Arizonan Charles D. Poston, a company director and field representative, solicited Brunckow's expertise. He thus found himself

[4]Ormsby, *Butterfield Overland Mail,* 84-85. For more on the San Pedro section of this historic route, see Conkling and Conkling, *Butterfield Overland Mail,* 2:147-154; and Ahnert, *Retracing the Butterfield Overland Trail,* 31-47.

[5]For a useful summary of this venture, see North, *Samuel Peter Heintzelman and the Sonora Exploring and Mining Company,* 19-45.

with Poston's party at Tucson in late August, arriving in time for the Fiesta de San Agustín. That fall the men settled at Tubac, then a rather rough and violent place.

The Sonora Exploring and Mining Company was responsible for opening several properties after leasing the Arivaca Ranch and obtaining mineral titles in the Santa Rita Mountains. These acquisitions included several silver mining sites in the Altar Valley south of Tucson, known collectively as the Cerro Colorado. They had been worked before, but now became paying propositions even if never fulfilling the dreams of overly optimistic investors. Indian troubles as well as difficulties with native Mexican laborers (one such incident resulting in the murder of Poston's brother John) slowed development.

Poston remembered Frederick Brunckow not only as a talented engineer but also as "a keen sportsman, fond of the chase, and added to his accomplishments the pleasing quality of being an excellent dresser of wild game. The mustang horse frequently formed a favorite dinner at Tubac."[6] Leaving behind the mess hall, his surveying instruments, and engineering notes, Brunckow traveled east with Herman Ehrenberg and others in 1858 to try and promote Arizona mining. The German engineer also ingratiated himself among interested capitalists. Soon afterward, even though earning the then-substantial sum of $1,800 a year, he resigned to join the St. Louis Mining Company. Brunckow had become interested in a claim east of the later Charleston site, reportedly discovered by Mexican miners in 1855.

In another version an early Arizona traveler noted on March 27, 1860, that the mine "was discovered some weeks since by several gentlemen connected with Capt. [Charles P.] Stone's commission for the survey of Sonora. . . . A block of the ore estimated to weigh five hundred pounds has been taken out solid, as a contribution to the Washington Monument. . . . Specimens chipped at random from the block were assayed by Mr. Brunckow and found to yield from $450 to $600 per ton. . . . The owners of the mine have already placed workmen on the ground for the purposes of holding their claim."[7]

Backed by the new company, Brunckow got started by hiring some

---

[6]Poston, "A Piece of History," *Arizona Daily Star*, February 17, 1880.

[7]Altshuler, ed., *Latest from Arizona!* 56.

Mexicans to build several small adobes. Then on July 23, 1860, eleven of these laborers murdered the German engineer, his machinist James Williams, and chemist John C. Moss. William M. Williams, identified at one point as general superintendent of the mine but actually a printer by trade, was away contracting supplies at Fort Buchanan, thirty-five miles to the west. Another employee, laborer and cook David Bontrager, also escaped death because he was Catholic. His captors released him some distance away.

That evening W. M. Williams returned to the darkened camp with two boys, William and Charles Ake. Their father, a farmer along Sonoita Creek, had sent them to help freight provisions. Williams discovered the body of his cousin near the storeroom door. Stumbling through the unlighted buildings, he tripped over another body, then felt a pool of drying blood. Williams tried striking a light with shaking fingers but could not find the candle. Rushing out in panic, he rounded up the Ake brothers and fled the grisly scene.

At Fort Buchanan the terrified trio blurted out their story. Officers ordered troops to investigate. Soldiers found the bodies of James Williams and J. C. Moss. Their search ended with the discovery of Brunckow at the bottom of a shallow mine shaft. A rock drill had been driven through his body. All three victims were buried in unmarked graves. After freed hostage David Bontrager finally showed up, he found himself under suspicion as a participant in the triple homicide. Luckily, information developed across the border supported his story of capture, and officers released him. Mexican authorities also recovered some of the stolen equipment, including a disassembled steam engine, but the perpetrators avoided arrest. Thus began the violent legacy of the Brunckow Mine, the scene of other deaths and long rumored a smugglers' rendezvous.

Mining activity slowed along the San Pedro after Brunckow's burial. But prospectors feared roving bands of Mexicans far less than falling into the hands of Apaches. In 1861 the Civil War erupted, and Commanding General Winfield Scott, the aged and bloated hero of the Mexican War, ordered federal troops in southern Arizona withdrawn to protect New Mexico from its seceded neighbor, Confederate Texas. Even as nameless Washington bureaucrats contemplated this decision, the Chiricahuas under Cochise launched hostilities against all remaining American settlers. These attacks followed an encounter between the Indians and the army at Apache Pass. The upheaval lasted eleven years before General O. O. Howard negotiated a fragile peace at the base of the Dragoon Mountains.

When officials carried out the organizational census of the Arizona Territory in 1864, the only persons found near the Brunckow Mine were three Hispanic laborers from New Mexico.[8] Deputy Census Marshal Gilbert Hopkins claimed this tally incomplete and promised to file a supplement, but he apparently failed to do so.

Two years later the California Column established Camp Wallen on the site of an old Spanish ranch house overlooking the Babocomari. Viewed as temporary by the government, it served its purpose in maintaining a military presence against the Apaches, even though some Indians succeeded in driving off a hundred head of cattle and horses on a single raid. Wallen closed in late 1869, its duties transferred to Camp Bowie at Apache Pass. Other posts would later tighten the gap, but the military's withdrawal from the valley failed to dampen interest in the Brunckow Mine.

Milton B. Duffield became its next owner to meet violence. A California pioneer in the days of '49, real estate speculator, Indian fighter, noted shootist, and the first United States marshal of the Arizona Territory, Duffield added the Brunckow to other scattered claims he and his family owned in southern Arizona. Trying to avoid creditors, he transferred this newest acquisition to his housekeeper, Mrs. Mary E. Vaughn. Milton Duffield was a strange character, an ill-tempered giant of a man who took delight in parading his physical strength and skill with deadly weapons. The hard-drinking ex-marshal was murdered at the Brunckow on June 5, 1874, by one Joseph Holmes.

Holmes worked for Edward N. Fish, another California gold seeker turned desert dweller. By then a monied Tucson wholesale and retail merchant, Fish, along with other investors, also claimed the mine. Riding up with a man named Joseph Oligher, Duffield—strangely unarmed for the occasion—forced entry into the adobe Holmes occupied. Aware of the former U.S. marshal's violent reputation, and his promise to kill

[8]*Federal Census—Territory of New Mexico and Territory of Arizona, Excerpts . . . 1864*, 58.

anyone found at the site, Holmes pulled up a double-barreled shotgun declaring that he, too, stood prepared to kill trespassers. Duffield ignored the threat and was shot to death. He thus ended his existence by filling an unmarked grave near Frederick Brunckow and the others. After two trials a jury found Holmes guilty of voluntary manslaughter. He only served seventeen days of a three-year sentence before escaping from the county jail, in the process freeing three other convicted killers. He was never recaptured.

With Milton Duffield out of the way, Mrs. Vaughn tried desperately to hold the Brunckow by pointing out official entries filed at the Pima County Courthouse. She published a trespass notice, but eventually succumbed to pressure and abandoned her claim. Bowing to negative public opinion, Edward Fish and his associates also turned their backs on the site after Duffield's death.[9]

Sidney R. DeLong and his two partners acquired title to the property by filing a relocation notice in late 1875. DeLong was no stranger to Arizona. Coming first to Tucson as an officer with the California Column, he eventually published a newspaper, became the town's mayor, and joined the prominent freighting-mercantile firm Tully, Ochoa & Company. For years he was post trader at Camp Bowie, and also worked as an engineer and surveyor. In the process he became one of the wealthiest men in the territory. DeLong served as a legislative councilman from Pima County at the time he took over the Brunckow. The old pioneer's reputation has suffered somewhat in recent years, primarily due to his participation in the infamous Camp Grant Massacre of 1871.[10]

Of DeLong's partners in the abandoned Brunckow Mine, Tom Jeffords is best remembered today for his association with Cochise, a story fictionalized in the novel *Blood Brother.* The other partner, Nicholas M. Rogers, worked as a station keeper in the Sulphur Springs Valley. Brutally murdered five months later by a roving band of Apaches, he was soon forgotten.

DeLong reported that the killing of Rogers, together with a man named Spence at Sulphur Springs and the similar death along the river

[9]For more details on this sordid episode, see Sacks, *Arizona's Angry Man*, 52-56.

[10]McClintock, *Arizona*, 2:600, 3:373-374; and Orton, comp., *Records of California Men in the War of the Rebellion*, 342, 347, 354, 385, 390, 409. For Sidney DeLong's own version of the Camp Grant affair and other Arizona observations, see DeLong, *History of Arizona*.

of someone named Johnson, caused federal authorities to request the removal of the three to five men regularly employed at the Brunckow. He and Jeffords ignored the advice. Having become the most important man for miles around, Sidney R. DeLong wanted it no other way.

But soon the relentless surge of history overwhelmed him. An explosive new era of mining, money, and adventure was about to shatter that isolated world of the San Pedro Valley.

Chapter 2

THE DISCOVERY

Following the footsteps of Frederick Brunckow and Sidney R. DeLong, other solitary prospectors and reckless adventurers crisscrossed the valley, all desperately dreaming of a chance at fortune. Yet the credit for discovering the Tombstone Mining District belongs to Edward Lawrence Schieffelin, a man as rugged as the rest but far more determined.

Ed came from a long line of strong, independent people. His grandfather, Jacob, had joined the New York State Militia as a teenager and within five years rose in rank to colonel. After studying medicine with private physicians, he finished the course at Columbia. Graduating in 1822, he traveled the eastern and southern states, Cuba, and Mexico in connection with a family-owned pharmaceutical business. Jacob Schieffelin eventually settled in Pennsylvania, raising eight children while living off real estate and other investments. He died in late 1880, already aware of his grandson's good fortune in Arizona.[1]

Edward L. Schieffelin was born October 8, 1847, in Tioga County, Pennsylvania. Nine years later, with his mother, younger brother Albert, and four-year-old sister Elizabeth, he traveled to San Francisco to join his father. Clinton Schieffelin, drawn by the Gold Rush, had sailed to California aboard the steamer *Pioneer.* Damaged hitting a reef at Talcahuano, Chile, that vessel eventually beached on the California coast at San Simeon Bay after emergency stops at Panama, Nicaragua, and Mexico failed to save its seaworthiness. Schieffelin scrounged space on the *Sea Bird,* coming north from San Diego, and finally steamed through the Golden Gate on August 20, 1852.[2]

---

[1] *History of Tioga County, Pennsylvania,* 682, 1004-1005; and Barrett, *Old Merchants of New York City,* 5:108-119.

[2] Although the *Sea Bird*'s emergency-expanded manifest is garbled, it does list a passenger named "Schiefflin" or "Shieffin." Rasmussen, *San Francisco Ship Passenger Lists,* 4:86-88. Also see Wiltsee, *Gold Rush Steamers,* 83-85.

Clinton crowded in with other miners on the Feather River. A year later he rushed to southern Oregon to examine rumors of new strikes. Buying land near Jewett's Ferry, he prospected and farmed before finally returning to California to meet his family. Together they left San Francisco and lived briefly at Crescent City. Then, summoned by the again-wandering Clinton, the Schieffelins abandoned California. Struggling through damp winter weather, they reached Clinton's Oregon farm, located outside a new community in Jackson County called Rock Point.

The older man submerged himself in civic and political affairs. Twice local Republicans nominated him for a seat in the state legislature. Clinton suffered defeat both times since Democrats dominated Jackson County politics.[3] He returned to his farm and helped raise a family that eventually numbered eleven children, only four of whom lived to see the turn of the century.

Young Edward showed no interest in agriculture, growing even more restless after learning about mining. Overcome by wild dreams of prospecting the Salmon River, he ran away at age twelve. Hardly beyond the shadow of hearth and home, the boy was persuaded by a thoughtful neighbor to give it up. As a teenager Ed worked at several Jackson County placer sites, including Forest and Jackson creeks, Sterling Gulch, the Applegate and Umpqua rivers, before finally striking out on his own in 1864.

Schieffelin struggled for years as a teamster and common laborer to finance prospecting trips into Nevada's Pioche country and Surprise Valley, as well as to the Salt Lake District in Utah. He made a winter trip into Idaho, visiting the Snake River and Boise Basin. Ed then joined others prospecting around Irataba, Mineral Park, and Prescott, Arizona. Discouraged after one of the group drowned in the Grand Canyon, Schieffelin returned to Nevada. Finding no work at Eureka, he spent the winter of 1874–1875 in the barren mountains above the Reese River Valley, chopping cordwood with his brother Albert at the old silver camp of Austin. Although earning enough for yet another Arizona trip, an attack of mountain fever delayed his plans.

Slow in recovering, Schieffelin found work with a stage company leading horses between Winnemucca, Nevada, and Silver City, Idaho.

[3] *Illustrated History of the State of Oregon*, 566.

The bustling atmosphere of that lively camp kept his dreams alive. A later trip home through northern California helped solve his lingering health problems. After just three weeks visiting his family, Ed left Oregon with $100 borrowed from his father. He traveled to Tuscarora and Cornucopia, Nevada, then on to other camps before ending up at Ivanpah in the southern California desert. On foot and again virtually destitute, Schieffelin found work and slowly earned enough for more prospecting gear. Buying two mules and outfitting at San Bernardino, Ed crossed the Colorado into Mohave County, Arizona, in February 1877.

Weeks later, while riding toward Prescott after once again searching sites near the Grand Canyon, Schieffelin, well equipped with rations, miner's kit, weapons, and mules, caught the eye of veteran civilian army scout Dan O'Leary. He and 2d Lieutenant Robert Hanna of the Sixth Cavalry had just enlisted a company of Hualapai Indians near Beale Spring to assist the military in controlling the Chiricahua Apaches in the southeastern corner of the territory. Schieffelin impressed O'Leary with his ruggedness and knowledge of frontier life. He was, after all, a powerfully built man with a piercing gaze, a feature remembered as "expressive of resoluteness."[4] When describing himself in 1892, Ed stated he was five-foot-eleven, with dark brown hair, blue eyes, medium complexion, and a mole on his right cheek.[5] He also carried scars from an encounter with a grizzly bear.

O'Leary invited Schieffelin to join his party. Ed jumped at the chance, seeing it as an excellent way to explore the country. They reached the San Pedro Valley in time to enjoy its mild spring weather.[6] Stopping at a small military cantonment founded just weeks before by Captain Samuel M. Whitside and named for the Huachuca Mountains looming above the camps, Schieffelin began making short, solitary trips into the surrounding country. He then accepted Lieutenant Hanna's invitation to accompany him and some soldiers on a scouting expedition.

The persistent prospector was in good hands. Although having just

[4]Maj. R. M. Hall to the *National Republican*, as quoted in *Arizona Daily Star*, August 12, 1879.

[5]Great Register of the County of Alameda (1894), Alameda, Calif., Precinct No. 7, Voting No. 324, Register No. 14890, Ed Schieffelin. Date of Registration: September 12, 1892.

[6]The group arrived in Tucson on May 21 and reached Camp Huachuca eight days later. Post Returns, Camp Huachuca, Arizona Territory, May 1877, National Archives and Records Service; and *Weekly Arizona Citizen*, May 26, 1877.

taken command of Huachuca's Indian scouts, Robert Hanna had been posted on the frontier since graduating from West Point in 1872: at Forts Riley, Hays, and Dodge in Kansas; and at Camp Supply in the Indian Territory. Between November 1875 and February 1877 he served at Fort Lowell near Tucson and learned something of Apache ways in both peace and war.

Schieffelin now rode with that officer through the Sonoita Valley and Patagonia Mountains. Skirting the southern end of the Huachucas, they arrived at the spot where the San Pedro River crosses into Arizona from Mexico. From there they followed the stream north nearly twenty miles. By then Ed decided to go off alone, as he could do no prospecting traveling with the soldiers. They wished him well and rode off, returning to Camp Huachuca several days later. Encouraged by stories of Frederick Brunckow's early activities, Schieffelin kept eyeing a series of conical-shaped hills more than a dozen miles east of the post. On that isolated landscape, he tried desperately to convince himself, he would find what had thus far eluded him.

Others were just as intrigued. That summer cattleman Henry C. Hooker and W. F. Leon, described as a practical miner with experience on the coast, toured several locations including the Brunckow. Afterward Leon declared southern Arizona "as favorable to successful mining as any other in the country, save the Comstock." He then added confidently, "at all points silver and gold ores are found and that where they are so generally found there must be some very large veins."[7]

Among Schieffelin's new acquaintances were many who warned him against traveling alone. Apache warriors posed a serious threat. As he prospected Ed could see an old Cochise lookout atop the Dragoon Mountains, across the valley from his ever-changing campsites. That formidable man had died in 1874, but his name and memory lived on, and with it fear of the Apaches. Just weeks before the Tucson press carried a story, attributed to a "messenger from Wallen," embellishing a report from two Mexicans at Davidson's Spring claiming to have buried four bodies at the mouth of the Babocomari. Many assumed Indians responsible and the incident helped heighten local apprehension: "fears are entertained for the safety of J. McKinney and a party who went

[7] *Weekly Arizona Citizen*, July 28, 1877.

prospecting through that part of the country a short time since and who have not been heard from."[8]

Other reports and rumors, together with repeated warnings from those at Huachuca, did not dissuade Schieffelin. Instead, it all helped provide the mining district he was on the verge of discovering with its name. "Whenever I went into Camp Huachuca for supplies on one of my trips," Ed recalled,

> some of the soldiers would frequently ask me if I had found anything. The answer was always the same, that I had not found anything yet, but that I would strike it one of these days in that country. The Indians at that time were very troublesome . . . and many settlers were killed previous to and during that year. Several times, in reply to my remark that I would eventually find something . . . the soldiers said, "Yes, you'll find your tombstone," and repeated that several times, the word lingered in my mind. . . .[9]

Of course there is more to discovering a rich mining district than simply choosing a colorful name. Schieffelin knew what it would take, explaining, "The ledges in Tombstone were pretty hard to find; they did not crop boldly out of the ground, and it was pretty slow work prospecting until you once learned the country, then it went fast enough."[10]

Running low on provisions, Ed stopped at the camp of George Woolfolk, a new arrival who planned starting a ranch. William Griffith and Alvah Smith then rode in from Tucson, hired by Sidney DeLong to carry out assessment work at the Brunckow (that together with other specified improvements was needed to satisfy federal law protecting claims on public land from relocation). Impressed by Schieffelin, Griffith suggested the prospector stand guard while Smith and he did their work, which should take no longer than a fortnight. Needing the money, Ed agreed, making good use of his time: "While standing guard, I could see the Tombstone hills very plainly with my field glasses, and I took a fancy to them, noticing that there were quite a number of ledges in the neighborhood of this Broncho [*sic*] Mine, all running in the same direction, about northwest and southeast."[11]

---

[8]*Arizona Weekly Star*, May 3, 1877.

[9]"History of the Discovery of Tombstone, Arizona, as Told by the Discoverer, Edward Schieffelin," typescript, Bancroft Library, University of California, Berkeley, 1-2. (Hereafter cited as "Schieffelin's account.")

[10]Ibid., 4.

[11]Ibid., 2.

Finishing the assessment work, Griffith and Smith looked over a promising ranch site before leaving to collect their pay. Finding some promising float, Schieffelin joined them at their new camp. Griffith carefully examined the samples and offered Ed a grubstake in exchange for a joint location. Returning later from Tucson, Griffith told Schieffelin he had changed his mind. Ed tried to dissuade him by pointing out two likely sites. Pleased by Schieffelin's forthrightness in offering him his choice, Griffith explained that while at the county seat he had met James Lee, a flour mill operator who agreed to help him claim 640 acres near the river, some four miles outside present-day Benson, under provisions of the new Desert Land Act. Although disappointed, Ed told him to use his own judgment.

Other newcomers arrived. Besides Woolfolk, Griffith, and Alvah Smith, Schieffelin now met Charles and John Bullard, there to check out stock grazing possibilities. Two others, William Sampson and Jacob Landers, started a ranch where the Babocomari joins the San Pedro. Schieffelin spent time with all of them, recalling that while these men "all camped together on the river, I made trips into the Tombstone country and back to the camp, leaving camp early in the morning of one day, remaining at Tombstone the rest of that day and night, and returning the next day to camp. I made several trips in that way. . . ."[12]

Ed discovered the Tombstone Mine on August 1, 1877, then found another he christened the Grave Yard. By Schieffelin's own account, he and Griffith traveled to Tucson to record that first claim on either August 25 or 26, as he recalled the town enjoying the "feast of San Juan." If his recollection of a holiday is correct, as is likely, the date had to be after August 28, the first full day of Tucson's Catholic community celebrating the Fiesta de San Agustín. Although surrounded by merrymakers, some coming from as far away as Sonora, Mexico, Schieffelin's thoughts were elsewhere: "I [had] found enough . . . to satisfy myself that there were good ores there, or ought to be. I was well satisfied of this, but could not go there and stay long enough to do anything, on account of the Indians. . . . The place was rather dangerous on this account."[13]

---

[12]Ibid., 4.

[13]Ibid.

Carrying assorted samples from the Tombstone and Grave Yard sites, Griffith and Schieffelin made the rounds. Whether the festivities dampened interest, or they felt the rocks not worth the trouble, no assayer offered to test them. More familiar with the town than Schieffelin, Griffith then showed the ore to various mine owners. But no one, not even the Brunckow's Sidney R. DeLong, seemed to care. DeLong brushed them aside, claiming he owned all the mines he wanted.

The near-penniless prospector again tried to persuade his new friend about mining, but William Griffith had already convinced James Lee. Those two filed on the Bunkam, an entirely different site, before Griffith returned to his ranch. Standing alone, Schieffelin recorded the Tombstone claim for himself on September 3.[14] Needing supplies, he visited Tully, Ochoa & Company. There a friendly clerk patiently explained that for ranchers the firm could extend credit, but for a lowly prospector it remained cash-and-carry. Ed ignored the advice and went elsewhere, using much of his remaining $6 to buy bacon and a sack of Mexican flour, cheaper than the comparable American product. Passing Fort Lowell, he persuaded the post blacksmith to shoe his mules, saving half the rate demanded by civilian liverymen.

On his return to the river country Schieffelin stopped at Sampson and Landers's camp. There had been many changes in the past two months. Back in July, Alvah Smith, Griffith, and the two Bullard brothers claimed twenty-five people clustered around the fledgling community of Whitside, named in honor of Camp Huachuca's Canadian-born commander. Everyone, so they said, felt safe from Indians. Yet something went amiss, and those assured declarations of July gave way to fear and defensive preparations in August. Although Smith, the Bullards, and George Woolfolk spoke of an exaggerated "forty settlers where a year ago no one lived," they now hastily formed a militia company to defend themselves against Apaches.[15]

The law provided for such contingencies, after filing a direct application with the governor, "Whenever thirty or more men shall associate together for the purpose of forming a volunteer company. . . ."[16] Failing

[14]Records of Mines, Pima County Recorder's Office, Book B, 696; and *Arizona Weekly Star*, September 6, 1877.

[15]*Arizona Weekly Star*, September 6, 1877.

[16]Hoyt, *Compiled Laws of the Territory of Arizona*, (1242) Sec. 7, 204.

to qualify with that number, a point not rigidly enforced, the river men organized the Upper San Pedro Rangers on August 28 during a meeting at Smith's ranch. Schieffelin signed on with nearly a dozen others.[17] This date proves he arrived in Tucson after the beginning of the Fiesta de San Agustín, the same day he became a volunteer, rather than August 25 or 26 as he remembered it, and before filing his Tombstone claim on September 3.

Respecting the Apache threat, Schieffelin now camped with Sampson and Landers. After Landers left for Tucson he joined Sampson on a three-week search for placer claims in the surrounding mountains. "When we returned to the river," Ed later recalled, "it was entirely deserted; everybody had left, and taken their outfits with them. Two adobe cabins that had been built on the river the preceding summer were deserted and everything was taken out of them. This was about the last of September, or the forepart of October, and here Sampson and I separated, he going to Huachuca, while I remained on the river a few days."[18]

Alone again, "then reduced to the last extremity; without provisions, almost without clothing, and with but 30 cents in money," Ed decided to find his brother Albert.[19] A family letter had mentioned his working at the Silver King in Pinal County, thirty miles northeast of Florence. There, to his growing dismay, Ed discovered that Albert had already left for McCracken Hill in the Owen Mining District on the Big Sandy in southern Mohave County. Schieffelin felt personally frustrated and physically stranded, explaining: "My mules were without shoes, and their feet were very tender; I had no money, or provisions, and had then got into a country where game was not plentiful, and it would be a day sometimes before I could get anything to eat. I could go no further, and the only thing to do was to find work, and get my mules shod."[20]

Schieffelin at first begged the foreman of the Stonewall Jackson Mine at McMillenville for a job. Turned away, he finally found work in the Richmond Basin, a dozen miles northeast of Globe, exposed to the bitter cold on the windlass of J. D. Wilson & Company's Champion Mine. For two grueling weeks he struggled through ten-hour shifts at $3,

[17] *Arizona Weekly Star*, September 6, 1877.

[18] Schieffelin's account, 7.

[19] Ibid.

[20] Ibid., 8.

before resupplying himself and shoeing his mules for the long journey across the interior Arizona wilderness. In Mohave County he found Albert working at the Signal Mill and Mining Company, a major extension of the McCracken complex. Ed hired on filling ore buckets at $4 a day to salvage his own sagging finances.

The brothers had not seen each other since their wood-chopping days in Nevada. One evening over dinner Ed showed Albert his few remaining samples, explaining in whispered tones about prospecting "in the lower country." Discouraged during his visit to Tucson in late August, Schieffelin had tossed most of them away on the road from Globe. In a final desperate plunge, he hoped to have the rest assayed, "and if they were not good ore, I would give it up and acknowledge that I did not know anything about mineral."[21] Albert offered no encouragement. Instead, Ed remembered, "For three or four weeks we worked in the mine, and I helped my brother build a stone cabin."[22]

Albert finally got around to introducing his brother to civil engineer Richard Gird. As the two cautiously shook hands, neither realized that this meeting represented the turning point for Schieffelin's persistence along the San Pedro. Gird, who had spent his whole life grasping at fortune, would come to see in this rugged Pennsylvanian the culmination of all his wild schemes. Both desperately needed the other to achieve that one bold, irreversible success.

Gird took on the world as a self-trained man. Sharing the confidence of those with similar backgrounds, he never felt cheated by opportunity. If one plan failed he adjusted quickly and began searching for another. In his own way Dick Gird was as tough and determined as Ed Schieffelin. Looking back from the perspective of age, he confessed to a mischievous but productive adolescence: "I was known as a 'bad boy' and was switched every day. . . . Yet at sixteen I had gained a fair knowledge of trigonometry and other higher branches and had read all the books I could get hold of in my father's and the school district library and elsewhere."[23]

In late 1877 he was "engaged in the building of the Signal Mill, under White & Allen, probably a couple of months," recalled Superin-

---

[21]Ibid., *9*.

[22]Ibid., *10*.

[23]*Ingersoll's Century Annals of San Bernardino County*, 600.

tendent Thomas Ewing. "Gird got either six or seven dollars a day while working for the Signal Company. He might have got $10.00 a day while setting the engine."[24] By the time Gird met the Schieffelin brothers he was no stranger to either mining or Arizona.

Richard Gird was born March 29, 1836, at Cedar Lake, New York. His father owned a farm and operated the local sawmill.[25] At sixteen Richard declared he wanted to see California. John Gird asked how he planned to do so. Confidently, the boy spoke up, "I expect you to advance the money." Next came the dreaded reply, "Then you will never get there. . . ."[26] Rebuffed, the disappointed youngster appealed to his mother. Finally convinced by his wife, John Gird relented and gave his son $1,500 on his promise not to encourage his other brothers to join him. Richard sailed for California with two of his father's Irish farm hands. Both deserted him in Panama. Crossing the Isthmus, he boarded the steamer *Oregon* at Taboga. After sixteen days and a short stop at Acapulco, he entered San Francisco Bay on November 20, 1852.[27]

Exuberant family correspondence from his older brother Henry had sparked Richard's imagination, filled as they were with legends of the Mother Lode. Arriving in Sacramento so soon after fire had destroyed two-thirds of its buildings, he was "compelled to pay two dollars and a half for the privilege of standing up all night in a tent."[28] At Coloma the brothers worked adjoining claims on the Greenwood placers, above the South Fork of the American River in El Dorado County. They did not get rich.

Dick Gird had contracted malaria—then called Panama Fever—while crossing the Isthmus. Doctors advised a lower altitude. Terribly weakened, Richard barely found the strength to walk alongside his brother's ox team pulling their meager belongings to Healdsburg in Sonoma County. After Gird's health improved they began a stock farm in the nearby Russian River Valley. For the next four years it was fairly profitable as their land holdings increased. But Dick Gird had not left

[24]Deposition of Thomas Ewing, San Francisco, September 1882, W. B. Garner, as Administrator of the Estate of Thomas J. Bidwell, Deceased, vs. Richard Gird, *Transcript on Appeal*, In the Supreme Court of the State of California, April 24, 1885, 144. (Hereafter cited as "*Garner v Gird.*")

[25]Hardin, ed., *History of Herkimer County*, 179, 385.

[26]Bancroft, *Chronicles of the Builders*, 3:88.

[27]Rasmussen, op. cit., 183, 185.

[28]Bancroft, *Chronicles of the Builders*, 3:89.

his father's farm in New York to work another one in California. In his spare time he studied assaying and other skills associated with mining and geology, training himself as a fairly competent civil engineer.

At Healdsburg he met Thomas J. Bidwell, a Missourian three years his senior who would play a major role in the early history of Tombstone. In 1858 the two sailed to Valparaiso, Chile, in search of adventure and opportunity. "We didn't go on any particular business," Gird acknowledged, "we went to see what we could find to do."[29] Carrying his gear strapped to his back, Gird prospected in the Andes foothills. Failing to find anything of value, he began working for Henry Meiggs, a San Francisco fugitive whom the younger man later characterized as "one of the greatest of modern financiers."[30] Under Meiggs's direction Gird helped supervise the construction of a small section of South America's first railroad, the southern route from Santiago. The two Sonoma County friends then suddenly left Chile, Gird explaining, "Bidwell was taken sick and had to get out of the country and I came with him."[31]

They stayed with Gird's family at Cedar Lake while Bidwell recuperated. Richard then suffered another malaria attack, forcing the pair to remain in New York longer than planned. Nursed back to health, they returned to California with Gird's sisters Mary and Ellen. Back in Sonoma County the two men became partners in a farm. They only worked together about a year, after which, as Gird recalled, "I just drew out and left things just as they were, and went off about other business."[32] The farm had never provided an adequate living. Gird became a surveyor before entering Arizona for the first time in 1862, traveling to La Paz on the Colorado River with Isaac Bradshaw. Continuing to prospect, he suggested Bidwell come and try his luck. They met at a place then called Weaverville on Martinez Creek in the Bradshaw Mountains.

Gird teamed up next with the two Tennessee-born Bradshaw brothers, Isaac and William, operating a ferry across the Colorado at Olive City south of La Paz. During an expedition into the interior, he opened a

[29]Deposition of Richard Gird; San Francisco, July 10, 1882, *Garner v Gird*, 77.

[30]Bancroft, *Chronicles of the Builders*, 3:89. For more on this volatile early hero of Mr. Gird's, see Stewart, *Henry Meiggs*.

[31]Deposition of Riehard Gird, San Francisco, July 10, 1882, *Garner v Gird*, 78.

[32]Ibid., 77.

freighting road by way of Granite Wash. Cited on early maps as "Gird's Wagon Road," this route eventually extended from the river to Prescott. During his travels, including a prospecting tour through Apache County, Gird began filling notebooks with geographic and geological details. Using various surveying instruments, he took accurate readings of mountain peaks and mapped vast areas. Through necessity Gird quickly expanded his knowledge of frontier life, its hardships and individualistic economic system. Nor did he escape trouble with Indians in protecting his fledgling business interests.

Gird and J. R. Simmons accompanied John Goodwin, Arizona's first governor, to Fort Whipple where the territory was formally established in late 1863. The first legislature convened nearby at the new town of Prescott the following September. That spring Gird and Simmons had prospected at Antelope Creek. Later the two joined Thomas K. "Pap" Thompson and others trying their luck along the Hassayampa River. In May 1864, after Gird returned from a quick trip to San Francisco, he and Simmons helped deliver provisions from La Paz for King S. Woolsey's expedition against hostile Indians. Gird ended this eighty-seven-day adventure suffering with fever. He slowly recovered in the back room of a Prescott saloon owned by John Roundtree and Dr. J. T. Alsap.

While Dick Gird prospected, his friend Thomas Bidwell started a restaurant at La Paz before moving on to Olive City in 1863. But Bidwell was interested in more than slinging hash. A year later he got himself elected to the First Legislative Assembly from Castle Dome City, a small mining town on the river fifty miles below Ehrenberg. With Bidwell in the legislature, Gird worked as a surveyor at Olive City.

Thomas Bidwell stayed in public service in one capacity or another for the rest of his life. Elected to the legislature again in 1868, this time from La Paz, he became Speaker of the House. From Ehrenberg in 1871 he sat with the Sixth House of Representatives, and three years later was a member of the Seventh Legislative Council. Bidwell also served as a deputy tax collector, probate judge of Yuma County, a notary public at Yuma, and later a justice of the peace at Ehrenberg, all without any formal legal training. Amid the busyness of his public life, he married and fathered two children, a daughter in Arizona in 1874 and a son born two years later at Gold Hill, Nevada. His wife died there eleven days after giving birth.

When he first entered the legislature, Bidwell did not forget Gird, who later recalled that with his friend's help, "I got an appointment as a Commissioner to make a map of the Territory of Arizona. . . ."[33] Thus Gird and Josiah D. Whitney, acting state geologist of California, drew and hand-colored a topographical map that General Irvin McDowell used in selecting several possible military sites. Bidwell's position in the legislature also helped Gird, along with a man named Henry Sage, win authorization to run a railroad between Castle Dome City and the Castle Dome Mine. Unfortunately there was no railroad for them to operate, nor were there funds forthcoming to build one.

For a time Gird returned to California, working first as a civil engineer in San Francisco. By the end of the decade he ran a real estate, employment, and collection agency that advertised widely in Arizona. He then joined the firm of H. B. Martin & Company, helping its founder (his brother-in-law) develop and manufacture steam engines and hydraulic pumps for mining and other commercial uses.

Although they installed a dozen of San Francisco's first hydraulic elevators, the company failed. Horace Martin skipped out for Chicago, where, among other things, he worked as a foundry engineer, patentee, and became involved again in the manufacturing of elevators. The old firm had borrowed heavily from John Currey, a former state supreme court justice, who obtained judgments against Gird in California and later in Arizona. With developments at Tombstone, Currey concluded, "I had no doubt about Mr. Gird's capacity to pay me eventually," adding, "I think the first payments made by Mr. Gird on this judgment were made in 1879."[34]

In still another scheme to make money, Gird tried having his Arizona map privately printed for commercial distribution. He borrowed needed cash from Bay Area real estate speculator Samuel A. Morrison. The plan did not go well, and at the time of the Tombstone strike Gird still owed Morrison several thousand dollars. He did not repay the debt until 1880.

A less-than-humble Dick Gird returned to Arizona in 1872, carrying just $16 in his pocket along with personal obligations topping $25,000. Undeterred, he got help from Thomas Bidwell, then operating a small

[33]Ibid., 80.

[34]Deposition of John Currey, San Francisco, September 1882, *Garner v Gird,* 218, 219.

vegetable farm and raising some cattle with B. F. Snyder and Pap Thompson on the California side of the Colorado River opposite Ehrenberg. Borrowing upwards of $100 to buy supplies, Gird outfitted himself and two others from San Francisco for what would be an unsuccessful prospecting trip into Mohave County. Two years later J. R. Simmons recalled seeing his friend at Cerbat and Piute Hill. Then, once again needing money, Richard Gird found work as a laborer before prospecting around Mineral Park.

To ease his financial woes, Gird supervised the building of reduction mills and put a smelter into operation for the Senator Mine at McCracken Hill. In 1876 he did the same at Ridenour and Crozier's Hackberry Mine on the east side of the Peacock Mountains, then prospected at Mountain Springs with Ed Clark and Matt Horner. They located four disappointing claims. Gird worked briefly as a laborer at Rawhide and McCracken Hill before returning to the McCracken Company as a surveyor in 1877. He also found time to invent a calculating machine and water meter (years before he and Bidwell had developed a gopher trap). That spring Gird helped install the Signal Mill, shipped from Aubrey Landing on the Colorado River, before moving on to bigger things.

Along with his cousin Thomas E. Walker, Gird surveyed the Signal townsite on a small mesa to the east of McCracken Hill, between the Signal and Senator Consolidated mill sites. The new community was about four miles below Greenwood, another mill town on the Big Sandy associated with the McCracken works. It created a small boom: "Town-lots here, sold by Richard Gird and T. E. Walker, are being disposed of rapidly and houses are going up, not one at a time but by whole streets."[35] After they sold most of the lots, Tom Bidwell joined them in selling the remaining property. By late 1877 Walker, who would follow his cousin to Tombstone, served as Signal's first postmaster. Bidwell held that office five months later, while also becoming a justice of the peace and notary. By now everyone simply called him "Judge."

Besides selling town lots and working as chief engineer at the Signal Mill, Gird acted as deputy mineral land surveyor. As was his custom in constantly searching for opportunity, he also conducted free assays for

[35]Signal, Arizona, September 23, 1877, unsigned letter to the editor, *Weekly Arizona Citizen*, October 6, 1877.

many a hopeful prospector. Thus, by the time he met Edward Schieffelin, Richard Gird enjoyed a solid reputation in territorial mining circles.

Gird had no trouble describing the formations that produced the ore Al Schieffelin now showed him from his brother's samples, including one nearly the size of a hen's egg that the older man "at once recognized as rich."[36] Agreeing to assay three pieces, Gird reported the first at $600 a ton while the second yielded a meager $40. That sample came from the Tombstone claim. The next day the third assay, from the Grave Yard site, showed $2,000 a ton, twice Ed's estimate, further convincing Dick Gird of the worth of Schieffelin's discovery. He now anxiously connived to become part of any expedition.

Ed hesitated and put him off, being more interested in convincing his brother to join this newest adventure. Al showed reluctance, claiming he had a good job paying $4 for an eight-hour shift and doubted he could find a better one anywhere in the territory. Al continued to speak highly of Gird, describing him as a good mill man and assayer, feeling he would make his brother a trustworthy partner. Within days Gird returned to the Schieffelins' cabin, finding it hard to conceal his enthusiasm to be accepted. He promised to stay at the site as long as necessary. Convinced of Gird's worth to any expedition, Ed decided to take him in. But Schieffelin also made it clear he would not wait until spring, as Dick suggested, and planned on leaving as soon as possible. Gird asked only for time to settle some private business.

Since the Schieffelin brothers had very little money, Gird, though still heavily in debt from his California failures, agreed to shoulder the burden of outfitting and paying general traveling expenses. To do so he borrowed $100 from Arthur Henry at Greenwood. Tom Walker came up with another $25 from Matt Horner, a sum never repaid. Gird had some money: "I earned it and got it out of the town site, some of it. I was getting at that time, when I worked, from six to eight and sometimes more dollars a day—ten dollars sometimes."[37]

Albert Schieffelin still felt everyone might be better served if he stayed behind to earn extra money. But Ed's determination to succeed proved contagious and Al soon changed his mind, pleasing Gird, who

[36]Gird, "True Story of the Discovery of Tombstone," 40.

[37]Deposition of Richard Gird, San Francisco, July 10, 1882, *Garner v Gird*, 80-81.

was particularly fond of him. Years later he wrote, "a better, more honest, truthful and naturally good man it would be hard to find than Al Schieffelin." Not that he disliked the older brother. Gird admitted that "Ed in all affairs of life was honorable and true," but characterized him as "somewhat self-conscious, given to personal display, ever worrying what others might be thinking of his goings and comings."[38] But the three were now partners come what may. On the strength of a handshake they agreed to share and share alike.

Preparing to leave Signal, Gird took time to buy out Pap Thompson's interest in a stock ranch he held with Tom Bidwell some four miles south of Skull Valley on Kirkland Creek near the Prescott road.[39] Bidwell and Thompson had bought the ranch three years before after dissolving their partnership with Snyder along the Colorado. For this new venture Bidwell had borrowed $500 from his friend Coles Bashford, one of the leading political and business figures in the territory. In buying Thompson's interest Gird covered his bets in case Schieffelin's dreams turned to dust.

The trio picked up supplies from Asa and Newell Kimball as well as from Jose M. Castañeda, to whom Ed had shown his samples. Gird charged items to his standing account with Kimball Bros. Their books show forty-seven such entries in the three months prior to his departure. Such generosity troubled William Corey, one of the firm's clerks, who later said of Gird, "I didn't consider him as a man responsible for his debts."[40] Gird arranged for more provisions from butcher Appolinar Bauer, another of the Signal crowd who later showed up in Tombstone, then bought a mule he named Mollie and added it to one Ed already owned. With all the necessary supplies—including assay equipment and surveying tools—piled into the bed of an old blue spring wagon parked behind the post office, they were ready to leave. Al Schieffelin rode his saddle horse. Ed and Dick sat together on the wagon, with Schieffelin's other mule, named Buck, tied behind. A mine whistle signaled noon as the three men pulled out. It was Thursday, February 14, 1878.

They started in haste. Gird's friendship with the brothers and his sud-

[38]Gird, op. cit., 49.

[39]*Weekly Arizona Miner,* February 1, 1878; and Testimony of T. K. Thompson, *Garner v Gird,* 388-390.

[40]Deposition of William Corey, Tombstone, August 10, 1882, *Garner v Gird,* 152.

den resignation from the Signal works, where he had been offered the superintendency, had not gone unnoticed. People around Signal accepted the trio heading out on a prospecting trip. Gird told Bidwell and others that he "was going off towards the southern part of the territory below Tucson," and later confessed his friend "was a great talker."[41] Ed showed reluctance in telling anyone, including his brother, the exact location of his strike. But they soon discovered that a party from San Francisco was already on the road ahead of them.

The Schieffelins and Gird moved down the Big Sandy to Bill Williams's Fork and on to Date Creek and Dripping's Spring, near the old Santa Maria toll road. From there they followed the Prescott road to Gleason's ranch, then on to Wickenburg by way of Martinez and the Hassayampa River. The three drove their wagon through the Salt River Valley to Phoenix (then a community of less than fifteen hundred). Concerned with dwindling foodstuffs while crossing the river at Hayden's Ferry, they bought fresh pork and then some eggs from a nearby ranch before pushing on toward Maricopa Wells and the old Overland Stage Road.

Covering thirty to forty miles a day, they reached Tucson in short order. Camping that night in a corral, Gird found time to visit friends, including John S. Vosburg, a local gunsmith and major investor in the Signal Mine. Restocking supplies with the help of a $300 credit arranged by Vosburg at Lord and Williams, the three men continued on. By noon the next day they approached William Ohnesorgen's crossing on the San Pedro.

Whatever concern they still felt over the San Franciscans quickly disappeared, Ed explaining: "We saw their wagons and mules in the corral, but there was only one of the party, named White, in the house, who was sick, and though he heard us as we went by, did not come to the door. Parsons, who was one of the party, had gone off towards the Whetstone Mountains to prospect, and he told me afterwards, if he had learned that Gird was with me, he would have followed us."[42] Instead, excited by rumors of a gold strike,[43] Parsons and White rushed on to Apache Pass and then into Guadalupe Canyon.

---

[41]Deposition of Richard Gird, San Francisco, July 10, 1882, *Garner v Gird*, 83, 93.

[42]Schieffelin's account, 14.

[43]*Arizona Weekly Star*, February 21, 1878.

From Ohnesorgen's Gird and the Schieffelins turned south along the east bank of the San Pedro. They passed the Mormon settlement of St. David, founded just two months before and occupied by a handful of people busily building houses of stone for protection against the Apaches. The prospecting trio, however, stopped for the night nearer the Babocomari. The next day all three were startled when Gird's rifle accidentally discharged, the bullet grazing the skin under his right arm as it passed through coat, vest, and shirt.

That evening they set up a dry camp near the Brunckow Mine. Unimpressed, Gird dismissed the location as worthless, describing it instead as "a mere blind to cover and shield a gang of smugglers. . . ."[44] With renewed apprehension Ed spotted others in the distance: "we saw them as they were riding across the country, into the Tombstone hills; we went into the old adobe and saw where they had camped, and had written their names on the walls."[45] The three partners waited until morning before unloading the wagon. The Brunckow adobe, with its nearly ten-foot-high walls and dirt floor, became their headquarters. To a small corner fireplace Gird added several bricks from an old wall and together with grate-bars and a sheet-iron cover converted it into a workable assay furnace.

With their camp established, the three men ignored the foul weather and hastened to the site of Ed's initial discovery. What they found disappointed Gird, who remarked that rather than the bonanza he expected, "It then seemed to be the graveyard of our hopes."[46] The claim proved to be nothing more than a small stringer. Shaking off the gloom, Gird decided they should go ahead and prospect anyway. At dawn he began trailing the gulches. Albert stayed behind most of the time to protect their supplies and equipment from marauders. He also served as camp cook, well known for his bread. Ed and Dick went out together every other day, Gird interrupting these trips sweating over his assay furnace while Schieffelin continued prospecting.

Still wondering about those they had followed from Signal, Gird finally suggested, "If we do not find something in a day or two, let us pack up and go and see where those fellows have gone." The normally

---

[44]Gird, op. cit., 43.

[45]Schieffelin's account, 14.

[46]Gird, op. cit., 45.

patient Albert, already losing his earlier enthusiasm, voiced indifference on whether they stayed or not. Thinking things through during an overnight ride to the Dragoons, Schieffelin decided, "if Gird was still in the notion to go away, he might go, and if Al. wanted to go with him, he might also, but I would stay there until I was satisfied,—I would not go now, whatever happened."[47]

As it turned out no one left. Instead, two of their mules and Albert's horse turned up missing, leaving Ed's saddle mule their only animal. At first they suspected Apaches, but three days later, finding tracks near an abandoned Indian trail along the river, Ed recovered the animals a dozen miles away on the road to Tucson. There was no more talk of leaving.

One afternoon Ed brought in a promising water-worn sample. Assays showed the value at about $2,000 a ton. The next morning, March 15, they began working the gulch. At one spot, convinced of value underfoot (some assays ran $9,000 and another registered an astonishing $1,500 in gold and $15,000 in silver), Ed exclaimed he was a "lucky cuss." The name stuck, and the Lucky Cuss Mine took its place in the lore of the Tombstone strike. With that discovery everything began falling into place.

Farther north they found slopes of quartzite, silicious limestone and porphyry, all honeycombed with possibilities. The trio now faced the difficult task of staking claims to conform with the law. Under provisions of the 1872 General Mining Law, boundaries of lode claims had been increased to 1,500 feet in length and 600 feet in width. Each claim needed to be carefully positioned to the locator's best advantage. On March 22 one layout proved so difficult they christened it the Tough Nut.

That same day they nearly ran out of provisions. Al Schieffelin and Gird headed for Tucson, returning with a Mexican boy to help with chores. Ed put aside his prospector's hammer and joined the others moving their camp from the Brunckow to a small cabin built at the Lucky Cuss. With the partners thus occupied, Henry D. Williams and Oliver Boyer, who had joined the Schieffelin-Gird camp for protection, began searching an area nearby.

Those two had ridden into the Tombstone Hills hunting stray burros.

[47]Schieffelin's account, 16.

They changed their minds after seeing assay equipment. Gird gave Williams, a man with some mining experience in Colorado, the benefit of his knowledge and agreed to assay their samples on the condition he and the Schieffelins share anything they found. Five days after the Tough Nut discovery Williams and Boyer located their own claim and ignored the promise. Gird angrily confronted Williams, "Haven't I kept my part of the agreement, made your assays . . . and here you have made your location and without saying a word."[48] Williams nodded in reply but said he planned on keeping the claim anyway.

Strapping on his revolver, Richard Gird stomped over and destroyed their markers and resurveyed. Strained negotiations continued before Williams finally surrendered the northern portion of his discovery. Because of all the trouble the trio named their new site the Contention. Williams and Boyer called theirs the Grand Central. It proved the best producer, with the Contention second and the Tough Nut third. Years later Ed was still upset that the best location in the district had gone to an ungrateful outsider.

Others taken in by Gird and the Schieffelins, although never as partners, included Gird's cousin Thomas Walker, who recalled, "I reached Tombstone about April 1st, 1878. I prospected . . . then went to work for Schieffelin and Gird, and took charge of the assay office, and kept their books and attended their mines."[49] The partners located the Old Guard for Walker after Bidwell's delay in coming down. For some reason Gird refused to pay the $2.50 filing fee for his friend. They would eventually file for others at Signal, finally including Bidwell and John McCluskey.

All this was no simple act of philanthropy. They did so with the understanding that these third parties would deed back half the claim. When Walker sold out "for ten or fifteen thousand dollars," Al remembered, "He kept one half of the money and Mr. Gird and my brother and myself kept the other half. I don't think that either McCluskey, or Walker, or Bidwell would have been located there on either of these mines, if they had not been extensions. There was a good deal of trou-

[48]Ibid., 19-20.

[49]Testimony of Thomas E. Walker, *Garner v Gird,* 394. Years later Walker reminisced about his Tombstone days: "Oxnard Citizen Writes Interesting Historical Article on Founding of Town of Tombstone, Ariz., and Early Days of Gold," Oxnard (Calif.) *Daily Courier,* October 31, 1921.

ble about people holding more than one claim in a location, and that was the reason why these claims were located for them."[50]

Finally, on April 5, 1878, the founders formally organized the Tombstone Mining District. The two Schieffelins, Gird, Thomas Walker, and Oliver Boyer signed the document.[51] Still angry with Henry Williams, a name conspicuously absent from the district creation notice, the partners continued prospecting. For the next several months county officials saw the names of the two Schieffelins and Dick Gird with regularity as they began recording their discoveries, including the Lucky Cuss, Tough Nut, Contention, West Side, Ground Hog, East Side, Defense, Owl's Nest, Good Enough, Tribute, and Contract.

Meanwhile, Williams loudly boasted that it was he who had discovered the Tombstone mines. Lazy editors at Tucson accepted his word and from time to time printed such items as: "To the indefatigable exertions, untiring attention and consummate skill of Mr. Henry D. Williams, the territory of Arizona is indebted for the discovery and development of this treasure." Or later: "Hank Williams discovered . . . the Grand Central. On that day there wasn't a human being in that country except the two or three prospectors at Williams' Spring."[52] Others complained, and an attempt was soon under way to clear the air: "There is some little dispute as to the first discoverer. . . . Some claim Schiflin [*sic*] and Gird; others H. D. Williams; and old settlers say Professor Bronkow [*sic*]. We would like to arrive at the fact, as it is likely to be a little bit of mining history for the future, which ought to be settled and chronicled now."[53]

One correspondent gave a fairly accurate account of developments, clearly showing that the Schieffelins and Gird had begun their exploration before Williams and Boyer stumbled upon their camp. Although its editor had not taken time to verify dates of entry at the recorder's office, the paper concluded such remarks were "A Satisfactory Solution of This Much Hooted Question."[54]

---

[50]Deposition of A. E. Schieffelin, San Francisco, July 1882, *Garner v Gird*, 338.

[51]Miscellaneous Records, Pima County Recorder's Office, Book 1, 526; and *Arizona Weekly Star*, April 11, 1878.

[52]*Arizona Weekly Star*, May 2, 1878; and *Daily Arizona Citizen*, April 23, 1878.

[53]*Arizona Weekly Star*, June 26, 1879.

[54]D. B. Rea to the editor, *Arizona Weekly Star*, July 10, 1879. Others had written to correctly point out that Schieffelin's discoveries predated both the Grand Central and Contention. See *Daily Arizona Citizen*, April 30, 1879.

But now those fortunate men on the San Pedro faced an even bigger problem than assigning credit for the discovery—how to develop their holdings into a paying enterprise.

Chapter 3

DEVELOPMENT

Richard Gird and the Schieffelins may have owned several promising mining claims, but without investment capital their dreams remained confined to a bleak forest of stone monuments. Success demanded more than common prayer, vain hope, and a few shallow holes in the ground. To raise badly needed cash they planned selling the Contention and Tough Nut to Massachusetts-born Josiah Howe White. White and his partner William C. Parsons, a man interested in Arizona mines for nearly a decade, were two of those who had caused the trio so much anxiety on that long road from Signal.

Josiah White had waited years for an opportunity like this. In 1859 the twenty-one-year-old railroad construction engineer got himself elected—by a slender four-vote margin—surveyor of Jersey County, Illinois, but soon fell victim to the lure of California. White helped build Sacramento's levee system, along with lifting his own reputation, after the massive winter floods of 1861–1862. Then, until leaving for Arizona fourteen years later, he became a San Francisco-based contractor and surveyor.[1] Gird and the Schieffelins now offered White his chance of a lifetime, asking $20,000 for the Contention—a site only casually examined—and $50,000 for the Tough Nut.

Earlier Gird predicted they would get at least $50,000 for all their discoveries. Ed ridiculed the whole idea, dismissing it as a fanciful daydream. But the chances for profit were clearly improving, even though White offered only $10,000 for the Contention and nothing for the Tough Nut. The partners accepted, and on May 17, 1878, they surrendered control over the second-richest property in the district.[2]

---

[1]*History of Greene and Jersey Counties*, 106, 107; *Illustrated History of Sonoma County*, 673-674; *History of Sonoma County*, 859-861; and *Langley's San Francisco Directory*, 1865-1878.

[2]Deeds of Mines, Pima County Recorder's Office, Book 2, 487-490; and Schieffelin's account, 21.

White filed the transaction with the recorder's office on June 7. Having no development capital himself, he immediately transferred title for the original $10,000, plus stock in a new company, to Walter E. Dean.[3] That thirty-eight-year-old San Francisco capitalist, who lived in style at the Palace Hotel, had learned the art of business at banking houses in his native New York. Hungering for adventure, he traveled west and by 1860 sailed for China. Turmoil on the Asian mainland, however, convinced the young speculator to retreat. Returning to California, Dean became associated with many famous mining companies there and in Nevada. Of his sojourn into the Tombstone Mining District it was later said that he did so when "cautious mining men were adverse to touching anything in Arizona."[4] This conservative outlook collapsed after the Contention began paying tens of thousands in dividends and bonuses.

On October 29, 1878, Walter Dean incorporated the Tombstone property as the Western Mining Company.[5] Both Josiah White and William Parsons stayed on; White as a major stockholder and superintendent of the Contention, with Parsons overseeing operations at other acquired properties such as the Contentment and Tranquility. The company kept its main office in San Francisco.

After gaining control of other Tombstone mines, Dean incorporated the Contention Consolidated Mining Company in late 1881.[6] He then moved the offices of the Western and the new company into the Nevada Block. That prestigious address represented the hub of empire in western mining and financial circles. It housed the Nevada Bank of San Francisco, controlled by surviving Comstock Silver Kings John W. Mackey, James G. Fair, and James C. Flood.

With cash always short in Arizona, capitalists such as Walter Dean instinctively understood the advantages. Within two years the *New Haven Journal* praised the territory's mineral possibilities by openly characterizing its citizens as "impecuniens," while at the same time citing the annoying practice of speculation in mining shares on the San Francisco

[3]Deeds of Mines, Pima County Recorder's Office, Book 2, 490-493.

[4]*History of Nevada*, three unnumbered pages between 124 and 125; and *Langley's San Francisco Directory*, 1878.

[5]Articles of Incorporation of the Western Mining Company, San Francisco, October 30, 1878; filed with the Secretary of State, No. 12109, October 31, 1878, Book 30, 335, California State Archives.

[6]Articles of Incorporation of the Contention Consolidated Mining Company, San Francisco, October 19, 1881; filed with the Secretary of State, No. 13423, October 19, 1881, Book 40, 96, California State Archives.

Stock and Exchange Board as a barrier to buying up valuable properties at low prices.

Others had shown interest in the new Tombstone mines even before the Contention sale. Charles Tozer, a San Francisco-based mining superintendent and investor who managed the Aztec Mine on the western slope of the Santa Rita Mountains south of Tucson, sent his associate Oliver Jacobs to examine the diggings. Jacobs asked Gird and the Schieffelins what they wanted for all their holdings. As Ed recalled, "We put the Tough Nut and two or three locations surrounding it, at $50,000, and the Lucky Cuss, with several claims near it, at $40,000. . . ."[7]

A delighted Oliver Jacobs carried away ore samples for his boss. Impressed, Tozer rushed to the San Pedro to see things for himself. He then sent for Nevada mining expert Isaac E. James. James, who had surveyed the boundary line between Nevada and Utah, supervised the Yellow Jacket Silver Mining Company at Gold Hill. He came to examine not only the Tombstone property but also mines in the Harshaw District, west of the Huachucas in the Patagonia Mountains. James wrote a detailed and glowing report on the Schieffelin-Gird holdings.

On May 17, 1878, the same day Josiah White grabbed the Contention, the Schieffelins and Gird bonded the Lucky Cuss, Owl's Nest, East Side, Tribute, Contract, Tough Nut, Good Enough, Defense, and West Side mining claims, together with certain mill sites and water rights, to Charles Tozer for the agreed upon $90,000. Dreaming of fortune, he gave the partners $1,000 with the understanding that the balance would be paid by July 2.

Tozer now desperately needed the help of friendly California investors to raise the rest. On May 20 he assigned his bonding agreement over to James Ben Ali Haggin, an erstwhile San Francisco attorney and major West Coast capitalist. Haggin was already busy in Arizona. His representative John Sevenoaks had just bought the Washington Mine in the Harshaw District, while negotiating Haggin's purchase of machinery for a ten-stamp mill from Lord and Williams at Tucson.

Interesting a man like James Haggin pleased Charles Tozer. Haggin excelled as a mining speculator. Together with his law partner and Kentucky boyhood friend Lloyd Tevis, the twenty-year president of Wells

[7]Schieffelin's account, 22.

Fargo, Haggin, Marcus Daly, and George Hearst often stunned their competition: "The syndicate registered, among many other achievements, ownership of the greatest silver mine, the Ontario; the greatest gold mine, the Homestake; and the greatest copper mine, the Anaconda."[8] Haggin then amassed another fortune in California real estate.[9]

Yet the cold and taciturn Mr. Haggin remained a cautious multimillionaire. He sent Joseph Clark from San Francisco to carefully examine Tombstone. Clark, it was reported, "considers $90,000 a large sum of money to pay for prospects, and to purchase at such a figure . . . rather hazardous."[10] Six weeks later he changed his mind and called it "the most brilliant gamble he ever saw."[11]

As Haggin vacillated in his spacious San Francisco office, "The bond expired on July 2d," Ed Schieffelin remembered, "and we extended the time to August 1st, with $100 a day forfeit. The 1st of August came, and they had not raised the money, so the mines fell to us again, and we went to work on them. We had a little money then, about $15,000. . . ."[12] Gambling on their own self-confidence, they wisely turned down another extension. But the Haggin-Tevis syndicate did not lose out completely. That September it took control of the Survey Mine north of the Tough Nut along the same lode. After picking up other sites they incorporated in California as the San Pedro Mining Company.[13]

George Hearst, father of future newspaper tycoon William Randolph Hearst, was one of the better-known capitalists involved with the Tombstone District. His activities on the Comstock, as well as his association with Haggin, made him a legend among western mining men. A self-educated man, Hearst's strengths lay with his innate abilities, natural boldness, and single-minded determination. Government geologist Clarence King, who was not fond of him, once joked of the rugged investor: "You remember, that . . . Hearst was bitten on the privates by a scorpion; the latter fell dead."[14]

---

[8]Marcosson, *Anaconda*, 35.

[9]Phelps, *Contemporary Biography of California's Representative Men*, 1:325-328; Leonard, ed., *Men of America*, 1064; and *New York Times*, September 13, 15, and 18, 1914.

[10]*Arizona Weekly Star*, June 27, 1878.

[11]*Weekly Arizona Citizen*, August 9, 1878.

[12]Schieffelin's account, 22.

[13]Articles of Incorporation of the San Pedro Mining Company, San Francisco, August 31, 1878; filed with the Secretary of State, No. 11500, September 2, 1878, Book 27, 324, California State Archives.

[14]Wilkins, *Clarence King*, 174.

Over the years Hearst made several trips into southern Arizona searching for valuable properties. At least twice Wyatt Earp served as his bodyguard. Tombstone benefited from Hearst's activities. His awesome reputation as a mining entrepreneur helped legitimize the district in the eyes of less confident speculators. But the formation of the San Pedro Mining Company remained some weeks away and did not help Charles Tozer, who now helplessly watched the extension agreement slip away. As compensation he bought the original Tombstone claim for $2,000.

Despite Haggin's reluctance others began showing interest in the new district. Several West Coast moneymen approached Signal superintendent Thomas Ewing about possible investment at Tombstone. James G. Fair introduced these gentlemen at William Sharon's Bank of California office in San Francisco. Following his long association with the Comstock, Sharon, then a United States senator from Nevada, stood tall among mining and financial leaders. As Ewing remembered, "Mr. Fair requested me to look at [the Tombstone mines] for Mr. Sharon."[15]

But Gird and the Schieffelins now planned to develop the property themselves to drive up the asking price. Besides, Ewing concluded, "When I looked at the property, they asked more than I could think of giving at the time from the development they had made. They were only down about 14 feet in three places, and had a little tunnel running just in the face. I declined to negotiate. . . ."[16]

As Ewing poked around, another of the Mohave County crowd, James H. Wooley, the middle-aged chief engineer of the McCracken Company's twenty-stamp Signal Mill, toured the mines with his eighteen-year-old English-born wife, Clara. She thus became the first woman to visit the district. Whether Mrs. Wooley viewed this honor with distinction or distress was not recorded. Thinking of the future, one observer remarked, "We extend a hearty welcome to this female pioneer and hope the day is not far distant when many more ladies will enjoy the beautiful scenery and delightful climate of the Tombstone." The Wooleys moved there in the fall.[17]

Unlike Gird and the Schieffelins, Henry Williams chose cash over

[15]Deposition of Thomas Ewing, San Francisco, September 1882, *Garner v Gird*, 143.

[16] Ibid., 144.

[17] *Weekly Arizona Citizen*, August 30, 1878; and Pima County Great Register, No. 1628, J. H. Wooley, Precinct No. 17, October 22, 1878.

bullion and sold his shares in the Grand Central to St. Louis-based mining speculator W. F. Witherell (who held interests at Arivaca and Oro Blanco) and Tucson lawyers Fred Stanford and Louis C. Hughes, known locally as "Pin-head." Two additional shareholders were Captain Samuel M. Whitside and 1st Lieutenant Hiram F. Winchester, a cavalry officer who later drank himself to death. The military had scoffed at Schieffelin's dream of finding mineral wealth, but now that it had happened everyone scrambled for a seat on the silver bandwagon.

Pockets stuffed with cash, Williams was soon living the good life on New York's Fifth Avenue. He was back within eighteen months, however, delivering temperance lectures on Tucson street corners (while under the influence some said) and boasting plans of discovering another great mine. His partner Oliver Boyer decided to hold on to his Grand Central shares, but he would never enjoy its riches.

Boyer was widely known as a disagreeable drunk. Once he threatened a man named Pat Cannon, but seeing his adversary heavily armed, he declined to do more than talk tough. Then, on June 23, 1878, Martin A. Sweeney and Joseph Price returned to the district after business at Tres Alamos. They stopped for a meal at Edwin C. Merrill's Grand Central Store near the Babocomari. In that canvas-covered saloon and all-around supply emporium, Sweeney spotted Oliver Boyer (who also used the name Jack Friday) and introduced him to Price. The three ate together before wasting the rest of that Sunday afternoon with shots of whiskey. Boyer finally called Sweeney aside for a private conversation, asking his help in composing a legal notice claiming 160 acres he hoped to develop as a townsite.

Soon they began quarreling. Boyer hotly accused his red-haired Irish friend of partiality toward the Grand Central's other owners, primarily Tucson lawyers Stanford and Hughes. Sweeney, a former prizefighter with fifteen years' experience as a cavalry sergeant, denied the charge and claimed he could pound the irate mine owner into jelly. Insulted, Boyer reached for a revolver. Merrill, the proprietor, quickly stepped between the two and disarmed him. Exploding into a bloody rage, Oliver Boyer pulled out a hidden pocket pistol and deliberately shot his seated and unarmed foe in the chest.

Price yelled at Sweeney to get a pistol from his horse. The wounded man staggered outside. A second bullet struck him in the throat. Three other shots went wild. Joe Price ran for help, calling for Merrill to hold Boyer. The storekeeper busied himself instead with the wounded man and sending a rider to Camp Huachuca for a doctor. Sweeney died within forty minutes. Amid the confusion Boyer also rushed to Huachuca, where post trader Frederick L. Austin spotted an opportunity and hastily bought up the killer's shares in the Grand Central for $1,000. The trader soon resigned—"outside business calls him away," it was said. "Austin is now largely interested in mining matters, and we expect to hear of his untold wealth at no distant day."[18] Boyer fled into Mexico. No one tried to stop him.

Angry miners organized an inquest and forwarded their findings to Sheriff Charles A. Shibell at Tucson. A grand jury ordered Boyer's arrest, but Mexican authorities shielded the fugitive from Arizona justice, at least until his money ran out. Finally returned for trial, Boyer was convicted and ordered hanged. This sentence was later reduced to life in prison.[19] In those days Arizona felons with more than two years to serve were transferred to Detroit. There they survived on a meal allowance of twenty-five cents a day. As he scraped food off a tin plate, Oliver Boyer had plenty of time to ponder the serious consequences of mixing bad temper with alcoholic overindulgence.

The decision to develop rather than sell out proved a wise move for Dick Gird and the Schieffelins. In late August, alerted by his longtime associate and political ally John S. Vosburg, former Arizona governor Anson Peacely-Killen Safford returned to Tucson after a nine-month absence with mineral speculation in mind. Daniel Gillette, another of Gird's one-time employers in Mohave County, traveled with Safford hoping to invest in some mines himself. They were joined in Tucson by Charles Tozer's boss, John Graham.

Soon others arrived, including Indiana State Geologist Edward Travers Cox and a couple of surgeons with money to burn. Cox had exam-

---

[18] *Arizona Weekly Star*, August 1, 1878.

[19] Ibid., June 27, July 4, August 1, 29, September 26, October 10, 1878; *Weekly Arizona Citizen*, (Florence, Ariz.) June 28, July 19, 1878; *Weekly Arizona Citizen*, (Tucson) September 28, October 19, 1878; and Reeve, ed., *Albert Franklin Banta*, 78-79.

ined mines in New Mexico fifteen years earlier. The government enhanced his reputation by publishing the report.[20] Also on hand was a thirty-five-year-old Philadelphia insurance agent named Elbert A. Corbin. The Corbin family, based in New Britain, Connecticut, and headed by the older brother Philip—a wealthy hardware manufacturer—already owned Arizona mining shares, primarily in the Patagonia District.

Safford, Gillette, Corbin, Tozer, Graham, Professor Cox, and the others visited the Aztec as well as sites at Arivaca, Cerro Colorado, and Oro Blanco. They made another swing through the Santa Rita, Patagonia, and Mule Pass mining districts before reaching Camp Huachuca and riding over to the San Pedro. Although exhausted, "They then visited the Gird mines, and all seem to be of one opinion, that it is the most wonderful sight they ever witnessed in mining."[21]

That trip turned the corner for Gird and the Schieffelins. Gillette offered $150,000 for everything but was just as quickly rebuffed. Then, on September 27, 1878, John S. Vosburg made an agreement with the trio to build a ten-stamp quartz mill and other essentials on property near the San Pedro.[22] For this Vosburg was to receive an undivided one-fourth interest in the mill site, all the water rights connected with it, and in the mining claims Lucky Cuss, Owl's Nest, Contract, Ground Hog, East Side, Tribute, West Side, Defense, Tough Nut, and Good Enough.

Vosburg ostensibly worked as a gunsmith and ammunition peddler at Tucson, but he and Safford had been associates for nearly a decade. As governor, Safford had appointed Vosburg, who represented Pima County in the Seventh Legislative Assembly, adjutant general in 1873 and again in 1875. Two years later Safford named his friend territorial auditor. The two had invested heavily in the Signal Mine but lost money. Tombstone would be different.

Anson Safford had been interested in mining for thirty years. He and Dick Gird first met in San Francisco in 1869. Afterward they saw each other occasionally in Arizona, including the time Gird worked for Daniel Gillette overseeing the furnace for a company at Rawhide. After

[20]Wilson and Fiske, ed., *American Biography*, 1:757; and Dunn, *Indiana and Indianans*, 1:35-37.

[21]*Arizona Weekly Star*, October 3, 1878.

[22]Miscellaneous Records, Pima County Recorder's Office, Book 1, 667; filed, June 18, 1879.

Safford became governor, his predecessor, Richard McCormick, provided a list of "intelligent pioneers and sound counsellors."[23] These men actually formed the core of McCormick's political ring, and included Richard Gird's old friend Thomas Bidwell. Safford was automatically a part of Vosburg's San Pedro arrangements. The two had formalized an agreement in late 1876 to divide all their mining investments. With the Vosburg deal signed, Safford interested Elbert Corbin in the Tombstone project.

On October 1, 1878, Corbin entered into an agreement with Vosburg—for a one-half interest in his share—to construct the ten-stamp mill, a ditch and river dam, together with a wagon road to the mines. Corbin agreed to furnish nearly $80,000, to be drawn at Gird's discretion from various bank accounts. All the money spent in excess of $40,000 was to be repaid with unspecified interest from the first net proceeds.[24] With the infusion of Connecticut capital the history of the Tombstone Mining District entered a new phase. Covering his bets, Elbert Corbin invested in other properties, including the Grand Central and Brother Jonathan, soon after his deal with Vosburg.

On October 26, Safford, the Schieffelins, Gird, Corbin, and Vosburg organized the Tombstone Gold and Silver Mill and Mining Company as an Arizona corporation, issuing half-a-million shares of stock. As Ed recalled: "The mill was to be built under Gird's supervision, and Gird was to be Superintendent. Safford went East and obtained the money, placing it to Gird's credit in San Francisco, and Gird went [there] and superintended the building of the mill at the Fulton Iron Works. . . ."[25]

That company earned a solid reputation supplying milling machinery, concentrators, hoisting and pumping apparatus, as well as sawmill and cable railway machinery to some of the largest operations on the coast.[26] Gird also arranged for the Fulton company to build a lumber mill for Tombstone. It would be set up in the Huachuca Mountains, the closest source of timber for mining or construction purposes. Richard

[23]Congressional Delegate Richard C. McCormick to Anson P. K. Safford, April 22, 1869, Arizona Secretary of State, Territorial Records, as quoted in Wagoner, *Arizona Territory*, 99.

[24]Miscellaneous Records, Pima County Recorder's Office, Book 1, 268-269; filed, June 18, 1879.

[25]Schieffelin's account, 22-23.

[26]Hittell, *Commerce and Industries of the Pacific Coast*, 662-663; *Bay of San Francisco*, 1:507; and "Fulton Iron Works," *Mining and Scientific Press* 39 (September 13, 1879).

Richard Gird, Albert and Edward Schieffelin. The partners seem bored by the photographic process. Preserving an image for history is one thing, but there was serious money to be made elsewhere.
Courtesy, Arizona State Museum, Tombstone.

transferred the lumber company to his younger brother William and to Al Schieffelin's friend from Signal, John McCluskey. He soon withdrew, but the site became known as McCluskey Canyon.

Reports out of Tombstone described several cases of claim jumping as people crowded in at dozens of sites. But most news satisfied the anxious: "The quick, sharp report of mining blasts at various distances tells you where the prospector is at work. Stray over the hills and ravines and every now and then you will pass some shaft or hole where men are busily at work."[27] That summer the Grand Central's first shaft had gone down twenty feet; the second reached fifty; and a third, forty-three. Foreman George Reynolds shipped ore to the San Pedro for testing on Captain Whitside's small Mexican smelter. The first sample ran $400 a ton, equaling early estimates from the rival Tough Nut.

[27] *Arizona Weekly Star*, October 24, 1878.

That September Fred Stanford and Louis C. Hughes surprised everyone by selling their Grand Central shares for a paltry $24,000. Stanford hung around, before long investing in the Lioness and Pectoral Mines. Despite these changes the Grand Central remained secure. Colonel Witherell, a man with wide experience in Colorado and Arizona, became more heavily involved. (Everyone called him "Colonel" even though he failed to graduate with his 1865 West Point class.) Professor John P. Arey, already closely associated with the colonel's other projects, was named superintendent of the Grand Central. He was eventually replaced by E. B. Gage, who rose as a major figure in the Tombstone Mining District over the next several decades.

Walter Dean, Josiah White, and others involved with the Contention drove their men just as hard as those employed at the Grand Central. That fall of 1878 they dropped a shaft 113 feet. With costs of only $5,000, miners extracted an estimated $80,000 in ore. One vein proved wider than the shaft itself, drawing the inevitable comparisons with the legendary Comstock Lode.

Such optimism proved unfounded at Tombstone, but in those days all rich silver strikes brought with them visions of another Virginia City. Nevada's famous bonanza remained the one great discovery against which investors compared all others, and Tombstone's reputation was spreading. An estimated three hundred miners had already abandoned the faltering McCracken works. Several months later one new arrival, describing Tombstone as "one of the best camps on the American continent," reflected on its isolation by suggesting, "If this camp was located in Nevada there would be 10,000 people here, but capital and population seems to avoid this Territory."[28]

At the San Pedro mill site, in the weeks following Gird's return from San Francisco, work continued, but not without problems. The company had contracted with the Mormons at St. David to build the dam and ditch to divert water to the mill. Delays at McCluskey Canyon, however, threw the schedule off.

Then, while supervising his army of laborers and ditch diggers, John Vosburg brought up the thorny question of liquor. As early merchant John B. Allen explained, "He did not want no liquor brought on to the

[28] *Daily Arizona Citizen*, March 10, 1879.

mill site, wouldn't allow it, and he objected to the men going across the river and getting it. There was so much dissatisfaction that it became necessary to remove Mr. Vosburg."[29] With Vosburg a major shareholder, Gird, who also disapproved of liquor, asked Tom Bidwell to find a point of compromise. The Schieffelins ignored the controversy, seeing it all as an unnecessary waste of time and energy. For recreation they happily shared beers with John Allen.

Mill construction gave rise to other complaints, especially from nearby ranchers fearing water losses. Accommodations had to be made with these people after Gird started work on the dam. Negotiations proved complicated, but everyone avoided serious trouble. By year's end it could be said:

> Tombstone Mill Site is now the scene of activity. Houses, shanties and *jacals* are going up rapidly, and several families are now on the ground. A restaurant has been opened by Mr. Ike Clanton. The foundation of the mill site is being blasted and leveled off to receive the structure that is now being rapidly shaped up at the timber and saw mill camp in the Huachuca mountains. A splendid road is being graded from the mill up to the mines. The Mormons are progressing finely with the ditch and dam. Everything connected with this enterprise shows that Corbin & Co. mean business, and ere long we may expect to see a genuine silver brick.[30]

Despite press reports assuring everyone that the mines showed promise of creating great wealth, physical conditions remained primitive. As J. B. Allen described operations in early 1879: "I know that Mr. Gird worked right at the mill as a day laborer, and the Schieffelins worked right in the mine just the same as their hired men did. . . ."[31]

On January 21, 1879, Gird, the Schieffelins, Safford, Vosburg, and Elbert Corbin entered into an agreement with Frank Corbin to construct another mill. All these parties now formed a corporation called the Corbin Mill and Mining Company, again consisting of half-a-million shares, to work ores from the Lucky Cuss, East Side, Tribute, Owl's Nest, and Owl's Last Hoot. With the completion of the new mill those five claims would become part of the Corbin Company, with Frank Corbin retaining half the capital stock of the original divided one-quarter share.

[29]Deposition of John B. Allen, Tucson, August 18, 1882, *Garner v Gird,* 360.

[30]*Arizona Weekly Star,* December 12, 1878.

[31]Deposition of John B. Allen, Tucson, August 18, 1882, *Garner v Gird,* 359.

Earlier he had arranged to have the Fulton Iron Works supply the company with a fifteen-stamp steam-driven reduction mill. Frank also hired Colonel Martin P. Buffum, a Union army veteran then working as a bookkeeper for a San Francisco theatrical daily, to supervise its construction. Afterward Corbin traveled to Connecticut to discuss these developments with his brothers. He returned to Tombstone in early May.

To accommodate eastern investors, offices for the Tombstone Gold and Silver Mill and Mining Company were opened at 310 Stock Exchange Place in Philadelphia. Under provisions of a written agreement negotiated to reduce expenses, the company first shared space and stationery with the Orion Silver Mining Company, another Arizona enterprise. That decision disturbed Gird and the Schieffelins as well as some of the others. Simply put, these men did not think much of the Orion property. Gird feared that negative reports on those claims might well hinder sale of their own shares, then being quoted at $6.50 to $6.75. In contrast, the Orion pulled in bids of only $1 to $1.08. By September those quotes fell to 87½ cents.

A meeting was called in July at the Continental Hotel in Philadelphia. Among those attending were Albert Schieffelin, Anson Safford, Philip and Frank Corbin, as well as Hamilton Disston, a major Orion stockholder. As Frank Corbin recalled: "We told them what we came for; that the parties here felt dissatisfied; that they had opened an office with the Orion Mining Company when we knew out here they had no good mining property. We thought it was against our interest to have a letterhead with the Orion first and the Tombstone second, and that was one reason that brought us on; that we had come on to adjust that matter."[32]

Everyone finally agreed to move the Tombstone company's office. At first they considered New York but then chose to stay in Philadelphia, finding space at 432 Walnut. After more meetings at Philip Corbin's New York City office, some of the group enjoyed a short vacation on Martha's Vineyard. They picked up Gird's sister Emma in New York and returned to Arizona in late October. Emma planned to see her brother's fiancée, Nellie McCarty, at San Francisco before visiting the mines.

In Connecticut Philip, Andrew, and George S. Corbin had organized

---

[32]Deposition of Frank Corbin, San Francisco, October 1882, *Garner v Gird,* 170.

Philip Corbin, the Connecticut hardware mogul and early investor in the Tombstone mines. Speculative capitalists such as he did more to assure the area's success and initial well-being than all the famous shooters combined.

their family's San Pedro holdings into the Hartford Mining and Milling Company—soon renamed the Tombstone Mill and Mining Company, a designation already used informally in Arizona, even on stationery. Shareholders consolidated their stock under the new corporation. On May 20, 1880, the Corbin Mill and Mining Company and the Tombstone Gold and Silver Mill and Mining Company, along with all mill sites and other property, became part of the new Connecticut corporation. They also incorporated in Arizona.

Early western mining remained a dangerous undertaking. There was no compensation for injury. Anyone seriously hurt was simply paid off and discharged. During Tombstone's early years all efforts at organizing a labor union met stiff resistance. Still, the mines did provide employment at a time and place where steady jobs were not easy to find.

After miners brought up the ore, laborers filled caravans of wagons for the trip to the San Pedro. There loads were emptied onto dumps at the top of the mill to be hammered into pieces about the size of a softball. Shoveled into machines called jaw crushers or macadamizers, the rock was easily smashed into pea-sized bits as quickly as sweating workmen could fill the hoppers. Inside, steam-driven batteries (huge wooden-braced contraptions containing five stamps, each weighing three thousand pounds) pulverized the material into powder amid the roar of iron against stone. Extracting the silver and traces of gold from this creamy paste was filthy, complicated, and hazardous work. It involved amalgamating pans, revolving mullers, riffled separating tables, and other apparatus spawning its own industry and vocabulary.

In the retort room furnaces separated residual impurities before the final product emerged as bars of silver. Workmen were often exposed to unsafe levels of arsenic, mercury, and lead. Everything could be handled by a crew of seven or less. Gird started with twenty, for an around-the-clock schedule, before refinements of technique reduced the work force. Of the entire process Mark Twain once quipped, "It is a pity that Adam could not have gone straight out of Eden into a quartz mill, in order to understand the full force of his doom. . . ."[33]

The Gird Mill cranked up for operational tests on June 1, 1879, after a four-to five-hour preliminary dry run on May 20. The company wanted to test the machinery and fill its cracks and seams with low-grade pulp. Early runs of third-class ore, some fifteen tons a day, yielded results the owners claimed to be over 80 percent assay value. First-and second-class ores began traveling through the milling process after the installation of a roasting furnace. All this pleased Gird, the Schieffelins, and their moneyed partners, who early on expected at least $20,000 in bullion each week.

Tombstone Mill and Mining Company bookkeeper Daniel Fields recalled: "The first shipment of bullion was made June 16th, 1879, from Millville. It was sent to Tucson. I think there were 8 bars in the first shipment. Thomas Walker and Thomas J. Bidwell went to Tucson with this first shipment of bullion in a little old blue wagon, the same one that the two Schieffelins and Gird came down from Signal in." The press reported the cargo's value at $18,744.58.[34]

Ed Schieffelin rode along with the others, and their arrival caused quite a stir along Tucson streets. The men watched as employees carefully placed each glistening bar into the vault of Safford, Hudson & Company's bank, before arranging shipment to Philadelphia via Wells Fargo. There investors crowded in to congratulate themselves. Profits

[33]Twain, *Roughing It*, 253-254. For more on the milling process, see Young, *Western Mining*, 193-203.

[34]Testimony of D. C. Fields (recalled), February 14, 1883, *Garner v Gird*, 415; and *Arizona Weekly Star*, June 19, 1879.

seemed assured: "On the lowest grade of ore, the net result will be about $900 per day, and on the highest grade $7,000 to $10,000. Mr. Gird, the Messrs. Schieffelin, Gov. Safford, J. S. Vosburg and the Corbin Brothers have fortune within their immediate grasp."[35]

Mill sites often spawned towns of their own, but they all became mere satellites of Tombstone. On October 28, 1878, Amos W. Stowe had recorded a claim to 160 acres on the river's west bank opposite the Gird and Corbin mill sites. Ostensibly he hoped to use the land for stock grazing and agriculture. But Stowe, who already owned two businesses in the district as well as holding mining shares, primarily with the Empire, soon realized the need for a town.

Millville had already started on company property, but Vosburg's and Gird's resistance to liquor stalled its progress. Stowe saw no money in temperance. Besides, as longtime resident James C. Burnett explained, "Mr. Stowe and Gird were not on very good terms. I think there was some trouble about mining claims."[36] Stowe hired Alexander J. Mitchell to survey the site. That English-born engineer and hotel proprietor, who had arrived in Arizona by way of Australia, got caught up early in all the excitement. He sold the Palace Hotel at Tucson and headed for the San Pedro. Later he mapped the district and had the results lithographed at San Francisco.

Calling his new town Charleston, Stowe began leasing lots with the understanding that leaseholders pay for the property, with improvements, after three years. The budding townsite impresario finally felt secure enough to bring his family over from California, thus escaping the Goose Club, a euphemism used in Arizona to describe men whose families lived elsewhere. Amos Stowe's generous leasing policy, which helped bankrupt the plan within months, appealed to many. Charleston shuddered under a construction boom that packed its streets with newcomers, all frantically grasping for a small slice of the town's twenty-six blocks. Having saloons and plenty of water—water proving a rarity for most district townsites—helped immeasurably.

Almost everyone crossed over. General merchandiser Walter B. Scott explained his reasons:

---

[35]*Daily Arizona Citizen*, June 4, 1879.

[36]Testimony of James C. Burnett, *Garner v Gird*, 349.

> I purchased the business and store of J. B. Allen in Millville in March, 1879. . . . After I commenced business myself I found it was necessary to put up a permanent building, the one I had purchased being only a canvas house. And I went to Mr. Gird and asked him if I could get a lease to sufficient ground there to erect a building on, and he said he supposed that I could but stated, however, that it would be necessary for me to comply with the regulations of the company. He said the company would own the ground, and if at any time it was found necessary for me to move I would have to do so, and that I should not sell any liquor. . . . I built a house on the other side of the river at Charleston.[37]

Crossing the San Pedro normally posed no problems because of low water levels much of the year. But at other times, especially during the traditional winter and late summer rainy seasons, the river could become a raging deathtrap for those foolish enough to attempt a crossing. During August of 1879, W. B. Scott reported high water suspending communications for thirty hours, adding, "a man crossing with a four mule team narrowly escaped being swept away with his entire outfit."[38] A bridge strong enough to support heavy ore wagons eventually eased the tension of mill workers living at Charleston.

Soon after Mitchell finished his survey the town boasted forty buildings, nearly all of them adobe, and many more tents. Weeks later a touring government agent reported: "At Charleston I found a second Leadville; three months ago there was not a building in the place, now there are some two hundred, and many others in course of construction, and most of them good substantial buildings; with a . . . steady stream of people pouring in there constantly."[39]

Although established first as a small boom camp in its own right, Charleston never offered a serious challenge as the center of the district. Less than a month after Mitchell completed his survey for Amos Stowe, the new town of Tombstone began its shaky start at its present location.

---

[37]Deposition of W. B. Scott, San Francisco, *Garner v Gird*, 222-223.

[38]*Daily Arizona Citizen*, August 16, 1879.

[39]S. C. Slade, Collector's Office, Custom House, El Paso, Texas, to the Secretary of the Treasury, July 6, 1879, General Records of the Treasury Department, Letters from Collectors, Custom House Nominations, Record Group 56, National Archives and Records Service. For another early description of Charleston, see *Weekly Arizona Citizen*, March 28, 1879.

One of the earliest views of Tombstone, looking toward the northwest, taken by Tucson photographer Harry Buehman in late 1879. Comstock Hill forms the background.
Courtesy Special Collections, University of Arizona Library.

Chapter 4

THE TOMBSTONE TOWNSITE

With the mining phase successfully launched, the district clearly needed a commercial and social center to accommodate newcomers. It had been nearly two years since Ed Schieffelin's discovery. Now, seduced by grandiose visions guaranteeing success, planners imagined more dollars piling up from real estate sales than from trying to scratch out another silver mine on the already crowded hillsides. Competition exploded as schemers everywhere lunged for the chance to make their special location the beacon of the district.

At first Watervale seemed the best site close to the mines. Also called Lower Town and situated a couple miles northwest of Tombstone, it boasted reliable water at three cents a gallon. Other locations offered far less reliability on that score, especially during the hot summer months. The district's first mercantile firms of Amos Stowe, together with John Kratzmyer and Oscar Dupuy's Pioneer Supply Store, had already carved out commercial footholds at Watervale. Kratzmyer soon withdrew, but the firm of Cadwell and Stanford willingly joined the others stuffing battered display cases with firearms, ammunition, tobacco, cigars, liquor, and, boasted Dupuy, "a fine assortment of Groceries and Mining Tools always on hand, at prices within the reach of all."[1]

Typical of early merchants, Andrew Cadwell and James Stanford (son of the Tucson lawyer who briefly owned part of the Grand Central) were both young—twenty-four and seventeen, respectively. Similarly, John Kratzmyer and Oscar Dupuy were only twenty-three and twenty-five. Men on the hustle felt at home in the Tombstone Mining District regardless of age. When the Pima County Board of Supervisors set up new election precincts in early October 1878, they picked the Pioneer Supply Store as polling place for Precinct No. 17. Dupuy, a former Tuc-

[1]O. H. Dupuy's Pioneer Supply Store advertisement, *Arizona Weekly Star*, July 4, 1878.

son dry goods clerk at Lord and Williams, served as the inspector with Richard Gird and Amos Stowe both poll judges.

Ignoring Watervale's apparent edge, other townsites clamored for recognition. These included Richmond, Graniteville and Granite Town, Merrimac, and Happy Valley—the latter so named because no liquor could be sold. That alone doomed its chances. Dozens of other isolated miners' camps pockmarked the surrounding hills. The district sported so many townsite plans that the press joked, "Every one seems inclined to set up a town for himself."[2] The Tombstone townsite actually changed locations during this period, reflecting economics and personal whim over popular preference.

When the Schieffelins, Dick Gird, and John Vosburg signed their original agreement with Elbert Corbin to build the first San Pedro reduction mill, it included, "An undivided one half [of Vosburg's] interest in the townsite, called Tombstone. . . ."[3] This did not, however, refer to the present Tombstone location. Instead, that site, which all early residents called Upper Town or Gird Camp and later Old Tombstone, lay astride the founders' West Side claim, "about equidistant from the Contention and Grand Central, the Tough Nut and Lucky Cuss mines."[4]

At first only scattered tents, shacks, small businesses, and saloons dotted Upper Town. One observer remarked with hope in late September 1878, "The new townsite is progressing finely; two stores, one butcher shop and a restaurant under way already."[5] John B. Allen moved in, setting up a small store connected with a ramshackle boarding house where the Schieffelins often stayed. That sixty-year-old businessman came to the district after selling his hotel at Cerro Colorado. From nothing more than a wooden and canvas rectangle, Allen supplied basic necessities while dreaming of making a fortune.

But to the growing discomfort of Dick Gird and the others, Upper Town began looking a little too permanent: "Houses of stone, adobe and of less substantial material are going up while wells and shafts are going down." Someone reported tents "in every nook and corner." The site "is growing daily and is now a town of some importance; there are

---

[2]*Arizona Daily Star*, February 7, 1879.

[3]Miscellaneous Records, Pima County Recorder's Office, Book 1, 668-669; filed June 18, 1879.

[4]*Weekly Arizona Citizen*, August 30, 1878.

[5]Ibid., September 28, 1878.

The first San Pedro reduction mill, designed for the Tombstone Mill and Mining Co. by the Fulton Iron Works of San Francisco. The house beyond was used as an office, and also doubled as a honeymoon retreat for Dick Gird and Nellie McCarty.
Courtesy, Special Collections, University of Arizona Library.

three restaurants . . . so that one stands a chance of getting a square meal." Later it was said: "There was a report the other day that Gird & Company had prohibited the sale of liquors on their mining claims; if this is the case, the upper town will not grow so fast."[6]

That was exactly what Gird and the others wanted, especially after people started laying out town lots. The idea of outsiders squatting on their property did not interest him or the company. They saw too many legal problems arising from a town and mining claims occupying the same ground. As John Allen explained: "Mr. Gird told me, and the Schieffelins had told me that it was necessary for the people to go, they must leave that place, the old town, because it was on their mining property."[7]

Where should they go? Already a group of determined men had begun planning a town for Goose Flats. But Colonel Alpheus Lewis and his son Robert aggressively promoted their own site, called Richmond,

[6]*Arizona Weekly Star,* January 9, 1879; and *Arizona Daily Star,* January 21, 1879.

[7]Deposition of John B. Allen, Tucson, August 18, 1882, *Garner v Gird,* 361.

and they had plenty of money to work with. The colonel had recently sold thirteen claims for nearly $150,000. Allen recalled that Lewis "formed a company in Chicago, to bring water into Richmond, and induced a great many, most all of the people to go there," because nearby Goose Flats was not a particularly popular location. To try and change those opinions and stimulate interest, the Tombstone Townsite Company began giving free lots "to everyone who would go there . . . provided they would go over and build."[8] At first no amount of generosity, calculated or otherwise, seemed to dampen Richmond's inside track on becoming "the town of the district."

Yet others, led chiefly by Richard Gird, began to take Richmond's noisy rival more seriously. Gird strongly opposed the Lewises' efforts for two important reasons: first, they planned on using water from a source near the San Pedro, thus possibly reducing levels for milling purposes; and second, his longtime friend Thomas J. Bidwell had maneuvered himself a one-fifth interest in the new Tombstone plan. John Allen even claimed, "It was Bidwell's brains that you might say originated and made the townsite amount to anything."[9] Many thought Gird was a silent partner, but he always denied it.

Despite surprisingly heavy opposition, Richard Gird had been named postmaster for the mining district on December 2, 1878. At first he located the office on company property at Upper Town. Gird now moved it to the Tombstone site, finding temporary space in J. B. Allen's new store. He then named Bidwell assistant postmaster. One man characterized the whole affair as "postal pap."[10] In those early days before Western Union arrived, the post office remained the single source of communication with the outside world. Mine owners, businessmen, and common laborers all needed access to the mails. Government records show about five hundred people entitled to its services just prior to Gird's appointment.[11] The idea of building a town without a post office made little sense to those living in such a remote area, and there were simply not enough people in the district at that time to justify more than one office.

---

[8]Ibid., 361-362.

[9]Ibid., 361.

[10]*Arizona Weekly Star*, June 12, 1879.

[11]Richard Gird for Tombstone Postmaster; Topographer, Post Office Department, Appointment Office, October 25, 1878, National Archives and Records Service.

Townsite organizers appreciated Gird's efforts. They donated property to his brother William for a new post office on the northwest corner of Fourth and Fremont; built by Al Schieffelin and Tom Bidwell, with the account carried on the books of the mining company. Construction costs were charged to William and the account itself was not closed until late 1880. The lumber, of course, came from Dick Gird's own sawmill in the Huachuca Mountains. Some of those associated with the mining company, including Albert Schieffelin, used the post office, which had two small rooms behind the delivery window, as their residence. Gird even swore Schieffelin in as a temporary clerk. It must have surprised many down-on-their-luck prospectors to have their mail handed to them by one of the area's luckiest men.

Criticism deepened over the mining company's control of mail services. Within a year authorities sent in two investigators. H. J. McKusick, superintendent of the government's railway mail service, reported that those in charge "practiced favoritism toward what was called the 'Tough Nut crowd,'. . ."[12] Special Agent J. H. Mahoney claimed "the Tough Nut Mining Company was running the Post Office to suit themselves; that the Post Office was closed up and opened when they saw fit. . . ." Mahoney also recognized the more serious complaint "that it was not safe to trust any letters going through the office."[13]

Dissatisfaction ruled the day, one angry citizen charging: "When the fame of the first discoveries in Tombstone began to attract capital to the locality, a town-site was surveyed and inducements held out to parties to settle. . . . A post office was established with Richard Gird as postmaster, stores were opened, barkeepers 'set 'em up' in numerous saloons, and the place was rapidly assuming the appearance of a prosperous town. All of a sudden the original proprietors announced that the town must be abandoned. . . ."[14] This angry dissenter, signing his anonymous scribbling "Justitia," blamed the whole process on a group of greedy speculators, aided in their nefarious activities by Gird, Safford, and Ed Schieffelin.

Despite all the grumbling, the old town had seen its day. In less than a year it was said: "Old Tombstone sleeps the sleep that knows no waking. Bailey's saloon is the last to succumb. The Chinaman's hashhouse came over to the 'bright side' a few days since."[15]

---

[12]Testimony of H. J. McKusick, *Garner v Gird,* 385.

[13]Testimony of J. H. Mahoney, *Garner v Gird,* 385.

[14]*Arizona Weekly Star,* June 12, 1879.

[15]*Arizona Daily Star,* September 2, 1879.

Another 1879 photo of Tombstone by Buehman from a slightly different angle, with the German cameraman's wagon and mule, hauling his cumbersome equipment, captured for posterity. Photography in those days was no simple business.
Courtesy, Arizona Historical Society.

Tombstone suffered its share of growing pains. If only foolish legend is believed, all these problems found their origins with a band of outlaws shielded by a string of corrupt county officials. That is until the forces of law and order, represented by the three mustachioed Earp brothers, brought them to dust. But Tombstone's earliest crisis involved not somber lawmen of mixed reputation; rather, the seemingly mundane business of real estate acquisition and transfer. The chief culprit in this sordid drama turned out to be the Tombstone Townsite Company. The company itself was dominated by a rather colorful assortment of enterprising men, some with questionable backgrounds, who all had one thing in common—a driving desire to make lots of money fast, with little concern given ethics or other annoying questions of personal conduct.

A major player was Joseph C. Palmer, a small man with big ideas. Born in Nantucket, Massachusetts, Palmer and fellow townsman Charles Cook rushed to California soon after hearing of the gold dis-

Townsite organizer Joseph C. Palmer, a tough and uncompromising individual, as he appeared during his earlier San Francisco escapades. *Courtesy, California Historical Society.*

covery at Sutter's Mill. They bought out the San Francisco banking firm Thompson and Company. Renaming it Palmer, Cook & Company, they opened as an express agency in 1849 on the city's main plaza. Two years later they abandoned the express business and devoted themselves almost exclusively to banking, furnishing bonds for city and state officials. Early on, as chief depository for many old and respected Spanish families, the company earned itself an enviable reputation.[16]

The Massachusetts men unfortunately shifted to real estate speculation and political intrigue, historian Hubert Howe Bancroft noting:

> Speculative bankers, like Palmer, Cook & Co., contrived by becoming bondsmen of state officers to obtain the handling of the money which should have been in the state treasury. Crime became easy and natural on both sides. Palmer, Cook & Co., who had nearly ruined the state's credit in 1854 by withholding the interest due on its bonds in order to depreciate them for speculative purposes . . . had both the state and the city of San Francisco in their power.[17]

[16]Barry and Patten, *Men and Memories of San Francisco,* 246.

[17]Bancroft, *History of California,* 6:616.

That power did not last and Palmer soon found himself mired in other controversies. The federal court thwarted his claim to two leagues of San Francisco real estate under the Rancho Punta de Lobos land grant.[18] Then, supporting David C. Broderick as United States senator, Palmer was accused by Elisha Peck, a state senator from Butte County, of offering a $5,000 bribe for Peck's support in favoring a joint convention to elect a U.S. senator. As one California legal historian explained, "Palmer escaped through a single dexterous movement of his counsel."[19]

Palmer and his cohorts finally closed their doors in late July 1856, because, as Bancroft wrote: "The law . . . was disregarded, and Palmer, Cook, & Co. again became the holders without security of $88,520, interest money due in New York on the state's bonds, but which they retained for their own use, the firm failing, and most of its members and agents absconding."[20] Palmer tried to rescue some of the assets by forming another bank but was derailed by an attachment of nearly $37,000. He gave up banking in late 1857 and turned to real estate.

Joseph Palmer survived these reverses, although the same could not be said for all his associates. The company had backed John C. Frémont, thus seriously damaging the old pathfinder's 1856 campaign as the first presidential candidate of the new Republican Party. The opposition tried to link his name with Palmer's scheme to swindle the public over the Mariposa Mine venture in the early 1850s. Frémont lost the election, but Palmer continued various businesses in California and elsewhere. During the Civil War he traveled to Mobile, Alabama, and tried to get hold of the cotton trade but by 1869 was back in San Francisco, again as a real estate agent. Five years later he styled himself a capitalist. By the time he traveled to Arizona and jumped into Tombstone's townsite troubles, Joseph C. Palmer had become a viticulturist at Mission San Jose. There, on 125 acres, he lived with his wife, two sons, daughter-in-law, two grandsons, two servants, a French chef, and a live-in lawyer,

---

[18]Hoffman, *Reports of Land Cases,* Joseph C. Palmer et al., claiming the Rancho Punta de Lobos, Appellants, vs. The United States, 216-218, 227-229, 249-272.

[19]Shuck, *Bench and Bar in California,* 103. Also see *Illustrated History of Plumas, Lassen & Sierra Counties,* 195.

[20]Bancroft, *History of California,* 6:618. For more on Palmer's nefarious California activities, see Cross, *Financing an Empire,* 1:72-73, 173-175, 198, 201-202, 208-209.

growing wine grapes with the help of nine vineyard workers, including six Frenchmen and one Italian.[21]

Palmer came to Arizona in early 1879 with James S. Clark, a Gold Rush '49er from Michigan. Earlier Clark worked as a San Francisco saloonman and livery stable operator. At age twenty-six he got involved in that city's 1851 Committee of Vigilance before moving up into the merchant class.

During the Civil War, Clark also became a Gulf state speculator in southern cotton. He operated out of New Orleans, then under federal control, with a cotton factor and commission merchant named Edward Fulton. Clark's situation was not helped by his attempted bribery of the provost marshal. Then, in late summer 1865, military authorities confiscated nine hundred bales of cotton claimed jointly by James S. Clark & Company and fellow San Franciscan Joseph Palmer, still in Mobile looking for any profits that might slide his way as the result of sectional strife.

They had consigned their cotton to T. C. A. Dexter, the area's supervising special agent of the Treasury Department, to aid its shipment over the military-controlled railroad from New Orleans to Mobile. "Mr. Dexter having received orders from the Treasury Department to ship all the cotton received by him, shipped the said nine hundred bales to New York," noted the Supreme Court, "where it arrived and was sold by the United States, and the net proceeds . . . $127,350, were paid into the treasury."[22] Clark, Fulton, and Palmer later filed for that amount with the U.S. Court of Claims, only to watch helplessly as their case was tossed out because of the statute of limitations. They appealed, but the Supreme Court ruled against them. In later years Clark would attempt redress by a special act of Congress.[23]

In 1879 Clark and Palmer traveled to Arizona, reaching Prescott on January 29. The press, describing the pair as "mining gentlemen" and "capitalists," made much of their arrival: "They brought through four

[21]1880 United States Census, Mission San Jose, Alameda County, California, E.D. 25, Sheet 19, Lines 10-29; and *Langley's San Francisco Directory*, 1869, 1874.

[22]Otto, rep., *United States Reports*, 9, *Clark v. United States*, 494.

[23]Nott and Hopkins, rep., *Cases Decided in the Court of Claims . . . and the Decisions of the Supreme Court in the Appealed Cases*, 11:774. Also see: House of Rep., 47th Cong., 2d Sess., Report No. 2031, "James S. Clark & Co.," 1-8; 51st Cong., 1st Sess., Report No. 2380, "Estate of James S. Clark," 1-6; and Report No. 377, "Estate of James S. Clark, Deceased," 1-7.

fine horses and a thorough brace wagon particularly adapted for travel in Arizona. Just what their programme is we have not learned, but believe they intend to examine Arizona mines thoroughly with a view to invest capital."[24]

After visiting their old friend John C. Frémont, then sitting as Arizona's governor and in Prescott only because the legislature was in session, the two left on February 6 and headed south over the stage road through Black Canyon. Nine days later they arrived in Tucson and let it be known they planned to look into mining property in the southeastern corner of the territory and in northern Mexico.

Wasting no time, the two Californians visited the San Pedro, prompting an optimistic reporter to gush: "Messrs. J. C. Palmer and J. S. Clark returned Sunday morning from their trip to Tombstone and surrounding country. They express themselves much pleased . . . and will probably make some investments. We hope they will. It is just such men that Arizona stands in need of to bring her out. . . ."[25] It did not take long for many weary citizens to wish they had never come.

Both understood the need for friendly relations with the powerful while at the same time pacifying the general public. That soon became impossible at Tombstone, so Clark bribed as much goodwill at the county seat as he could by building a block-square ice plant. He then stood out as a savior to all suffering desert dwellers. Southern Arizona's political power sat firmly in Tucson, and Tombstone was then still part of Pima County. Months later the press complimented Palmer, saying, "We can't have too many of his stamp."[26]

Another man interested in the new Tombstone townsite, Tennessee-born Michael Gray, had preceded Palmer and Clark into southern Arizona. His father had moved the family to Texas in 1832 when the boy was only five. Michael joined the army at fourteen and saw action in the fighting that preserved Texas independence. During the Mexican War he served as a second lieutenant in John Coffee Hays's Texas Ranger Mounted Volunteers. Seeing no action, he mustered out after only forty-one days. Gray later followed Colonel Hays to California.[27]

---

[24] *Weekly Arizona Miner,* January 31 and February 7, 1879.

[25] *Arizona Weekly Star,* March 13, 1879.

[26] *Arizona Daily Star,* December 12, 1879.

[27] Military and Pension Records, Michael Gray, 2nd Lt., Capt. Gillaspie's Co., 1 Reg't (Hays's) Texas Mounted Vols., National Archives and Records Service.

Justice of the peace and townsite manipulator Mike Gray as he looked when entering the legislature from Benson in 1887. *Courtesy, Arizona State Library.*

Between 1852 and 1855 Mike Gray served as Yuba County sheriff and proved his mettle on more than one occasion; not restricted to shooting one heavily armed attacker stone-dead with a single-shot pocket Deringer.[28] Earlier in his tenure one defendant faced the gallows for looting two tons of merchandise from a warehouse—a single sack of potatoes tipping off authorities. The prisoner was found "guilty of grand larceny punishable with death." The case prompted the statement in 1879: "This was the only conviction and execution in the State under the then existing law, and it was repealed at the next session of the Legislature. This is a remarkable case on account of its being an instance of hanging for stealing, an unusual thing in the United States in modern times."[29]

Gray began speculating in mining properties soon after leaving office. He traveled throughout California and Nevada as well as into the Mexican states of Sinaloa and Sonora. There one of his sons was born in

[28] *Daily California Chronicle* (San Francisco), August 13, 1855. For other references to Gray's service as county sheriff and as a volunteer fireman, see *History of Yuba County,* 120; *Hale & Emory's Marysville City Directory,* 59; and *Colville's Marysville Directory,* 95. Gray's son John P., writing in 1940, mistakenly identified his father as "acting sheriff" of Yuba County. Records suggest instead that he replaced R. B. Buchanan in 1852 and was succeeded in late 1855 by William B. Thornburgh.

[29] *History of Yuba County,* 126.

1862. Mike Gray came to Tucson from San Francisco in late 1878 to look over the new Pima County mining districts. After a whirlwind trip to Tombstone, the press reported: "We are happy to note that he contemplates casting his lot with us; he is a genial and companionable gentleman of culture, with energy and foresight which makes him a good addition to Arizona."[30] Despite all the kind words, Gray quickly joined Palmer and Clark in promoting the Tombstone townsite scheme.

Both Thomas J. Bidwell and Anson Safford quietly maneuvered themselves into the townsite organization, fueling speculation that Gird and others associated with the mining company, particularly the two Schieffelin brothers, had some secret interest in its success. Officials at Millville repeatedly denied the charge, but many of them accepted free lots or paid ridiculously low prices for choice locations.

Tom Bidwell's connection is interesting as he had very little cash and no valuable mining claims. Instead he spent most of his time acting as lumber agent for Gird's sawmill and presiding as justice of the peace at John B. Allen's store at Old Tombstone. What attracted the other townsite members to Judge Bidwell was his friendship with the district's most influential men and his still-strong political ties at the territorial capital.

After recruiting Safford and Bidwell, Palmer and Clark finalized their plans. In trying to sort out the controversy a year later, someone signing himself "Squatter" described their efforts:

> The present townsite was located under the preemption laws of the United States by Samuel R. Calhoun, some-time in the month of February 1879, and in [March], A. P. K. Safford, James Clark, Joseph Palmer and Thomas J. Bidwell entered into a negotiation with Mr. Calhoun, and Mr. Calhoun in consideration that the . . . parties would survey and lay out a townsite and use their best efforts to build a town . . . did give them 4-5ths of his right. All the old settlers are familiar with the efforts, and by whom the success was consummated.[31]

The two Calhoun brothers, Samuel and Charles, had located the Mountain Maid mining claim. It covered a large portion of the proposed townsite, the most level piece of ground within reasonable walking distance of all the important mines. Charlie built the first structure

[30] *Arizona Weekly Star*, January 9, 1879. For a short biographical sketch of Mike Gray, see *Daily Nugget*, October 16, 1880.

[31] *Weekly Nugget*, March 25, 1880.

on what became Tombstone, described at one point as a "mansion . . . built of mud and carraja poles,"[32] on a spot later occupied by the Mount Hood Saloon and then by the Grand Hotel. Palmer, Clark, Safford, Bidwell, and Samuel Calhoun drew up the Tombstone Townsite Claim. Dated March 5, 1879, this document was filed with the United States Land Office at Florence on April 19, and with the Pima County Recorder's Office at Tucson three days later.[33]

Safford and Samuel Calhoun signed the document, but curiously, James S. Clark used the name of his young son Maurice E. Clark, then living in New Orleans before going to Germany to study mining at Frederick Brunckow's old school. Joseph C. Palmer substituted his twenty-three-year-old son Joseph Bruce Palmer, still in California looking after his father's vineyard interests. Thomas Bidwell used the name Phronia J. Bidwell (another time it shows as Pluonia). It is not clear whom this represents; it was not Bidwell's deceased wife or either of his two children. By the time he began selling his interest, however, Bidwell ignored the deception and used his own name. Since common knowledge correctly identified the townsite claimants, perhaps those three felt safer from potential legal disruptions by listing out-of-state relatives or a nonexistent party rather than themselves.

To satisfy Land Office requirements, the proprietors hired Solon M. Allis, who had worked for the Aztec and Toltec mining syndicates, to conduct a survey. At first the town occupied a 320-acre rectangle lying on a slightly northwest-to-southeast axis. From the west, and its conical-shaped landmark Comstock Hill, numbered streets ran roughly north to south with named streets intersecting to form 300 by 300 foot blocks—each divided into twenty-four lots—in the main business district and residential areas. An exception was a line of 240 by 300 foot blocks between Toughnut and Allen streets. As the town grew other exceptions were made. To the south Toughnut was the first east-west street, followed by Allen, Fremont, and Safford.

What eventually became Tombstone's main business thoroughfare was named for John B. Allen, who abandoned his canvas shack at Old Tombstone when offered free lots in exchange for his promise to build.

---

32 *Tombstone Epitaph*, August 12, 1882.

33 Tombstone Town-Site Claim, Pima County Recorder's Office, Land Claims, Book 1, 527.

He established himself on the southwest corner of Fourth and Allen streets, building a store for general merchandise with a blacksmith and stable adjacent.

Allen, born on the rocky coast of Maine in 1818, had been in Arizona since 1857. Remembered by Tucsonan Harry Drachman as tall and impressive, he enjoyed patrician features yet displayed an affable manner. With his Boston education, Allen easily got himself elected to the Fourth Legislative Assembly and was instrumental in having the capital moved from Prescott to Tucson for a time. He then served as territorial treasurer for several years. Experiencing a precarious boom-and-bust cycle in his own finances, however, Allen tried developing agricultural, ranching, mercantile, and other interests. Twice he was elected Tucson's mayor and briefly acted as postmaster in 1866. A dozen years later he worked as a jobber and general merchandiser before opening a small hotel in the Altar Valley. At Tombstone he prospered for a time, even becoming Millville's first and only postmaster. That office closed a year later when mail service for isolated river settlements and ranches shifted to Charleston. Allen would leave the district rather early.

Although seldom in Arizona, John Charles Frémont was territorial governor when Solon Allis platted the townsite. For whatever favors might result, the owners decided to name the main street in the governor's honor. Unfortunately, the tide of commerce soon made Allen Street the center of activity. The old explorer, always hopelessly dreaming of making a fortune in western mining, did not feel slighted. He saw himself as too important an individual to serve in some remote territory anyway and spent most of his time in New York City. Under pressure Frémont resigned from office in late 1881.

Safford Street was, of course, named for Anson Safford. He was a man of piercing gaze but diminutive stature, standing only five-foot-six. Raised in poverty, he left his Illinois childhood home for the California gold fields at age twenty. Self-educated, he served two terms in the legislature before moving on to Nevada. There his political interests continued to blossom. After a two-year tour of Europe he maneuvered an appointment as United States surveyor-general for Nevada. Afterward Safford petitioned President Grant and became Arizona's third governor. He served eight years and is best remembered as a champion of public education. He went on and helped found the banking house of

Safford, Hudson & Company at Tucson. Always with an eye open for opportunity, the feisty ex-governor later became president of the Arizona extension of the Southern Pacific Railroad.

As the town grew other streets would be added, mostly to the north and east. Tombstone had plenty of thoroughfares ready for businesses and housing, it had the post office and justice court, and sat on the most logical piece of real estate for miles around. It even had some of the most talented crooks in the area running the place. Yet one thing was missing. Tombstone had no water. Townsite organizers persuaded the Brunckow's Tom Jeffords to try and dig a series of artesian wells, such as those found near St. David, but the idea fizzled. They clearly needed something more than promises if the town was to survive.

This very point kept competition from Richmond alive. Colonel Lewis and his son planned to supply their site with water from Crystal Springs Ranch near the San Pedro River. Control of that location, earlier known as Fritz Springs and later renamed Lewis Springs, concerned Richard Gird, who even as late as October 1879 wrote Safford concerning a backup water source: "I am looking up the water question for the town of Tombstone, and Ed. has found what he thinks an ample supply in the Dragoons about 13 miles away and at an elevation that will reach all the mines. It is a splendid undertaking, and is forced on us by the action of Lewis and his company who propose to take the water from Crystal Spring [*sic*] that furnishes water for the mill."[34]

Mill worker S. W. Wood spoke about problems with Crystal Springs "that discharged its water into the river above our dam, [Lewis] diverted the water from the river, and run it over the flat [ostensibly used as a vegetable garden], and that slacked off the head of the water for us, so that we had to hang up our mill a portion of the time. . . ." Tom Bidwell tried bribing the man in charge with a $200 payoff. Colonel Lewis, whom it appears planned some sort of confrontation, squelched the deal after which Bidwell ordered Wood "to go and take the spring by force: first decoy the man who was holding the spring away, and then take some armed men and hold it."[35] A fairly sensible solution by the standards of 1879, but the plan failed anyway.

Dick Gird was still more interested in water for his mining company

[34]Richard Gird to Anson P. K. Safford, October 13, 1879, *Garner vs. Gird*, 206.

[35]Deposition of S. W. Wood, Tombstone, August 10, 1882, *Garner v Gird*, 244.

than in domestic consumption at Tombstone. Yet when it came to the town, James S. Clark and his associates kept most of the necessary details in mind. They grabbed control of Sycamore Springs, seven miles to the northeast. That place would supply the town through a piping system and reservoir by late 1879, thus doing away with daily wagon runs from Watervale. Walter Dean and the Western Mining Company had originally planned a reduction mill for the site. Luckily for Tombstone they settled instead on the San Pedro.

Even without settling the water question, the Townsite Company faced another serious problem. Organizers had failed to pay the necessary fees—a mere $1.25 an acre—when filing their initial claim with the Land Office. This delayed the granting of a townsite patent, a document necessary to guarantee real estate titles. At first the Townsite Company ignored the issue and started selling lots anyway. Customers resisted pressure to invest and instead reopened the debate on whether this new townsite would become the district's principal community or not.

Seeing all this as a godsend, Lewis pressed forward with Richmond and the choice of Crystal Springs as a water source. He hoped to convince settlers to abandon Tombstone, despite its control of the post office and stage stops. Lewis brushed aside complaints about uneven ground by pointing out that his town was closer to the mines. By mid-May his company had completed its line survey designed to raise water to a reservoir atop what was then called Treasure Hill, seventy-five feet above the townsite and adjoining mines. Lewis offered water to both, free to those buying choice lots. The colonel boasted there would be no need for water wagons at Richmond.

Gradually, however, public opinion accepted the Tombstone site mainly because of the post office. Since the physical location of a post office can not be moved without official approval (a point ignored by Gird), many residents convinced themselves that the Townsite Company had in some way received the blessings of the United States government. They felt certain the patent would arrive in good time and all questions would be resolved. Unfortunately many of these people turned out to be dreamers, the very trait that had drawn them to such a wild, undisciplined boom camp in the first place.

Besides clouded titles, the scarcity of building materials, primarily adobe and lumber (then selling on the open market for $100 per thousand feet), slowed Tombstone's growth. Yet once again, Richard Gird came to the rescue by allowing excess lumber from his sawmill to be sold commercially. At first Tom Bidwell served as mill agent, before stepping aside for his friend from Signal, William A. Harwood. Neither Bidwell nor Harwood intended any townsite other than Tombstone to benefit from the excess lumber coming out of Gird's mill.

Technically Richard's brother William owned the major interest. In reality Dick Gird pretty well controlled things: "My brother Richard told me in San Francisco that the company would furnish the mill and pay our running expenses, and that we might go on and run the mill and furnish them lumber at $50 a thousand, and all lumber outside that we could sell, we might have all we could get for it, provided that the money, the price of the lumber, should run through their hands until they were paid for the mill."[36]

After supplying needs for ore bins, hoisting works, mining timbers, boarding houses at Millville, and other demands of the company, lumber from Gird's sawmill was used to build the first houses in Tombstone. Attempts were made to satisfy customers with high-priced stock shipped from Tucson, but supplies always fell short. With profits guaranteed, the Mormons at St. David (then a community of about seventy) began planning a second sawmill at Miller's Canyon in the Huachucas. District residents, so desperate for lumber, happily read reports of mill machinery passing through Tucson in early July. Optimism faded, however, after Josiah White contracted most of the Mormon output for the Western Mining Company and its Contention Mill.

All this activity in the Huachuca Mountains eventually spawned a small vacation industry for those Tucsonans wishing to escape the summer's relentless heat. New Yorker Edward C. Burton opened a hotel, whose Japanese cook served a breakfast guaranteed to block the arteries of even the most hardened desert dweller—fried chicken, porterhouse steak, scrambled eggs, buttermilk, and coffee with cream. Captain Whitside planned on subdividing a portion of the valley east of Camp Huachuca for civilian homes. Richard Gird ignored tourism. Instead, he sent two prospectors into Ramsey Canyon searching for mining prop-

[36]Testimony of William K. Gird, *Garner v Gird*, 353.

erty. They christened one site the Discovery, but assaying at only $200 a ton, it never rivaled Gird's Tombstone holdings.

Others cared little or nothing about mining. The financial possibilities presented by Tombstone's lumber shortage stretched beyond Arizona, convincing San Diego lumberman Philip Morse to take advantage. Despite poor health Morse had traveled to San Francisco via Panama in 1865, finding work as a lumber salesman. He shifted his talents to San Diego four years later, managing McDonald & Company for his future father-in-law. Morse then became assistant cashier of the Commercial Bank and served two years as city treasurer before moving to Arizona in 1879.[37]

That summer Morse and Jacob Grunendike started a sawmill south of Fort Bowie. From the Southern Pacific's end of track freighters hauled the mill's machinery through Tucson. Within six weeks the Chiricahua canyons hummed with activity as teams hauled finished lumber to Bowie and Camp Rucker, at the upper end of White River, and to points as far distant as Sulphur Springs and Tucson. But Tombstone remained the primary market. William Harwood, as he had done earlier for Gird, acted as the new firm's local agent.

With Gird's lumber arriving, and Morse & Company's output not far behind, the sounds of construction rang sweet in the ears of the townsite proprietors. Amid this blooming optimism questions surrounding the nonexistent patent went unanswered. By September 1879 the Townsite Company was selling choice lots at healthy prices: "Corner lots advance 20 per cent every twenty-four hours, and some real estate owners will be sadly left out."[38] The Tombstone real estate boom had begun, and a sordid affair it soon turned out to be. Almost everyone in town got involved in this sticky business one way or another.

Not only were town lots being sold, but over the next few years so were whole sections of the townsite, subject to numerous verbal and behind-the-scenes agreements. On July 1, 1879, Samuel R. Calhoun and his brother Charles sold their one-fifth interest to Mike Gray for $650. That same day Gray bought half of Tom Bidwell's share for $500. Then, on January 7, 1880, Bidwell sold his remaining interest to Gray for

[37] *City and County of San Diego*, 133-136; and *Illustrated History of Southern California*, 194-195.

[38] *Arizona Daily Star*, September 2, 1879.

$750. Many of Bidwell's friends felt he sold out too fast and far too cheaply. But on February 2, 1880, Safford sold his one-fifth interest to Michael M. O'Gorman, another agent for Gird's sawmill, for $1. Twenty-seven days later O'Gorman, who jointly owned several town lots with William Gird, sold that same share to James W. Locker for $750. On March 2 Locker turned around and sold his one-fifth interest to John D. Rouse for $2,000.

Rouse, a partner in the New Orleans law firm of Rouse and Grant, knew James S. Clark from his many difficulties with federal authorities. Rouse and his associate William Grant enjoyed a wide practice handling claims against the government.

On June 20, 1880, Joseph C. Palmer and his son sold their one-fifth interest to James S. Clark for $1. John J. Anderson bought a one-fifth interest from Clark for $5,000 on November 9, 1880. On March 4, 1881, Mike Gray sold one-fifth to Maurice E. Clark for $3,000. In turn young Clark, then in Dresden, Germany, sold all his holdings in Tombstone back to his father for $5 on January 24, 1882. Two years later James S. Clark sold a three-fifths interest to attorney George G. Berry—who had known Mike Gray since 1857—for $5,000.[39] It was all quite adventurous and rather shady.

[39] *Tombstone Townsite Title Abstract*, 2-19.

Chapter 5

Tombstone on the Rise

As news of Tombstone's richness swept the territory, scores abandoned home and property in their mad rush to the San Pedro. It was no easy journey. Contrary to popular belief, most townspeople and laboring men on the frontier did not own horses. Many, carrying what they could strapped to their backs, trudged the nearly eighty miles from Tucson. More than one anxious prospector pushed a wheelbarrow over the rutted roadways, piled high with belongings. Those with cash sometimes found space on freight wagons or unscheduled bucklines. Others traveled established routes to Tres Alamos or Ohnesorgen's before walking south or begging transportation. During that first hectic summer of 1878 young James Branson, taking up some of the slack while hurriedly stuffing dollars into his own pockets, opened a one-man express line between Tres Alamos and the Tombstone mines.

Outsiders watched in amazement as this strange menagerie, gripped by fear of arriving too late for immediate gain, plodded on relentlessly. These were quintessential frontier types, people who dismissed danger, hardship, and starkness of surroundings. Their dreams overshadowed any sense of reality, leaving behind only sparkling visions of silver.

By November 1878 John D. Kinnear started a stage line called the Tucson and Tombstone Express. At first the company's four-horse, twice-weekly runs averaged seventeen hours and cost $10, round trips $15. Andrew Cadwell confidently challenged Kinnear by opening the Pioneer Tombstone Stage Line from his and Jim Stanford's store at Watervale. They also left twice a week, charging $7. Cheaper, yes, but their route was longer, with passengers forced to spend the night at Turner's ranch. Kinnear matched the lower fare and shortened the trip to thirteen hours by using a cutoff to his ranch at Ash Spring, a water source on the northeast side of the Whetstones. He soon forced Cadwell out of the transportation business.

Of early service it was said: "Stage riding in Arizona is usually regarded as a synonym for inexpressible torture, but a noteworthy exception to the rule may be found in a ride to Tombstone on J. D. Kinnear's line."[1] Despite all this prearranged journalistic enthusiasm, stage travel remained hazardous. Horses suffered as each company tried to outdo the other in travel time. Many animals died even though Kinnear used sixty horses in rotation, for three changes of stock, on his many trips between the two towns. Drivers pushed their teams to the limit, and coaches sometimes toppled over. Frequent travelers, buying space with injury or death a daily possibility, quietly prayed for the advance of the Southern Pacific Railroad.

Suffering queasy stomachs and shaken nerves, most stage passengers during those early months reached Tombstone around midnight. Those without accommodations guaranteed in advance faced the prospect of sleeping on cold ground. Luckier ones bought or begged space between banquet-sized restaurant tables, one traveler remarking with dry humor, "a dining room is very convenient when breakfast is considered; and the landlords are very generous in that respect."[2]

Wide-eyed newcomers discovered a tiny town with ramshackle written all over it, yet one offering predictable amusements. The liquor business flourished, helped by a disproportionate number of young bachelors. Consumption was not confined to beer and shots of red-eye. Wine, champagne, and a wide variety of mixed drinks also delighted thirsty patrons. Accomplished bartenders, respectfully called "mixologists," won loyal followings and found themselves in demand among rival saloon owners. With the price of draft beer eventually stabilizing at a nickel a glass and stronger drinks selling at two for a quarter, business boomed throughout Tombstone's heyday.

Trouble from liquor took many forms, not restricted to fights, hangovers, or liver damage. For one unfortunate fellow branding empty whiskey barrels "an explosion took place, bursting a hole through the head of the barrel and driving the branding tongs with such force into the man's face as to break his nose, fracture his jaw and bruise his face badly. The injured man lay insensible for several hours. The mystery is

[1] *Arizona Weekly Star*, January 30, 1879.

[2] *Weekly Arizona Citizen*, August 15, 1879.

what caused the explosion. If it was the remains of nitroglycerine or lighting whisky, what became of the men who drank it."[3]

None of this discouraged anyone. Later that year it could be said with pride: "Parties visiting Tombstone now will have no occasion to thirst for the oil of life. Mr. H. Horton shipped to-day to Messrs. Vogan & Flynn 6300 pounds of liquors, the greater part being the now favorite brands of whiskey. . . ."[4] These included Amazon Bourbon, O. K. Cutter, Miller's Extra, Argonaut, Log Cabin No. 1, Old Kentucky, and Chicken Cock & Rye. Hennessey and Martell brandy, gin, rum, and other drinks were also on hand, ordered across the polished bars of ornate saloons or from hole-in-the-wall joints called sampling rooms.

The district's chief activity, however, was still mining, not loitering in saloons. One sinister story swirling over the hills and along the river in early 1879 mistakenly claimed that old Spanish and Mexican land grants, awarding their owners all mineral rights, blanketed the district. Congress, worried prospectors now insisted, had recognized these ancient titles, which had been purchased by predatory capitalists, thought to be Joseph C. Palmer and his cohorts in the Tombstone Townsite Company. Anxious miners "were prepared to 'make Rome howl.'"[5] Fortunately nothing came of it, and residents quietly settled back into normal routine as quickly as aroused.

As work pushed forward relics were found from earlier days. Ed Schieffelin picked up a stone hammer on open ground. Sweating workmen at the Corbin mill site unearthed a Spanish spur with its distinctive oversized rowel. Others discovered the barrel of an ancient blunderbuss, all viewed as further evidence of men moving through the area in less crowded times.

Not all the excitement concerned mines, mill sites, and archaeological treasures. In early April rancher and part-time miner Frank Patterson rode in from Charleston to sample the new town's distractions. While touring saloons in a drunken stupor, he quarreled with local stockhand Charles Snow. Both were high-strung men in their mid-twenties. This fact, mixed with alcohol, sparked the trouble. Patterson punctuated his

[3] *Arizona Daily Star*, August 21, 1879.
[4] *Weekly Arizona Citizen*, December 27, 1879.
[5] *Arizona Weekly Star*, March 27, 1879.

argument with gunfire, superficially wounding his fellow Texan. Justice of the Peace Thomas Bidwell set bail at $500. Patterson avoided any serious difficulties but his name would surface again, embroiled in later controversies involving the Earp brothers.

That same month Albert Smith killed John D. Boardman. Again Judge Bidwell heard the case. Boardman, an early volunteer with Ed Schieffelin in the Upper San Pedro Rangers, had repeatedly threatened Smith for reasons long since forgotten. Watching him approach, Smith stepped outside his house and leaned a rifle against the wall. He ordered Boardman to stop but was answered by a bullet that missed its mark. Smith grabbed his gun and fired. Boardman turned and ran about twenty steps before tumbling into the dirt. He died within minutes. Bidwell ruled the killing justified and discharged the prisoner.

Nervous outsiders began wondering if the district was even safe to visit. R. M. Hargrove, Charleston's newly appointed deputy sheriff, arrested one American and two Mexicans involved in a shooting affray, stopping the former with a warning shot. He then arrested another man in an unrelated shooting incident. The harried deputy must have questioned his decision to enter law enforcement as he now held four violent offenders in a place with no jail. Officers "kept prisoners in their own houses or guard them in other inconvenient and insecure ways," reported the press, before transportation could be arranged to Tucson.[6] Citizens begged the board of supervisors for a jail. They declined to consider the request, pointing out that no town in the district had yet been incorporated.

Toward the end of May tempers rose twice to the point of gunfire. No one was killed, but it caused one disgruntled observer to remark: "The effect upon our peaceable element is somewhat irritating, and causes them to speak quite savagely about hemp, limbs, etc., and in a manner that would indicate an intention to suspend such proceedings rather summarily. In fact, it is a question in my mind whether they would allow the perpetrators the right of counsel, and whether it would not result in the general good."[7] Still, despite the rawness of the place, rough nature of the work, and trouble from young hotheads, there had only been three deaths reported in the district—two shootings and one from natural causes.

[6] *Daily Arizona Citizen*, April 26, 1879.

[7] *Weekly Arizona Citizen*, May 30, 1879.

As the town struggled into existence the Tombstone Mining District paused to catch its breath: "The first hot fever of prospecting has passed . . . and the consequent period of dullness is being experienced."[8] Yet no one's enthusiasm lagged as workmen stockpiled tons of ore from such well-known properties as the Lucky Cuss, Tough Nut, and Contention. More than a ton from the Grand Dipper—including amongst its shareholders the county sheriff—was hauled to Tucson for processing at Warner's newly opened two-stamp mill. Owners of the Three Brothers, another mine showing early promise, shipped their richest ore to San Francisco. Low cash reserves remained a problem for many potential mining tycoons, a condition further whetting the appetite of persistent speculators.

It all proved to a smaller number of skeptics that Tombstone was for real; the deposits ran deep, not simply surface teasers called blanket lodes. With most of the ground already claimed, prospectors and would-be prospectors (mostly laborers unable to find work in a glutted market) drifted into the Dragoon and Huachuca mountains. Some gold was discovered, but most of these men returned to Tombstone and took regular jobs after more heavy machinery arrived.

Reduction mills remained the key to prosperity, for as one Tucson journalist noted: "Ore on the dump, however rich it may be, cannot pass as currency from hand to hand. . . ." The Gird and Corbin mills were designed only to work ore for the Tombstone Mill and Mining Company. All other major producers, such as the Contention and Grand Central, likewise planned just for their own use. Smaller companies required independent custom mills willing to contract out. Rumors of the Golden Star Mining Company's machinery being hauled in from Cave Creek delighted cash-poor owners. While they waited nervously, the press concluded, "Many of the best looking prospects . . . have no more than the assessment work done on them, and though they may be worth thousands in the near future, the pauper millionaires who own them are . . . for the time 'rustling for grub.'"[9]

In May 1879 Captain Whitside, Lieutenant Winchester, and F. L. Austin, the Huachuca trader who had so cheaply snagged Oliver Boyer's

[8] *Arizona Weekly Star*, May 22, 1879.

[9] Ibid., March 27, 1879.

interest, negotiated their shares in the Grand Central for more than $100,000 to unseen eastern investors including Chicago businessman Nathaniel K. Fairbank. They congratulated each other, but with the Grand Central eventually recognized as the richest mine in the district, many questioned their judgment in selling out so soon.

Tom Jeffords, Tucson physician John C. Handy, and other partners then sold the promising Head Center and Yellow Jacket to San Francisco commission merchants William Moody and Adam Farish for the seemingly ridiculous sum of $30,000. Many shook their heads in disbelief, while at the Contention owners turned down a million-dollar offer. Wild estimates claimed more than that was already piled on company dumps.

Riches below ground had seduced swarms of pick-and-shovel men. Shafts at all the major companies were down well over a hundred feet. The Lucky Cuss, Contention, and a few other sites employed round-the-clock shifts. So much ore stood at the Tough Nut in early 1879, that with the mill not yet ready, Gird reduced its work force. A thousand tons awaited processing at the Contention. More and more the district shifted from its period of speculation to one of serious production. As work rushed forward at the Gird Mill, machinery for the Corbin site kept rumbling through Tucson under the gaze of appreciative onlookers: "This morning came a dozen more wagons with two immense boilers, smoke-stack, pumps, engine, pans, stamps. . . ."[10] It was an impressive spectacle. After the Gird Mill went on line that June, everyone sensed a new era.

Amidst all the material progress the district's death toll rose a few notches that summer. Near Charleston, Dan Keho and Dennis Consadine, two drunken prospectors who worked a claim in the Huachucas, quarreled after leaving town. Keho staggered back bragging about a killing. Having heard their share of boastful chatter, the saloon crowd dismissed it all as whiskey talk. That is until a man's body was discovered with its face crushed and a bullet hole through the heart, fired close enough to burn his clothes. Justice of the Peace James Burnett held a makeshift inquest and ordered the prisoner jailed at Tucson to await a grand jury. Keho was eventually sentenced to ten years.

[10]*Daily Arizona Citizen*, May 2, 1879.

Some of the major sites associated with the Tombstone Mill and Mining Co., located just beyond the gully south of town. The Grand Central can be seen to the left, in the distance against the skyline. Below that point is the Contention.
Courtesy, Arizona Historical Society.

They buried Consadine on the last day of June 1879. Six days later a double killing rocked Old Tombstone. A man named Quinn argued with Lucky Cuss superintendent Jeremiah McCormick over the affections of a prostitute. The twenty-seven-year-old McCormick settled the question with his fists in front of Danner and Owens's saloon. A bleeding Quinn complained to his friend John Hicks, who left for the new town to awaken his brother. After more whiskey fortified their courage, both cautiously approached the superintendent's house. Five pistol shots from McCormick and a man named Jackson settled the argument. John Hicks fell dead. His brother, one bullet striking near the eye and tearing through his left cheek, died later.

Deputy Sheriff Newton J. Babcock arrested the shooters. Witnesses claimed bystanders had disarmed one of the brothers before any shots were fired. An inquest held at Fatty Smith's saloon swept the matter aside, concluding John Hicks "came to his death by a pistol ball shot by the hand of an unknown person." The system thus protected McCormick, a man of some importance to the Tombstone Mill and Mining Company. The press simply concluded, "This sad affair has cast a gloom of sorrow over the friends of both, and will no doubt stand as a warning against bad women and bad whisky."[11]

With court cases crowding his day, Thomas Bidwell resigned to accompany Frank Corbin on a trip to San Francisco. He recommended William Harwood as his replacement. Supporting that choice, forty-nine citizens signed a petition on July 3. The next day, however, a slightly larger number organized behind Mike Gray. Accepting Bidwell's resignation, the board of supervisors appointed the controversial townsite functionary.[12] Either he or Harwood would have assured the court's continued presence at the new town, further annoying Colonel Lewis and other residents of Richmond. Gray relocated his wife and family from California, while building a small hotel and hash house to make a few extra bucks helping ease the burden of newcomers.

Edward Travers Cox, the Indiana mining expert who had visited southern Arizona in 1878, joining Elbert Corbin and the other investors, returned in July for another look. Traveling with Henry C. Hooker of the Sierra Bonita Ranch, Cox first examined claims at Dos Cabezas and the Dragoons before pushing on to Tombstone:

> It was nearly one year since my first visit to this district. Then there were only a few miners' camps to be seen, made of ocotillo stalks, and a small assay office made of plank owned by Richard Gird and the pioneer locators, Ed. and Al. Schieffelin. Now there are three small villages in sight and the slopes are covered with miners' camps. The new town of Tombstone is a flourishing place, contains quite a number of buildings, mostly adobe, a hotel, a number of restaurants, saloons and a large store, owned by Gen. Allen.[13]

Everyone agreed the future looked bright, filled with progress and good times. Even farmers, many of them Chinese—distrusted and isolated along the San Pedro—found ready markets for their produce

---

[11]Ibid., July 9, 1879.

[12]Resignation of Thomas J. Bidwell, June 30, 1879; Petition for William A. Harwood, July 3, 1879; and Petition for Mike Gray, July 4, 1879. Records of the Pima County Board of Supervisors. For Gray's appointment as justice of the peace, see "Board of Supervisors," *Arizona Daily Star*, July 8, 1879.

[13]E. T. Cox to the editor of the *Los Angeles Commercial*, August 2, 1879, as quoted in *Daily Arizona Citizen*, August 16, 1879.

among those hopeful souls pouring into the district. Field hands earning $2 a day and board tended a wide variety of fruits, vegetables, and grains. Potatoes and onions sold for ten cents a pound and corn at half that price to people less interested in diet than the glitter of bullion.

That summer and fall the three major producers hummed with activity both above and below ground. Swarms of men struggled to build elaborate hoisting works, operated by huge upright steam engines hauled in from California and assembled on site. All this impressive mechanical apparatus increased production without expanding the work force. Other mines kept production and profits high by reducing payrolls. This not only satisfied distant stockholders, but all the new machinery swelled a sense of pride buried deep within nineteenth-century values: "the steam whistle [on the Contention hoisting works] will in a day or two, for the first time, send its shrill notes over the district, as at once the final, everlasting requiem of the Apache, dearth, and death; and the clarion harbinger of civilization, life, and prosperity."[14]

Colonel Lewis watched helplessly as his dreams for Richmond slowly faded away. Tombstone was clearly winning the race as the district's principal townsite. Earlier contenders had long since dropped out. Residents of Watervale moved en masse, including A. J. Mitchell, who piled his surveying tools and assay equipment into a tent beside a lot on which he planned building a house. Many residents of Charleston shifted their allegiance, raising the clamor for legally secured real estate titles. With every detail enviously noted by outsiders, construction of new stores, hotels, and residences continued, despite persistent shortages of lumber and adobes. Palmer, Clark, and Gray's gamble was paying off. Lewis could do little more than fume.

One short September news item reflected, in its own unique fashion, the town's growing financial stability: "An enterprising sportsman has just exported a complete outfit of 'Keno' from Tucson to Tombstone. This indicates that money is beginning to circulate out there, and that checks and silver bricks are no longer the only currency of that district."[15]

Stage service also expanded. William Ohnesorgen and J. D. Kinnear's former agent, Henry C. Walker, started the Tucson and Tombstone Stage

[14] *Arizona Daily Star*, August 23, 1879.

[15] *Daily Arizona Citizen*, September 1, 1879.

Line. They announced three trips a week, a $7 fare, and four-horse Concord coaches. In time, both the new company and Kinnear added some six-horse teams. To attract attention they painted their coaches in decorative fashion and even named some. Ohnesorgen and Walker called one of theirs the "Tough Nut," while Kinnear chose "Old Reliable," and for another the "Grand Central."

Ohnesorgen and Walker papered the trail to Tucson with four hundred posters advertising their service. All but a few were torn down or defaced. The owners offered a $10 reward for the conviction of those responsible. Their agent, Marshall Williams, later reported a more serious prank. Someone had removed axle nuts from one wheel. Had it gone unnoticed, the incident could have led to serious injury or even death. In another episode a team was poisoned before its Tucson departure. Authorities failed to identify and arrest the culprits. John Kinnear ignored accusing eyes as the mad rush for transportation dollars continued without pause.

For a time Kinnear, who had reduced his three trips a week to twelve hours, dropped his fare to $4, and travelers rejoiced. Later the rate stabilized at $7—still a sizable sum for masons and carpenters earning $6 a day, miners averaging $4 for a ten-hour shift, and laborers, some of whom earned no more than a plate of food and fifty cents a day. By year's end it was noted at Tucson: "Two lines of four and six horse stages go out six times a week, loaded down with passengers and express freight, and the road is lined with private conveyances. Women and children are numerous among those going."[16] The two companies had agreed to the alternate schedule in order to give Tombstone more reliable service. Kinnear soon scrapped the plan and began daily runs himself.

Freighters had more to worry about than simply long rides and uncomfortable roadbeds. In August 1879 they welcomed news of Ohnesorgen's private toll bridge opening near his old overland stage stop on the upper crossing of the San Pedro. Although north of the main road, the bridge easily handled twenty-ton wagons and proved no inconvenience to teamsters fearful of the river's soft bottom and frequent high-water delays.

Near the river routes used by freighting companies and stage lines,

[16]*Arizona Weekly Star*, August 7, 1879.

gangs of workers prepared the Contention mill site. In the process they spawned a third major town. That September John McDermott and D. T. Smith surveyed the boundaries of Contention City on sloping ground east of the river and north of the mill. Town lots sold fast enough that within a week there was "one store, one hotel and restaurant, one saloon, one meat market, one Chinese laundry, and some dozen or more shanties which are buildings in embryo. . . ."[17]

Meanwhile caravans from freighters Buckalew and Ochoa passed through Tucson with loads for the Contention easily topping a hundred thousand pounds, some of it the largest milling machinery yet seen in the territory. At the same time a Corliss engine, one of the technological wonders of the nineteenth century, was being built in Rhode Island for the new mill site.[18] Owners hoped it could be installed with the rest of the equipment by early 1880. Six mule trains waited on site, with two more ready to move south from the Southern Pacific's stalled railhead at Casa Grande. Owners needed the extra items since they planned a wet milling process but wanted to be able to run less productive dry crushings if water shortages made it necessary.

To see his handiwork at the Gird, Corbin, and Contention mills firsthand, James Spiers of the Fulton Iron Works traveled from San Francisco to Charleston with Frank Corbin, T. J. Bidwell and Gird's sister. For the seventy-mile run between Casa Grande and Tucson they were stuffed into a coach carrying ten other passengers, including J. D. Kinnear and Wells Fargo detective Robert H. Paul. Everyone endured a punishing ride bathed in sweat and dust. Once free to roam the San Pedro, Spiers quietly solicited new business.

Eventually superintendents of the Head Center and Sunset constructed mills north of Contention, with their own small settlement called Bullionville. Surrounded by mill sites and astride the main corridor into the district, Contention City became an important commercial hub. The Grand Central set up its facilities two miles south. Five miles beyond stood the Boston-Empire Mill of the Boston and Arizona Smelting and Reduction Company. A nearby collection of saloons and retailers, shacks and boarding houses was christened Emery City.

---

[17]*Arizona Daily Star*, September 23, 1879.

[18]For more details concerning this rather remarkable piece of machinery, see Ingram, *Centennial Exposition*, 696-699.

In Tombstone business boomed and lawyers found more work than they could handle, prompting one man to reflect, "They have to carry their offices around in their pockets or hats."[19] Along with real estate troubles and mining disputes, it seemed as if ordinary citizens were determined to keep the lawyers busy. Miner Charles Foley and a chop-house proprietor exchanged harsh words and gunshots, one of which hit Foley in the shoulder. Such criminal cases were seldom as lucrative as mining litigation, but hungry lawyers scooped up whatever they could find.

The excitement of silver proved so strong that soldiers began deserting surrounding military camps to get in on the action. James Abbott and two others slipped out of Fort Bowie with seven army horses, including the personal mount of Captain Daniel Madden. Within days Abbott was arrested in Tombstone. He had unwisely introduced himself first as James Grier and then as Charles Burns; using two names in such a small place was a clear giveaway. The unfortunate Mr. Abbott ended his dreams of riches in an army stockade.

Tombstone showed little interest in military rustlers. Confronting questions of law and order, the town had trouble enough trying to deal with fist fights, robberies, stabbings, and occasional gunfire. Ignoring earlier rebuffs from the board of supervisors, townspeople took it upon themselves to construct a small city jail. And none too soon. Fred Bartholomew attacked W. C. Cullen with a shotgun. Cullen stood his ground and blasted away with his six-shooter. That night's revelry ended with no injuries—all the shots went wild—but witnesses congratulated themselves on their decision to build a jail.

Amid all the gunsmoke real estate kept selling, even without assurances of a townsite patent. Buyers, acutely aware of the dangers of investing under those circumstances, prepared to hold on with force of arms if necessary. During that fall of 1879, and for many months thereafter, it was not uncommon to see heavily armed men camping on lots in the prime business district. Quarrels erupted between those contesting the same plot of ground. Considering all the tension it remains a miracle that no blood flowed along those crowded streets. Legal questions dominated the atmosphere as Tombstone, which really had no for-

[19] *Arizona Daily Star*, September 26, 1879.

mal right to exist, grew amid ever-rising and highly speculative real estate prices.

Before long a fresh problem arose: "A conflict has also grown up between the town authorities, and Mr. Field, owner of the Gilded Age mine, which it is claimed was located about four months prior to any town association."[20] Early on, six mines intersected the townsite, including the Mountain Maid, a goodly portion of the Vizina and Way Up, the northern third of the Empire, and a small portion of the Good Enough. Except for its southeastern tip, the entire Gilded Age cut a wide swath into the bottom center of town. Edward Field and W. S. Sanford & Company of New York had openly claimed the property since December 1878. Complicating matters, they had yet to discover anything of value. Instead, one sarcastic report noted, "The Gilded Age is taking out large quantities of lime cement."[21]

With Field coming up empty, and before he could start cross-cutting a new ledge, Mike Gray of the Townsite Company saw his chance and swore out a complaint for criminal trespass. Gray had found an unexpected ally in section twenty-three of the 1872 federal mining law: "no location of a mining claim shall be made until the discovery of the vein or lode within the limits of the claim located."[22] The irate owner of the Gilded Age was transferred to Charleston for trial. Authorities there knew a hot potato when they saw one and refused to act, claiming the case belonged in the district court.

Field, a tough customer pushing fifty, refused to be thus intimidated and returned to work. He stubbornly ignored all community pretensions, but not before filing suit against Gray and the Townsite Company. This highly charged case, involving claims and counterclaims questioning the status of mining locations and a townsite occupying the same ground, meandered through the courts for years. The Townsite Company and local lot owners had not heard the last of the man from Massachusetts. It proved Richard Gird had been right all along when he kicked everybody out of Old Tombstone.

Despite all the tension there were signs of progress. In early October editor and publisher Artemus Emmett Fay started Tombstone's first

[20] *Weekly Arizona Citizen*, September 19, 1879.

[21] *Arizona Daily Star*, September 17, 1879.

[22] Leshy, *The Mining Law*, 372.

newspaper. Before turning twenty Fay worked as co-editor of a paper in his native New York. Eventually he moved south into Pennsylvania, his first wife's home. In the booming oil center of Titusville and surrounding Crawford County, he gained more experience with the *Herald* and the *Oil City Times,* before becoming city editor of the *Titusville Daily Courier.* In 1875–1876, as clerk of the Democrat-controlled Pennsylvania House of Representatives, Fay observed countless examples of behind-the-scenes political maneuvering. Watching it all unfold served as a private education for the young journalist.

In late spring 1877, surprised by rumors that the *Daily Courier* was for sale, Fay decided to try his luck in the Southwest. At Tucson he joined one-time Grand Central part-owner Louis C. Hughes as co-editor and publisher of the *Arizona Star.* That August, Hughes announced his retirement from journalism to practice law full time. Fay ran the paper himself until December 1878 when Hughes reluctantly returned as a partner. Weeks earlier Fay had been elected one of five men to represent Pima County in the Tenth Legislative Assembly. That body also included such later Tombstone personalities as John H. Behan, attorney Thomas Fitch, and southern Arizona political figure Samuel Purdy—an old acquaintance of Mike Gray's.[23]

Fay sold his interest to Hughes in early March. He helped edit the *Star* during this latest transition, even while working on plans—encouraged by the trader Carlos Tully—of starting his own newspaper at Tombstone. Finally, on October 2, 1879, Fay peeled the first issue of the *Weekly Nugget* off an old Washington hand press. In doing so he carried on a colorful southwestern tradition. That particular press had been used to publish Arizona's first newspaper in 1859, the oddly spelled *Weekly Arizonian.* Shipped round South America the year before and deposited on the waterfront at Guaymas, Mexico, the press traveled north to Tubac lashed unceremoniously to an oxcart.[24] Artemus Fay now made good use of this vintage apparatus to chronicle the exploits of Arizona's legendary boom camp.

The town's pioneer editor originally planned to call his creation the *Silver Nugget,* but changed the masthead just before press time. The whole

[23]Wagoner, *Arizona Territory,* 90, 514.

[24]Hattich, "Highlights on Arizona's First Printing Press," 68, 70.

operation crowded a tent on the north side of Fremont west of Fifth. "We print an edition of fifteen hundred copies," Fay cautiously informed his readers, "that friends may mail a paper to their acquaintances in the states and elsewhere." With copies priced rather high at twenty-five cents, the new editor tried defusing complaints with the comparison, "only the same as a drink or a cigar."

There was much to write about. Besides trumpeting the town's bright future, a traditional strategy, Fay refused to gloss over its blemishes. In that first issue he reported: "We would very much dislike to see another such disgrace upon our camp as that which occurred on Sunday night last—the promiscuous firing of revolvers by lewd women, and men who seem to spend most of their time with them. . . . At one time it was dangerous to pass along the streets."

In his opening salutation, Fay admitted somewhat humbly, "We make our bow this morning heels over head in debt." He then outlined his purpose:

> It shall be our aim to make the *Nugget* a purely local journal. . . . Independent in all things, but neutral in nothing that interests the public, we shall fearlessly expose fraud, bogus mining speculations, and all schemes which might result in gain to individuals, but disastrously to the district. . . . During our journalistic career we expect to make enemies as well as friends, but will unflinchingly perform our duty to the public . . . let the chips fall where they may.

Having a newspaper was important, of course, but Tombstone welcomed all signs of progress whether business or social. On the night of October 26, 1879, Lorra Fuller, a blacksmith's wife, gave birth to the first child born in the camp, a daughter named Inez. In typical fashion, Fay reported, everyone celebrated the happy event by getting drunk.

Another correspondent scribbled this picturesque view:

> It is now only about two months since [Tombstone] began to assume the proportions of a town—it is now a city, of the western style. Of course, there is much of canvas and rough, unplanned boards about it, but still the enterprise, spirit and thrift is there shown and is daily accumulating like a snow-ball rolling down hill. Daily, and almost nightly, Sunday and all, the sound of the axe, hammer and saw is heard, and teams are pouring in with all kinds of goods. . . . Hotels, boarding houses, lodging houses, and shops of all branches of mechanical industry are already here. The professions are well

> represented by several M.D.s and half a dozen or more lawyers, with more coming. The town is about to incorporate, when police courts . . . and all the paraphernalia, grandeur and eclat of a full-fledged city will be born to life.[25]

Hoping to forestall problems such as those arising from the Gilded Age case, general lapses in law and order, and the dangerous practice of lot jumping, a group of forty-one citizens had signed a petition in August asking the county supervisors to incorporate their town.[26] The board ordered the sheriff to conduct a census. In late September Charlie Shibell enumerated a community of 474.[27] Weeks later estimates increased that figure three-fold, and within eight months it officially topped two thousand. While Shibell counted heads in Tombstone, calculations for the entire district conservatively suggested "the population is now estimated at 1,500, where four months ago scarcely 100 people were to be found."[28]

In early October the supervisors finally agreed with the locals, and in three successive issues of the *Weekly Nugget* there appeared a "Notice of Incorporation," setting a hearing date of November 1 to consider the petition. At that meeting the board approved the application and scheduled municipal elections to take place in twenty-four days. They also rejected a plea to change the town's name; some found the sepulchral reference offensive, "but the board thought inasmuch as the district, post-office, stage lines, and everything in and about the city on the hill was called Tombstone, they would not change the name [given] it by the discoverer of the district."[29]

The decision to incorporate came none too soon. The town was growing rapidly and lot jumping remained common: "This evening about dusk Charley Calhoun took it into his head that some one was going to lay claim to one of his lots on Allen street, and went out and took position with his 'shooter' on the lot, and after a rather emphatic address and firing two shots was taken into custody by the deputy sheriff. A little too much stimulants was the only trouble."[30]

---

[25] *Arizona Daily Star,* October 16, 1879.

[26] "Petition to Incorporate Town of Tombstone," August 25, 1879; filed November 1, 1879 by William S. Oury, Clerk of the Board of Supervisors, Miscellaneous Records, Pima County Board of Supervisors.

[27] "1879 Census of the Town of Tombstone," together with Shibell's letters of transmittal, October 9, 20, 1879, Miscellaneous Records, Pima County Board of Supervisors.

[28] "Arizona Mining Districts," *Mining and Scientific Press* 39 (October 18, 1879).

[29] *Arizona Daily Star,* November 2, 1879.

[30] *Arizona Weekly Star,* September 17, 1879.

An unsettling sense of danger, violence, and moral decay worried many. One visitor with wide experience on the Comstock remembered Tombstone in 1879 as "a lawless, terrifying town."[31] A later world traveler, by contrast, recalled the place as "a wild and romantic camp."[32]

County officials and ordinary citizens alike wondered about Tombstone:

> The incorporation is very timely. . . . Saloons of the lowest class and dance houses where men and women of the vilest sort herd together and make the night hideous are becoming well settled institutions, and unless suppressed or strictly regulated they will breed crime and death. Shooting has been a frequent thing, though happily the marksmanship has been very bad and but little damage has [been] done. A vigilance committee has been talked of, of course, but it is probable that the better sense of the community will try a new course.[33]

Not everyone favored incorporation. Always hoping for more immediate gain, Tom Bidwell complained to Anson Safford: "The town is still slowly improving and growing, and we should have made some money out of it had not Gray been a fool. . . . Gray and others have been d—d fools enough to incorporate the town and [I am] inclined to think the people will find it a very expensive luxury in the end."[34] But violence strengthened the arguments for incorporation. Locals would then assume the responsibility for controlling excesses and policing the unruly town on Goose Flats. "Let Tombstone incorporate," the press pleaded on the day the supervisors met, "that the town may become as famous for its virtues as the district is for its mines."[35]

On a morning in early November Orrin House passed the old Brunckow adobe on his way to the Evening Star claim. In a ravine to the north he stumbled upon the body of J. Van Houton, an older man at age fifty-five and the former owner of the Eureka Saloon at Tucson. The ex-saloonman, who had abandoned that trade for Tombstone mining property, had been badly beaten with stones about the head. Alive but insensible, Van Houton died that evening. He left behind, it was later disclosed, a wife and family destitute in British Columbia. This particularly brutal murder caused great excitement throughout the district.

[31]Upshur, *As I Recall Them,* 149. [32]Burnham, *Scouting On Two Continents,* 32.

[33]*Daily Arizona Citizen,* November 3, 1879.

[34]T. J. Bidwell to A. P. K. Safford, December 5, 1879, *Garner v Gird,* 198.

[35]*Arizona Daily Star,* November 1, 1879.

Investigations uncovered a desire by several parties, including the deceased, to relocate the Evening Star after the required assessment work had not been done. This, it was thought, provided a motive for the slaying. As details unfolded names began to surface. As it turned out, Charleston livery stable operator Frank Stilwell and his friend James Cassidy also had their eyes on the Evening Star. On November 7 both men camped there awaiting the lapse of the original filing so they could jump the claim after midnight.

They found others there for the same purpose, including the "well known and irrepressible genius" Frank Ames, called "The Tinker" by skeptical locals. Ames and his friends set up a tent to wait out the clock. An hour after midnight Cassidy finished his bottle, claimed possession, and gleefully awoke his friend with the news. Pleased at outmaneuvering their rivals, he and Stilwell returned at daybreak the four miles to Charleston for breakfast. In their absence Ames jumped the site, conveniently convincing himself Cassidy had failed to follow proper procedures. When Cassidy and Stilwell returned they found The Tinker hard at work. Van Houton then showed up to relocate the mine, explaining the expiration date was the eighth not the seventh. Several others, estimated from ten to a dozen men, had the same idea.

Those camping on the Evening Star that second night included one of the Gattrell brothers from Charleston (an original owner of the disputed property), former Texas Ranger Peter Spencer, Frank Patterson (the same man who had shot Charles Snow seven months before), and a man identified only as Holland.

In the darkness gunshots were exchanged between Stilwell, Cassidy, and Frank Ames over their earlier quarrel. No one was injured, but Van Houton decided to move his camp nearer the Brunckow. There, before morning, he was attacked. Authorities arrested Stilwell, Cassidy, Patterson, Spencer, Ames, and Holland before trying to sort out details. Emotions ran so high against the prisoners, following a hastily organized coroner's inquest at Charleston, that Justice of the Peace Burnett ordered Deputy Sheriffs Buttner and Babcock to transfer the shackled men to Mike Gray's Tombstone justice court. After a two-day hearing all were released except Stilwell and Cassidy. Attorney Webster Street represented the territory, with Wells Spicer defending the two remaining prisoners. Heavily armed guards escorted both men to the county jail at Tucson. There F. L. B. Goodwin took over as defense counsel.

Released on bond, they would eventually be discharged on the plea that prosecutors lacked sufficient evidence for a conviction. More than two years later someone signing themselves "S." absolved Frank Stilwell of blame, claiming to have heard the deathbed confession from the actual murderer. The mysterious Mr. "S." (Webster Street, Wells Spicer, or Peter Spencer?) then told his story to Mike Gray, who "expressed surprise that no suspicion of the real perpetrator should have occurred to him before."[36] The identity of this individual, unfortunately, was never publicly revealed.

Meanwhile, even before securing his freedom, Stilwell joined with E. T. Hardy and started a short-lived partnership to operate a stage line between Charleston and Patagonia. Frank was not about to allow a simple murder indictment to get in the way of scrounging for cash. But the names of Frank Stilwell, Peter Spencer, Frank Patterson, and even Wells Spicer would be heard again, associated in one way or another with troubles involving the Earp brothers. The Van Houton murder, more than most other events, convinced a growing number of skeptical citizens that what they needed was some law and order.

After counting the ballots on November 24, thirty-one-year-old William Arthur Harwood became Tombstone's first mayor. He easily defeated Dr. Henry M. Matthews by a margin of 116 to 59. Harwood had been a crony of both Dick Gird and Tom Bidwell at Signal. He and Bidwell briefly ran a bookstore there, and for seven months he served as constable for his friend's justice court. Both men came to the Tombstone District in late September 1878, and as Harwood recalled, "we lived in a little cactus cabin with a canvas cover. It was about twelve or fifteen feet from this cabin to where the boys had the assay office."[37] Later he became a deputy sheriff as well as working as Gird's lumber agent. Harwood now moved up into the mayor's chair.

Along with His Honor, the four newly elected councilmen were Alder M. Randall with 168 votes, Andrew J. Cadwell with 147, James M. Vizina with 142, and Albert Schieffelin with 129 votes. Losers

---

[36]"S." to the editor, reprinted from the *Tombstone Daily Nugget* by the *Arizona Daily Star*, March 25, 1882.

[37]Deposition of William A. Harwood, Tombstone, August 10, 1882, *Garner v Gird*, 160.

William A. Harwood, Gird's lumber agent and Tombstone's first mayor, owned the two lots on Fremont Street west of C. S. Fly's boarding house and photo gallery. *Courtesy, Arizona State Library.*

included such subsequently important Tombstone personalities as Artemus Fay with 84 votes, and Sylvester B. Comstock with a single vote—presumably his own. Voters also decided the important position of town marshal, electing thirty-year-old Fred White over his younger rival, Daniel S. Miller, by a vote of 107 to 62. Totals were low because many of those otherwise eligible to cast ballots had simply not bothered to register.

At least a city government of some sort was in place, although terms would be short. Tombstone's second city election, coinciding with normal county schedules, took place within six weeks. Thus, in early January 1880, Mayor Harwood stepped aside for Councilman Alder M. Randall, a former postmaster at Mineral Park, now a thirty-three-year-old co-proprietor of a small Allen Street saloon and billiard parlor. Randall defeated Benjamin A. Fickas, 104 to 99. The Townsite Company smiled with delight.

Daniel Miller tried again for the city marshal's office but was beaten a second time by Fred White, 113 to 101. The board of supervisors then appointed Miller a constable. Smith Gray and Sylvester Comstock—who had learned a thing or two in the weeks separating this triumph from his earlier defeat—were both elected councilmen by a ten-vote majority. Andrew J. Cadwell and attorney Harry B. Jones made up the rest of that four-man body, with Mike Gray acting as clerk and recorder.

Gray, who aside from his duties at the justice court ran a popular lodging house and restaurant that catered to as many as seventy-five people a day, actually enjoyed a better reputation—although mixed—than did other members of the Townsite Company. One report during that fall of 1879 characterized him as "the commandant of the town," and "the center of attention."[38] To help ease Gray's burdensome schedule, the board of supervisors granted a new petition and on January 5 reappointed Thomas J. Bidwell as a second justice of the peace.[39]

The establishment of a working city government did much to assure Tombstone's legitimacy in the eyes of the United States Land Office. Under provisions of federal law a patent could be issued to the mayor of an incorporated community and held in trust for the people by their duly elected representative—in this case, Mayor Alder M. Randall. Exactly how this arrangement unfolded was left to Arizona authorities. In theory it should have worked well, but with Tombstone nothing could be taken for granted.

---

[38] *Arizona Daily Star,* September 17, 1879.

[39] "Proceedings of the Board of Supervisors," *Arizona Daily Star,* January 8, 1880.

Tombstone in early 1880, soon after its second round of municipal elections, yet before the first great building boom. The sleepy, ramshackle appearance belies a raucous atmosphere that shocked its more conservative visitors.
Courtesy, Arizona Historical Society.

Chapter 6

More Growing Pains

Mark Twain once joked that civilization's march began with whiskey. Only then, he explained, did missionaries, immigrants and traders, real estate speculators, and the lawyer tribe follow along, dragging with them newspapers, politics, jails, churches, and other social refinements. The always-observant Mr. Twain had again stumbled on to something. One surprised revenue agent, visiting tiny Tombstone in late 1879, found seventeen licensed saloons. Earlier someone remarked, "as is usually the case in mining camps the saloons are numerous and gorgeous, while the churches are small, in fact microscopically small."[1]

Tombstone's sinful reputation easily convinced the self-righteous that it desperately needed some Bible-pounding. Itinerant evangelicals jumped at the challenge (a few left in disgust within hours), spreading their high-pitched message from street corners. In time, however, the Methodists, Catholics, Presbyterians, and Episcopalians all built permanent churches. The small Jewish community met in private homes or occasionally used the Masonic Hall or justice court.

Reverend G. H. Adams, Methodist superintendent of missions for Arizona, preached the first formal indoor service in early October. Men crowded in—only three women attended—forcing less aggressive worshipers to stand outside the storefront volunteered for the occasion. Only two hymnals surfaced amongst personal belongings, but everyone raised their voice on cue. Adams delivered his forty-five-minute exhortation by the light of one flickering candle and small lamp. Editor Fay, then a Methodist himself, praised the performance. Looking to improve its image, the Townsite Company donated property so Adams could build a church for his enthusiastic followers. Mike Gray responded by

[1] *Arizona Daily Star*, August 23, 1879.

giving two lots to the Catholics. Yet six months passed before religion formally organized in Tombstone.

Bankers got a jump on the preachers. At first these services were handled from Tucson. Every two weeks employees of the Pima County Bank stuffed canvas sacks with cash to meet payrolls and other demands. In time they arranged for coin and currency to be deducted from standing accounts with the Pacific Bank of San Francisco and delivered by Wells Fargo.

After bank president P. W. Smith moved to Tombstone in late 1879, taking over John B. Allen's debt-ridden store, he began offering some services. By early 1880 he advertised as a branch, then opened full time as the Agency Pima County Bank, with himself as manager and Heyman Solomon as cashier. By early summer, crowding out its corner of the mercantile store, the bank moved into separate quarters in Smith's building. It handled many valuable accounts, including those of the Grand Central Mine, whose superintendent, E. B. Gage, had been an early supporter of the Tucson house. The Pacific Bank of San Francisco, the Farmers and Merchants Bank of Los Angeles, the First National Bank of Chicago, and the Chemical National Bank of New York were among its major correspondents.

A second Tucson bank, Safford, Hudson & Company, opened a branch in mid-1880 with Anson Safford, John Vosburg, James Toole, and Charles Hudson its chief backers. As cashier they hired Milton B. Clapp, a man with banking experience at San Francisco and a former Brooklyn insurance and real estate agent.[2] The new bank, from which Clapp also ran his own insurance business, found space on Fifth Street at the rear of the Oriental Saloon.

Even with a large bachelor population Tombstone still needed a school. A. E. Fay, Ben Fickas, and E. M. Pomeroy served as trustees. Located behind P. W. Smith's in a one-room shack with mud roof and dirt floor, classes opened to nine students in February 1880. Many of those eligible simply failed to show up. Before year's end, however, attendance rose tenfold, giving rise to the less-than-subtle complaint: "The public school of Tombstone is at present held in a little bird-cage on Allen street, about large enough to accommodate a man and his wife, if

[2] *Brooklyn City and Business Directory*, 1874, 1875; and *Langley's San Francisco Directory*, 1875-1880.

they hadn't been married long. . . . This condition of things is an outrage on the pupils and a burning disgrace to the district."[3] By mid-summer Edward Field of the Gilded Age donated land for a suitable building.

Women found Tombstone particularly restrictive. A caste system was rigidly based on their husbands' occupations. Wives of superintendents seldom mingled with those of common miners, and wives of gamblers and saloonkeepers enjoyed even fewer contacts beyond their own circle. For most women the town offered few amusements. Barred by custom from saloons and gambling halls, one could attend only so many church socials. Despite a small public library intellectual stimulation was hard to find, complicated by the daily drudgery of nineteenth-century home life. Women formed clubs to discuss the latest eastern and California literary periodicals, helping break the cycle of work and boredom. For some there was no escape. Saloonman Tom Corrigan's wife twice attempted suicide. A badly aimed pistol shot to the head forced surgeons to remove her right eye. Whether the bleakness of frontier life or her husband's violent domestic habits drove her to despair was not reported.

With the town now recognized as the commercial and social center of the district, Artemus Fay boasted: "The Tombstone district now bids fair to surpass the Comstock in its balmy days. We are assured by men who have had a great deal of mining experience, that Tombstone stands an excellent chance of being recorded as the most wonderful mining camp ever discovered on the American continent."[4] That proved an exaggeration, but no one denied Tombstone's optimism. And with good reason; only the week before it was said: "There are now 480 places of business and private dwellings erected in this city, and 21 new buildings in process of erection, including several blocks of stores. No town on the Pacific coast is improving with the rapidity of Tombstone."[5]

Most accepted the crude living conditions, either crowded into town or huddled around open camps before building themselves shacks or buying small tents. Miners accepted accommodations no larger than packing crates and happily called them home. George W. Parsons made special mention of his first shelter: "Rough house—simply roof and

[3] *Tombstone Epitaph*, October 20, 1880.

[4] *Weekly Nugget*, March 25, 1880.

[5] Ibid., March 18, 1880.

sides with openings all over through which wind came freely—mile and ¼ from town." Sanitary conditions were poor, and as Parsons described: "Rats and mice made deuce of a racket last night around a fellows [*sic*] head on ground. Rolled over on one in the night and killed him—mashed him deader than a door nail." Cats, understandably, became luxury items. Early on Parsons caught one disagreeable feline "and hurried home anxious to get there before I was clawed or bitten to death. Peace tonight among the rats."[6]

George Parsons arrived in Tombstone on February 17, 1880, at the age of twenty-nine. His first impressions were: "One street of shanties some with canvas roofs. Hard crowd." The next day he saw Ed Schieffelin: "Rough looking customer."[7] But Parsons, whiling away his hours as a junior San Francisco bank clerk, had long dreamed of adventure. Resigning, he came to Arizona with his friend Milton Clapp and is best remembered now for keeping a daily journal describing Tombstone's formative years, even if many of his early entries mirror naïve exuberance over frontier expertise.

Years later, even with memory failing to keep pace with all the details, Parsons tried remembering those wild youthful days. He gloried at seeing himself described, however inaccurately, as being "for seven years . . . one of [Tombstone's] controlling spirits in the interest of law and order, being one of the council of ten when the first vigilance committee was formed, and always in the saddle with the first to drive the Apaches out of the country or assist a beleaguered ranch."[8]

Accepted into Tombstone's mining community, George Parsons developed promising claims not only in the district but also in northern Mexico. Working diligently for local Republicans, he quickly became a partisan of the Earp brothers. For George they always stood out as larger-than-life figures; but then there was much about the Earps he never understood.

While Parsons and other newcomers chased private success, bigger changes were taking place. The Schieffelin brothers decided to sell their 250,000 shares in the Tombstone Mill and Mining Company to a group of close-knit Philadelphians. These men, including an attorney, banker,

---

[6]Parsons, *Private Journal,* February 17, 1880, 92; February 21, 1880, 94; and March 7, 1880, 100.

[7]Ibid., February 17, 18, 1880, 92, 93.

[8]Guinn, *Historical and Biographical Record of Los Angeles,* 834.

mercantile appraiser, and other business types, were headed by Hamilton Disston, a one-time Orion shareholder whose father, Henry, owned the giant Keystone Saw-Works. Also on hand were representatives of the petroleum-rich Hulings family from Oil City, Pennsylvania, as well as a couple highly placed Boston moneymen.[9]

After playing tourist at the San Xavier Mission they met the Schieffelins at Tucson, along with George S. Corbin; T. Henry Asbury, president of a hardware company; and Benjamin F. Hart, a Philadelphia heavy-spring manufacturer. The Disston party traveled with two men they picked up in San Francisco: mining expert William A. Williams and Frank X. Cicott, the coiner from that city's United States Mint. Cicott knew some of the Arizona parties because he lived at the Baldwin Hotel where most of the Tombstone mining men, including Richard Gird and Anson Safford, invariably stayed when visiting the Bay City.

Waldron J. Cheyney attended the Tucson meetings, later admitting:

> I was an interested party . . . although my name did not appear in the contract. Mr. Gird seemed to be present and active all of the time, but he didn't have, as I understood, any personal interest in the sale. My understanding was that the Schieffelins' interest alone was being sold; that Mr. Gird agreed to . . . retain his interest in the stock of the company as part of the inducement held out to the Philadelphia men to buy the Schieffelin interest. . . . Mr. Gird told me that he would not sell his stock at anything like that price. . . .[10]

What was not generally known is that Richard Gird did have an interest, as Albert Schieffelin explained: "I and my brother got $600,000 for our stock. . . . It was understood that Mr. Gird was interested with us in that sale, between ourselves. The public generally did not so understand it. We didn't know that it interested the public any, and we sold all of the stock that stood in our names on the books."[11] Ed later described this arrangement himself: "He was interested with us . . . and I and my

[9]For some background on the Disston family, as well as on Marcus and Willis Hulings, see Scharf and Westcott, *History of Philadelphia*, 2267-2268; and *History of Venango County*, 349, 882-883.

[10]Deposition of Waldron J. Cheyney, Philadelphia, November 1, 1883, *Garner* v *Gird*, 229. For more on this important family, see *Portrait and Biographical Record of Arizona*, 113-114; and *Gopsill's Philadelphia City Directory*, 1879–1881.

[11]Deposition of Al. E. Schieffelin, San Francisco, July 1882, *Garner* v *Gird*, 229. The amount paid the Schieffelins was subject to much speculation in southern Arizona. On March 18, 1880, the *Weekly Nugget* gave a figure of $1 million. Wells Spicer, perhaps letting the excitement of the moment get the better of him, came up with $2.5 million. *Arizona Daily Star*, March 9, 1880.

brother were interested in what Mr. Gird got for his stock. We were to divide it equally each to have a third."[12]

After completing negotiations at Tucson on March 13, the actual transfer of money took place in Pennsylvania. The agreement called for payments of $100,000 a month, beginning in April and ending in September 1880. The Schieffelins deposited most of their proceeds in Philadelphia, where Albert lived for a time, afterward transferring some to their account at the Anglo-California Bank in San Francisco.

Even with Gird working feverishly as an intermediary, the negotiations were less than harmonious. A quarrel broke out between the Disston party and the Corbins over Elbert's earlier contacts with the Orion Company. Then, as longtime Tombstone Mill and Mining Company employee August Baron recalled: "Ed. didn't want to sell at the time. He was the only one that didn't want to sell. He hung back about eight hours before the sale went off. He wouldn't sign the papers until he got mad and signed them. He sold the mine cheaper than he calculated. . . . [After] Ed. came back from Tucson . . . he said, 'I wish I been out on a prospecting trip, so I was not home.'"[13]

Originally the plan called for the Corbin-Vosburg interest to join forces with the Philadelphians and buy out Gird and the Schieffelins. But as Frank Corbin admitted, "Some of the parties thought it was a little more than we could stand up under." They decided instead to ease out the brothers and retain Richard, after which Frank sensed "a little coolness on the part of the Schieffelins towards Gird."[14]

It was important to those buying the Schieffelin interest that Gird stay on as a major shareholder and superintendent of the mines and mill. Benjamin Hart explained: "Had I supposed Gird was selling his stock with the Schieffelin stock, I would have hesitated very much about taking it. We would have considered they had a motive in their desire to sell."[15]

Exactly why the brothers sold out when they did, and for a price which seemed below value, might be answered in part by Richard Gird himself. Writing Philip Corbin some months later, he acknowledged:

---

[12]Deposition of Ed. L. Schieffelin, San Francisco, June 8, 1880, *Garner* v *Gird,* 284.

[13]Deposition of August Barron [*sic*], Tombstone, August 11, 1882, *Garner* v *Gird,* 318, 319. For more on August Baron, see McClintock, *Arizona* 3:101-102.

[14]Deposition of Frank Corbin, San Francisco, October, 1882, *Garner* v *Gird,* 175.

[15]Deposition of B. F. Hart, Philadelphia, November 2, 1882, *Garner* v *Gird,* 237.

"You know that I have always favored the proposition that you should get your outlay back, that is my position still. I came near quarreling with the Schieffelins over that proposition, and that is the main reason why they sold out. I want to be kept right on the record as a man who would never do another an injustice."[16]

Thomas Bidwell, who had long argued with Gird and the others to sell out rather than develop the mines—while slipping between periods of praise and gloom over their mineral prospects—offered his own explanation to Safford: "I preferred and advocated . . . going on and make certain combinations and reap the benefit of it; but Ed. and Al. was [*sic*] afraid of a stock deal and the mine would be gutted out and [they] could not sell the stock for anything; and . . . they would not consent to running both mills and paying dividends, and trying our hands at a stock deal. Dick and Corbin were willing to do this under the circumstances."[17]

After the Schieffelin sale the Tombstone Mill and Mining Company reorganized. The new owners dominated nearly 200 acres of mining claims, 500 acres for mill sites, various water rights along the San Pedro, vast holdings of heavy mine and milling machinery, as well as the Millville townsite. Big-time capitalism had indeed come to the San Pedro Valley.

Capitalism of another sort was pushing closer. On March 20, 1880, the city of Tucson celebrated the arrival of the Southern Pacific. Two days before, the *Weekly Nugget* reported: "Mr. Ed. Schieffelin brought into our office yesterday, the silver spike to be driven in the last rail at Tucson. . . . It weighs about a pound, is of solid silver, and was made by Mr. Arnold at the Corbin mill, from Tough Nut bullion. It is a very handsome piece of workmanship."

Richard Gird planned on making the presentation to railroad president Charles Crocker himself, but a sudden case of pneumonia interrupted the process. Gird had hoped to curry Crocker's favor and persuade him to build a spur line to Tombstone. Instead the old railroad mogul slipped the silver spike into his pocket as a personal memento and promptly ignored the place. Although track eventually did reach Contention City, and finally Fairbank, Tombstone would not see service until 1903. At that point it no longer really mattered.

---

[16]Richard Gird to Philip Corbin, January 24, 1881, *Garner v Gird*, 215.

[17]T. J. Bidwell to Anson Safford, April 8, 1880, *Garner v Gird*, 204

The advance of the powerful Southern Pacific soured stagecoach profits. William Ohnesorgen bailed out, dissolving his partnership with Henry Walker. The firm became H. C. Walker & Company. Marshall Williams stayed on as its Tombstone representative even after becoming Wells Fargo's local agent. With the railroad pushing east, both old and new stage lines started shorter and cheaper runs north to its depot at Benson.

Even before Charlie Crocker turned his back on Tombstone, journalists reported no loose lumber lying around the unruly camp on Goose Flats. At the height of this new crisis people frantically began fencing in vacant property:

> Last Sunday it was rumored that the Land Office would not grant a patent to certain lands covered by town lots, and in the evening the jumping process commenced, and vigorously prosecuted. . . . All the vacant lots on Allen and Fremont streets were jumped, in fact the spirit of the acquisition of Territory was so great that the sacred precincts of the graveyard were encroached upon. . . . Some trouble is anticipated when the former owners make their appearance on the disputed premises.[18]

Defending himself against loud cries of mismanagement, Mike Gray wrote:

> Since I have been in charge of the business no man has ever applied to me for a lot upon which to build, whether he had the money or not, who has not received [one]. . . . I have given all my time to the care of this property, and the promotion and prosperity and development of this place, and it seems to me that it is an act of great injustice on the part of the jumpers to attempt to take from my associates and myself our property. We have complied with the law, expended money upon this property, and it in equity and justice belongs to us absolutely as any property belongs to any man.[19]

Less than satisfied by Gray's remarks—few missed the self-serving tone—residents faced a more serious challenge. With Mayor Randall safely in office the Townsite Company made arrangements with A. J. Mitchell to resurvey the site in late March. The surveyor general at Tucson approved the results.[20] In the name of the "corporate authorities of

[18]*Daily Arizona Citizen*, March 5, 1880.

[19]M. Gray to Editor *Nugget*, March 3, 1880, as reprinted in *Arizona Daily Star*, March 7, 1880.

[20]Plat of the Tombstone Townsite Claim, March 31, 1880, Federal Records Center, Laguna Niguel, California.

the Town of Tombstone" Alder Randall sent the new plat and required $400 filing fee, paid by James S. Clark, to the United States Land Office. It receipted the funds April 9. Then on May 22, unbeknownst to ordinary citizens, the mayor and council deeded 90 percent of the townsite to Joseph Bruce Palmer, Maurice E. Clark, Mike Gray, and John D. Rouse. Gray, in his capacity as village clerk and recorder, witnessed the agreement. Tombstone authorities recorded this grant deed two days later.[21]

When word of Mayor Randall's decision reached the streets, people began bombarding the General Land Office at Washington with complaints. Property owners and non-property owners alike signed petitions of protest, charging fraud and all sorts of chicanery between the mayor and council and the townsite proprietors. Critics vigorously condemned as unconstitutional an Arizona provision allowing "All persons who select and lay out a town site . . . shall be deemed occupants thereof. . . ."[22] They also warned that Randall's transfer would "result in confusion and perhaps bloodshed," and asked instead that the patent be issued to the county probate judge.[23]

Despite all the confusion over these shady real estate transactions, the boom continued. Established buildings and others under construction crowded both sides of busy Allen Street. Walking east to west that spring of 1880 one would have seen, among other burgeoning landmarks, M. Calisher & Company. That California firm had already tested the Arizona market, most notably at Globe. They willingly opened a Tombstone branch in 1879. Situated on the north side of Allen between Fifth and Sixth, the firm ignored its dirt floors while vigorously selling others finished lumber (at too high a profit to waste as flooring for themselves). David Calisher, the old man's teenage son, managed the store for his absent family, a heavy responsibility in such a rough and isolated place.

Across from Calisher's and a bit to the west, next to the Star Restaurant, stood pioneer general merchandisers Cadwell and Stanford. That

---

[21]Deeds of Real Estate, Pima County Recorder's Office, Book 7, 99-102; and *Tombstone Townsite Title Abstract*, 10.

[22]Hoyt, *Compiled Laws of the Territory of Arizona*, (3633) Sec. 11, 611.

[23]Petitions to James A. Williamson, Commissioner General Land Office, Washington; and Petition to United States Land Office, Florence, Arizona, July-August, 1880, Old Townsite Files, Tombstone, Records of Former General Land Office, Record Group 49, National Archives and Records Service.

Allen Street looking west in the spring of 1880. Adobes are piled on the corner of Fifth, across from the original Wehrfritz Building, for use on the Vizina & Cook Block. Work is nearly finished on the second story of Bilicke's Cosmopolitan. The last sign barely visible on the right marks the O.K. Corral. Across the street, behind the large rendition of a running horse, is the Dexter Livery, Feed & Sale Stable.
Courtesy, Huntington Library.

firm moved from Watervale to Fremont Street in late 1879, before it became clear Allen Street would be the new town's commercial center.

Nellie Cashman opened the Nevada Boot & Shoe somewhat west of Calisher's. This resourceful woman would operate a number of businesses in Tombstone. During those early months she divided her time between the new silver camp and a restaurant at Tucson called Delmonico's. In June 1880 Nellie and Jennie Swift reorganized the Allen Street boot shop into the Nevada Cash Store. They stocked provisions and furnishing goods, as well as fresh fruit they claimed was delivered daily from Los Angeles, while boasting "Choice Family Groceries a Specialty." Cashman eventually took over the Arcade Restaurant and Chop House, and even helped her widowed sister, Frances Cunningham, run a small hotel in the Vickers Building on Fremont Street. Later Nellie opened a hotel at Bisbee, before returning to Tombstone and becoming co-proprietor of the Russ House at Fifth and Toughnut.

Cashman enjoyed a charitable reputation, grubstaking many a prospector, donating to worthy causes, collecting money for the construction of Tombstone's Catholic church, and helping establish a baseball park for local boys. Still, with all her philanthropy, she harbored deep-rooted prejudices—especially against Jews and Chinese. Her Irish background led to her dislike of Englishmen as well. Despite everything, she employed a grateful Chinese cook who commissioned a painting of her from a photograph he carried on one of his visits to the Asian mainland.

A woman of independent spirit, Nellie Cashman led a life just as adventurous as did others she now encountered at Tombstone. Years earlier she and her mother traveled to Pioche, Nevada, when it was still a wild and dangerous place, to open a restaurant and miners' boarding house. By 1874 Nellie joined the Cassiar gold rush to British Columbia. There she financed a winter expedition that helped save hundreds during a scurvy outbreak, personally braving the ravages of weather to keep open a precarious lifeline. Such selflessness established her reputation. Returning to San Francisco, she visited the Comstock Lode at Virginia City and saw other Nevada boom camps. In late 1878 she left Los Angeles for Tucson. Within months she opened Delmonico's. But Nellie Cashman could not ignore the persistent rumblings along the San Pedro, and it is for her years in Tombstone that she is best remembered today.

Across Allen Street from Cashman's, near the corner of Fifth, Joseph Tasker and Jacob Hoke opened a general merchandise emporium dealing in home furnishings, clothing, building hardware, agricultural tools, as well as wines, liquors, and cigars. Both men came from San Diego, where they had once driven water wagons to cut the dust in that coastal city. Tasker served two terms on the county board of supervisors in the early

1870s. At Tombstone he and Hoke soon dissolved their partnership. Hoke went on to run a lumberyard while Tasker took on George Pridham as a partner.

During the spring of 1880 work began on the Vizina and Cook Block, eventually covering the three lots next to Nellie Cashman's on the northeast corner of Fifth and Allen. Proprietors of the Oriental Saloon would lease that corner, turning their place into an instant landmark.

Until then thirsty citizens could satisfy themselves across Fifth Street at Benjamin Wehrfritz and Siegfried Tribolet's Golden Eagle Brewery and Saloon. Commonly called the Wehrfritz Building or simply the Golden Eagle—and later the Fredericksburg Lager Beer Depot, before being renamed the Crystal Palace in the summer of 1882—it originally occupied a space only thirty feet wide and fifty feet deep. That portion housed a saloon and reading room while the brewery operated from a somewhat smaller building to the rear. From there twenty-five kegs of lager and bock beer flowed "like a Niagara cataract," claimed one charmed reporter. At first even bulk prices were steep—$7 for ten gallons and $3.75 for a five-gallon order. In the saloon bottled beer initially sold at three for $1, or $3 per dozen on outside orders. Pampering their customers, the partners offered German lunches featuring such delicacies as Swiss and Limburger cheese, sea trout, herring, smoked halibut, sardines, pig's feet, sheep tongues, head cheese, and other imported delights.

Two of Tribolet's brothers, Godfrey and Charles, the latter a one-time Los Angeles policeman, ran a butcher shop from an adobe building on Fifth Street. Described as "one of the Pioneer butchers," Godfrey built his own corral and slaughterhouse outside town. He kept busy purchasing cattle in the hundreds. The youngest brother, Robert, gained considerable notoriety selling liquor to the Apaches at a spot four hundred yards south of the Mexican border, causing one army officer to suggest the law should "have justified the hanging of the wretch Tribollet [*sic*] as a foe to human society."[24] Nothing came of it, and in time the offender was found residing quietly in Tombstone before moving on to Bisbee as a self-described workingman.[25]

---

[25]*Tucson and Tombstone General and Business Directory*, 197; and Cochise County Great Register, No. 2525, Robert Tribolet, Bisbee, August 18, 1884.

[24]Bourke, *On the Border With Crook*, 480; McClintock, *Arizona*, 1:254-255. For more on this family, see Adams, *History of Arizona*, 3:479-480.

South of the Wehrfritz Building, on the corner directly across Allen, Samuel Danner and Sylvester Owens opened the Bank Exchange Saloon in a tent fifty feet deep. They had moved over from Old Tombstone in late summer 1879. Their place served for many months as the new town's social and political center. With this in mind the proprietors added a large meeting hall behind the saloon, celebrating with a grand ball; cover charge $2, "admitting Gentlemen and Lady."

Davis S. Chamberlain, later a millionaire Iowa patent medicine manufacturer, came to Tombstone with $5 in his pocket; but was not so poor as to overlook an amusing incident at the Bank Exchange. The bar—planks supported by empty whiskey barrels—ran down the west side. Along Fifth Street a bench provided seating for patrons, their backs resting against the canvas wall. Following a run-in with a local tough, one victim carefully noted the man's position on the bench. Once outside he isolated the reflected shadow. Finding a two-by-four topped with a rusty nail, he "imbedded the spike into the soft parts of this man's rear end and then made his getaway."[26] The limping badman searched in vain for his attacker. It was noted with amusement thereafter that he never again sat on Danner and Owens's bench.

Chamberlain may even have sympathized with that man's discomfort. While crossing the plains as a teenager in 1865 his revolver exploded, forcing a nine-month convalescence at Salt Lake City to heal a shattered knee. He then struggled at various Montana placer sites for $6 a day. On May 10, 1869, after a brief trip home, Chamberlain watched the unfolding ceremony at Promontory Summit, Utah, celebrating the completion of the first transcontinental railroad. Next he hunted gold in Idaho before finding less strenuous work amongst Oregon retailers. Sailing for California, he organized a trip to Arizona. There he helped run a blacksmith and wagon shop, as well as earn extra dollars at the Vulture Mine south of Wickenburg. In late 1871 he witnessed the Chinese massacre at Los Angeles, an explosion of racial violence that, even when told secondhand, made a deep and lasting impression on the Earp family. After more travels and various economic ups and downs, including three years in Utah running a stage company and entering the grocery trade at

[26]Chamberlain, "Tombstone in 1879," 233. For more on Chamberlain's life before and after his time in Tombstone, see Harlan, *A Narrative History of The People of Iowa*, 4:267-268.

Second story construction nearly complete on Bilicke's Cosmopolitan Hotel. Sign marking Otto Geisenhofer's City Bakery can be seen to the right.
Courtesy, Arizona Historical Society.

Los Angeles, Chamberlain finally made his way to Tombstone. He became involved in a small water company, a general supply store, and, with the help of Tolley Ogden, began providing food services to the Contention Mine.

Next to Danner and Owens, New Yorkers James Vogan and James Flynn opened a wholesale liquor store and saloon in late 1879, described as "a rather elegant place on Allen street, where the drouthy public keep the proprietors busy dispensing Harry Horton's best brands. Jim [Vogan] hasn't lost any of his entertaining powers, and everybody from Tucson makes it a point of honor to interview him."[27] Vogan had given up man-

[27] *Arizona Daily Star,* January 17, 1880.

agement of Tucson's popular Congress Hall Saloon. Ironically, he died at the Congress Hall in 1885 at age forty-six.

West of the Wehrfritz Building, toward the corner of Fourth, workers busied themselves putting up the second story on the Cosmopolitan Hotel that spring of 1880. Owned and operated by Carl Gustav Frederick Bilicke and his nineteen-year-old son Albert, the Cosmopolitan boasted the glory of being the first two-story building in town. By November they bought an adjoining lot for a sizeable addition.

Bilicke had opened in a tent in July of 1879, offering the first beds (rather than cots) in any Tombstone hotel. Later eighty guests enjoyed the luxury of suites and parlors furnished with black walnut and rosewood bed sets with spring mattresses. In a pinch, cots and shake-downs could handle a hundred more. The office and restaurant—its facilities soon leased to separate managers—fronted Allen Street. These accommodations lured both stage companies to unload their passengers there. Tired and dirty, they stumbled off the coach to be greeted by the "evening zephyrs of Allen street" from the hotel's Steinway. "A year-old mining camp is getting along at a pretty good gait when it indulges in pianos," mused a local press release.[28]

Young Albert Bilicke was born in Oregon in 1861. The family then moved to Idaho before moving again in 1868 to San Francisco. His father had first seen that city sixteen years before. Albert attended public schools until age fifteen, after which he took a commercial course at Heald's Business College.

Carl Bilicke, having made a small fortune in the Pacific Northwest as a general merchandiser, lived the semi-retired life of a self-styled capitalist before losing much of his money during the 1875 run against the Bank of California. Looking for a way to change his luck with a change of scene, the family traveled to Arizona in early 1878. Carl opened the Miners' and Farmers' Store at Globe City. Still thinking of himself as a capitalist, he also handled banking transactions. Within five months the family moved to Florence and bought the old Elliott House. Finally they settled in Tombstone and became firmly entrenched in the hotel business.[29]

Just east of Bilicke's original site, Otto W. Geisenhofer opened the City

[28]Ibid.

[29]*Press Reference Library: Notables of the Southwest,* 59; McGroarty, *Los Angeles,* 2:104-105; and Spaulding, *History of Los Angeles,* 2:131-132.

Bakery. He had come to the United States from Bavaria in 1875. Hoping for a better life in the Southwest, and after first toying with the possibilities of central Arizona, the twenty-three-year-old Geisenhofer made his way to Tombstone. In October 1879, together with his older brother Michael, he started the bakery in a tent. By early spring they improved the location by rebuilding in adobe. The brothers maintained a steady business supplying families, hotels, and the surrounding mines and isolated camps with fresh baked goods. After selling the property to Bilicke, Otto moved his bakery up Allen Street and ran it in connection with what tradesmen then called a coffee saloon.

West of the Cosmopolitan early arrival Charles Brown had also opened a hotel. This Ohio native was one of the first businessmen to locate on the new townsite, owning two lots on the northeast corner of Fourth and Allen. There he "built and bossed the first hash house in Tombstone, for which many a weary tramp may thank the Lord."

Brown opened his Mohave Hotel in a tent in April 1879, but soon converted to a one-story wood-frame. Of Brown's enterprise it was noted: "This has from the first been one of the favorite places of the town, and its daily arrivals have averaged 20. He has now 16 bed-rooms, with 40 beds, and can accommodate 100 to 150 with beds by resort to cots and floors."[30] By the summer of 1880 Brown began rebuilding as a two-story adobe fronting Allen Street. Already renamed Brown's Hotel, the new accommodations occupied the entire upper floor, with most street-level space leased to other businesses.

Originally the ground floor corner was set aside for a branch of a Tucson stationery store. After that deal fell through, Colonel Roderick F. Hafford moved his saloon there. Hafford had originally opened in that portion of Brown's property between the old Mohave Hotel and Bilicke's Cosmopolitan. He now jumped at the chance to lease the refurbished corner location. He was not, however, its sole proprietor. Hafford acted as an associate of the San Francisco firm William B. Hooper & Company, one of the first wholesale liquor houses in southern Arizona. They operated under their own name in Tucson, Phoenix, and at Alexandra in Yavapai County, but in other towns linked their business with locals—in Globe with C. M. Coover; in Harshaw with William Kane; in Silver City, New Mexico, with the Luke brothers; and in Tombstone with Colonel Hafford.

[30] *Weekly Nugget*, June 10, 1880.

Directly south of Charlie Brown's new edifice, Missourian Michael Edwards opened the California Variety Store in a spacious adobe crowded with heavy shelving, counters, and glass showcases. From that busy corner he operated a wholesale and retail business concentrating on general merchandise, crockery, family supplies, and miners' outfitting materials. He catered to families by offering discounts for cash. As sole agent for the Vulcan Powder Works, Edwards enjoyed steady trade with the mining companies. Dick Gird and Al Schieffelin's old friend from Signal, Jose M. Castañeda, managed the place.

Across Fourth Street from the California Variety was P. W. Smith's—John B. Allen's old stand. West of Smith's large frame building was the Dexter Livery and Feed Stable, its location marked by a large sign showing a running horse. Across from the Dexter, slightly to the west at mid-block on the north side of Allen, stood Montgomery and Benson's O.K. Corral. The front covered two lots in width but narrowed through the block to Fremont Street, where a small rear entrance was added between the Union Meat & Poultry Market and Papago Cash Store.[31]

John Montgomery had been in the San Pedro Valley since 1874, having gotten there over a rather circuitous route. From his Ohio birthplace Montgomery traveled to San Francisco by way of Panama in 1852. Drawn west by the Gold Rush, he followed his dream for the next twenty-two years through mining camps in California, British Columbia, and Montana. He settled in New Mexico by the early 1870s, dividing his time between raising stock and prospecting. Then, in Arizona, Montgomery became associated with Thomas Dunbar at his station at Tres Alamos. He was soon named postmaster there and, along with Dunbar and others, also harvested large crops of barley, corn, and potatoes for the Tucson market.

Excited by Schieffelin's discovery, Montgomery moved south, joining up with Edward Benson, a man ten years his junior, to open the O.K. Corral at Old Tombstone in February 1879. Later they shifted their profitable business to the Allen Street property, facing the same battles over clear real estate titles plaguing everybody else. Within four years he bought out Benson's interest. A confirmed bachelor, Montgomery stayed in Tombstone

[31]Real Estate Deeds, Pima County Recorder's Office, Book 7, 470, June 8, 1880, Recorded September 28, 1880; and Assessment Roll of the City of Tombstone, Montgomery and Benson, S 20 ft. Lot 5, W 1/2 Lot 6, E 1/2 118 ft. Lot 6, Lots 17 & 18, Block 17, Paid December 10, 1881.

for the rest of his life, becoming deeply involved in politics as a member of the Republican Party.[32]

To help advertise the town's exploits, John Philip Clum and two partners founded another newspaper that spring. Clum would play a major role in Tombstone's early history, with some characterizing his contribution as less than honorable. Leaving Rutgers after his freshman year, Clum took a job as a weather observer for the government at Santa Fe in 1871. Within three years he managed to maneuver himself into heading the San Carlos Apache Reservation in Arizona during some rather troublesome times. He held that post until 1877. Under pressure, he resigned in a huff and turned to journalism.

At Tucson Clum bought the *Arizona Citizen* and promptly moved it to Florence. Simultaneously, he practiced law in Pinal County but with no formal training soon abandoned the bar. Within a year he returned the paper to Tucson where it remains today. In early 1880 Clum, not yet twenty-nine years old, sold out to R. C. Brown and together with new partners Thomas R. Sorin and Charles D. Reppy started Tombstone's second newspaper in a small canvas tent at 220 Fourth Street. Eight weeks later they moved up to Fremont, three lots west of the post office. The trio changed the paper's name from *Clarion* to *Tombstone Epitaph* just before the first issue of May 1, 1880.

Years later Clum took full credit for choosing this appropriately colorful name. Yet mining engineer John Hays Hammond, nephew of the celebrated Texas Ranger and California pioneer John Coffee Hays, claimed he had suggested it and that the "editor . . . immediately accepted the idea and in gratitude sent me free copies of the paper for many years thereafter."[33] Reacting to the *Epitaph's* first issue, whoever dreamed up its masthead, the *Tucson Record* prophesied, "Its name will either be a fortune to the paper or break it in six months."[34]

Clum joined numerous political and social organizations, a practice that had often made him an object of ridicule at Tucson:

> Elder Clum has been in Arizona five years. During most of the time he was agent at the San Carlos. After being dismissed he took unto the profession of

[32]1880 United States Census, Tombstone, Pima County, Arizona, E.D. 3, Sheet 14, Lines 13-14; and *Portrait and Biographical Record of Arizona*, 641-642.

[33]Hammond, *Autobiography*, 1:79.

[34]*Tucson Record*, as quoted in *Tombstone Epitaph*, July 2, 1880.

> law, taking a little hand in medicine at the same time. From these he stepped into a newspaper, and became an active member of the church, church socials and women's gossip clubs. During the last year he has taken somewhat to the stage, having made his appearance more than once in church concerts and "nigger" minstrels. . . . When it was known that an acrobatic club was to be formed in our city, his turning proclivities at once made him the available candidate for the presidency, which position he now holds with great credit. His last departure is for the pulpit. . . .[35]

Assuming "Parson Clum" would flop, reports noted with amusement that Presbyterian minister J. E. Anderson angrily characterized him as false and treacherous.

John Clum did not change much after hitting Tombstone. It started with him becoming postmaster on June 4.

Editors at both the *Nugget* and *Epitaph* found plenty of news. Attorney Wells Spicer, along with friends Amos Stowe and Thomas Ogden, laid out a whole new town—adding to the already serious problems surrounding lot jumping and murky real estate titles. Christened New Boston, "It adjoins the old town of Tombstone on the east, but will have no political connections with it—that is, no Mayor, no Town Council, no Marshal, no police—none of that kind of nonsense. . . ."[36]

Spicer ordered maps drawn showing five hundred lots: "There is already a great demand for lots in New Boston. There will be a rush and a hurrah for them; knowing ones, who claim to have good foresight, tell me that the old town of Tombstone will soon become to the new city of New Boston what the 'Barbary Coast' is to the city of San Francisco."[37] Spicer boasted of space set aside for a depot, failing to mention that no railroad showed any real interest in the place.

Many in Tombstone were not amused after Spicer and his pals blocked the eastern end of Allen Street, leading to a description of the enterprise as "a tent town, but the occupants of the lots have plenty of grub, evidently intending to stay. The Marshal is trying to keep Allen street clear, to the extremity of the town-site line, and locators are everywhere putting in a claim by actual occupation for town lots or county seats."[38] Some

---

[35] *Arizona Daily Star,* August 29, 1879. For less critical views of this man's life, see Walter, "Necrology-John P. Clum," 292-296; Parker, "John P. Clum," 32-37; and Clum, *Apache Agent.*

[36] *Arizona Daily Star,* June 3, 1880.

[37] Ibid.

[38] *Weekly Nugget,* August 12, 1880.

arrests were made, with prisoners hauled to Tucson to explain their troubles before an exasperated magistrate, but Spicer's real interests lay elsewhere.

Distancing himself from what he called "small-fry affairs," he concentrated instead on looking after mines, politics, and anything else showing promise of making a buck. And he did so with chameleon-like adaptability. At Tombstone he represented himself as a staunch Republican, in tune with most mine owners and many leading community figures. Years before, however, he claimed to be an equally staunch Democrat.

Wells Spicer was born in New York in 1833. Twenty years later he was a busy young man in Iowa. At Tipton he practiced law, helped publish the *Cedar County Advertizer* before becoming its sole proprietor, ran as a Democrat for prosecuting attorney, got himself elected county judge for two years, was appointed to something called "Notary Public and Commissioner of Deeds for the State of California," and became a charter member and first officer of a fraternal organization known as Siloam Chapter No. 19 (he would be tossed out of the Masonic Lodge at Tombstone in 1885 for failing to pay his dues[39]).

He married Abbie Gilbert in 1856, and they had one son. Six years later some of their property was sold at a sheriff's auction. Abandoning these reverses, Spicer left for Colorado with dreams of finding a gold mine. By 1874 he was in Salt Lake City advertising himself as an attorney and mining engineer. While there he served as Brigham Young's lawyer and helped defend John D. Lee in the celebrated Mountain Meadows Massacre case. In Wells Spicer's eyes Tombstone offered fresh possibilities, the switch in fortune he so craved. This he would not find, however, with his plans for another town.[40]

New Boston may have irritated the Tombstone townsite crowd, but those running the mines ignored the place: "The heavy ore teams from the Contention make their regular trips across the new site regardless of the regular angles of the streets and blocks. It is rumored that the old Tombstone Company will remove New Boston on the plea that it will obstruct their streets. No artillery will be used."[41]

---

[39]*Proceedings of the M. W. Grand Lodge,* 1:224.

[40]For some background material on Wells Spicer, see *History of Cedar County,* 401, 454-455, 461, 476, 551; Stout and Stout, *Cedar Land,* 2:216-217; and Sloan, comp., *Gazeteer [sic] of Utah, and Salt Lake City Directory,* 280.

[41]*Tombstone Epitaph,* July 19, 1880.

It was just as well. Spicer faced enough troubles over his new town without tangling with local freighters. They were all hard men. One of the most determined was a thirty-three-year-old Missourian named Julius E. Durkee, who had come into the San Pedro Valley with hardly a dollar to his name. Working in the mines, he dreamed of other ways of making a living besides slaving through a ten-hour shift for $4. Durkee decided instead to become the district's freighting impresario.

While visiting mining superintendents he solicited shipping contracts at $3 a ton, thereby undercutting his competition by fifty cents. Having no money to satisfy these hurriedly negotiated obligations, Durkee approached E. B. Gage, hoping the Grand Central's boss would finance his grandiose scheme. Impressed by the younger man's persistence and grit, Gage agreed to bankroll the enterprise. With his pockets stuffed with cash, Durkee rushed back to his native state. There he purchased a mule herd large enough to keep twenty or more teams hauling their thirty-ton loads down to the mill sites. The new firm eventually employed dozens of teamsters and laboring men. Unfortunately Tombstone's newest entrepreneur proved overly fond of gambling, drinking, champagne baths, and seasons of "riotous living and wild debauch." While an inmate of the insane asylum at Stockton, California, he died flat broke in 1896.[42]

Wells Spicer needed no confrontations with the likes of Julius Durkee. But Charles H. "Ham" Light, a local blacksmith and another major ore-freighting contractor, filed charges over the New Boston townsite. Mike Gray ordered Spicer's arrest. Everyone reached an understanding before trial, and Light withdrew his complaint. Despite repeated threats, forty-three lots had been sold and ten others reserved within the first month of the new town's existence. A number of buildings had been started, including family homes and an office for Spicer. Stowe and Ogden contracted sixty thousand adobe bricks to raise a two-story building—the upper floor to serve as a public hall—on the corner of B and Banner streets.

The center of New Boston lay astride the Silver Belt mining claim, the northern half of which occupied the Tombstone townsite roughly between Tenth and Twelfth streets nearly up to Fremont. Trouble with

[42]Pima County Great Register, No. 4852, J. E. Durkee, Precinct No. 17, October 18, 1880; *San Francisco Chronicle*, August 9, 1896; and In the Matter of the Estate of J. E. Durkee, No. 1662, August 28, September 30, 1896, Superior Court, County of Los Angeles. Durkee's estate was so small it required only a $300 bond. Even so, his old friend E. B. Gage, then living in Prescott, Arizona, served as co-executor, with former Tombstone lawyer Ben Goodrich handling the legal work.

overlapping claims persisted for weeks, and by mid-August the *Epitaph* reported: "Jumping lots in New Boston, is still all the rage. . . . The whole town has been jumped." Mayor Randall ordered Fred White to prevent any encroachment on the original Tombstone site. The marshal happily "tore down the fences at the intersection of Allen and Ninth streets. . . . Wind, of course, is cheap and plentiful."[43]

Worried about lot jumping, Mrs. M. L. Woods checked her property only to find someone hard at work fencing it in. After her complaints were ignored she pulled a revolver and scared off the startled interloper. Imagine her surprise when she was "shortly afterward informed by the Marshal that she had fenced in Tenth street. . . ."[44] Checking maps that supported White, Mrs. Woods stomped off to Spicer's office demanding another lot. One assumes he complied without argument; whether or not over the barrel of a Colt revolver is not known. Mrs. Woods started an eatery but within weeks gave up on New Boston and transferred her energies to Tombstone, opening the Melrose Restaurant on the south side of Fremont between Fourth and Fifth.

Partly from all the confusion, New Boston, however briefly, had seen its day, especially after a Tucson court ruled against its inhabitants. Wells Spicer moved on to other things. Besides his budding law practice he wrote dispatches to the *Arizona Star* at Tucson and worked with Thomas Gardiner as associate editor of the *Arizona Quarterly Illustrated,* a small newspaper-like journal devoting much of its space to Tombstone topics.

Some viewed with skepticism Spicer's role as a journalist. In early January he had written the *Star* suggesting the Corbin Mill had been sold to the Western Mining Company for the Contention. Other sources then claimed, no, the Farish brothers of San Francisco, who already owned the Head Center, Yellow Jacket, and Naumkeag, and had bonded the Mammoth and Rattlesnake mines, had bought the Corbin property. Frank Corbin returned from San Francisco just in time to be hounded by swarms of reporters. Corbin ridiculed the affair as he pushed through the unruly crowd, characterizing Spicer's report as a "wilful and malicious falsehood, uttered by some evil disposed person from a motive to injure our company to the advancement of others."[45]

---

[43] *Tombstone Epitaph,* August 14, 1880.

[44] Ibid.

[45] Frank Corbin to the editor, *Daily Arizona Citizen,* January 12, 1880. Also see Wells Spicer, "Tombstone Inscriptions," *Arizona Daily Star,* January 7, 1880.

Amidst all these charges bouncing back and forth through hostile newspaper columns, Richard Gird had married Nellie McCarty at San Francisco and brought her to Millville in time to chuckle over this latest controversy. Their large adobe home, with its wide veranda overlooking the San Pedro Valley, shared space with offices of the Tombstone Mill and Mining Company. There the couple honeymooned surrounded by the pounding stamps of the Gird and Corbin mills, a twenty-four-hour reminder of fortune within reach. Wild rumors concerned them not.

But Tombstone soon witnessed a more serious tragedy than Spicer's shortcomings as a member of the fourth estate or his persistent problems with the New Boston townsite. A troubled marriage, normally seen as a private matter, stunned residents as the final act unfolded before public gaze. Mr. and Mrs. Mike Killeen stood at the center of this drama. He tended bar at Lowery and Archer's Palace Saloon, just east of Tasker and Pridham. She had been working as a maid and living at Bilicke's hotel during their weeks of separation.

On the evening of June 22 May Killeen attended a dance at the Vizina and Cook building at Fifth and Allen. After the festivities she stepped outside to speak with Frank Leslie, a thirty-eight-year-old Texan commonly known as Buckskin Frank. Standing with Leslie was his young friend George M. Perine, then interested in mining and milling but who had come from San Francisco just months before representing his father, a roofing material manufacturer. Perine also served briefly as postmaster just prior to John Clum's appointment.

In contrast, Leslie carried a reputation as a desperate man, having scouted for the army on the Great Plains and again against the Apaches in Arizona until 1877. Traveling to San Francisco, he found work as a bartender at Thomas Boland's Pine Street saloon. By 1879 Leslie had become a bookkeeper and then bartender at Kerr and Jurado's billiard parlor,[46] before briefly returning to Texas. He came to Arizona from San Antonio and managed the bar at the Cosmopolitan with W. H. Knapp during that summer of 1880, despite a sketch in the *Arizona Quarterly Illustrated* characterizing him rather loosely as a mining man.

He now agreed to accompany Mrs. Killeen back to her lodgings. There

[46]*Langley's San Francisco Directory*, 1878-1880.

the two seated themselves on the balcony to enjoy the warm evening air. Mike Killeen followed his wife into the hotel. Climbing the stairs unnoticed by all but George Perine, he slipped quietly down the hallway and stepped outside. Hearing gunshots, startled witnesses across the street looked up and saw figures grappling in the shadows. Most claimed at least three men were involved. As the smoke cleared it was seen that Leslie suffered two superficial scalp wounds. Killeen had been hit in the right side, the bullet passing through his lung and lodging in his back just beneath the skin. A second shot shattered his jaw, coming out below the right eye.

As Mike Killeen lay suffering from his wounds, Artemus Fay concluded, after talking with witnesses, "The stories are very conflicting, and it is almost impossible to get at the real facts, but we do not believe the half has been told."[47] Fay did not yet know the answers, but his instincts were right. Even Clum correctly surmised, "The causes which led to these desperate results found their origins in domestic infelicity and the blind, resistless passions of love and jealousy."[48]

Justice of the Peace Mike Gray conducted a preliminary hearing. To the satisfaction of almost no one he called only two witnesses, Mrs. Killeen and Frank Leslie. She claimed her husband shouted an oath and fired at Buckskin Frank. Only then did Leslie rise. After a second shot the two men scuffled and a frightened Mrs. Killeen ran into the hotel looking for Albert Bilicke as her friend fell.

Frank Leslie testified that Mike Killeen appeared out of nowhere, called him a son of a bitch, and fired. The shot missed. As he rose a second bullet grazed his left temple. Leslie fell to his knees, reached for Killeen, and knocked him off balance. Seeing his own revolver lying unattended, and thinking it impossible to disarm his attacker, Leslie grabbed the weapon and began shooting. Killeen got off a third shot as Leslie backed into the hallway firing one last time. Frank rushed downstairs to the bar and grabbed another revolver before being disarmed and arrested.

Who knew what to believe when Buckskin Frank was involved. As one contemporary described him:

> This man Leslie is a peculiar case, one of a class not infrequently met on the frontier—apparently well educated, gentlemanly and liked by all who know him, with as much sand as the country he ranges—but a novelist who can make a little truth cover a large area. As much fact as you could pick up on a

[47] *Weekly Nugget,* June 24, 1880.

[48] *Tombstone Epitaph,* June 26, 1880.

> pin point would last him a year. But there is one thing about it, his prevarications are all harmless. He never lies to hurt anybody—least of all to hurt Frank Leslie.[49]

Judge Gray's decision to call only Buckskin Frank and his new lady friend, while ignoring numerous witnesses from the street below, cast doubt on the court's ability to discover the truth. The *Nugget* characterized the proceedings as "a very lame affair." As speculation grew and overshadowed the facts, others demanded to hear Mike Killeen's version. After all, they argued, the badly wounded man still clung to life, though no one could say for how long. Indeed, he would be dead within days. Gray relented early on the morning of June 24. Killeen's testimony suggested some interesting possibilities.

Struggling to get words past his shattered jaw, Mike Killeen said he went to the party to see his estranged wife but was told she had left with Buckskin Frank. Angered by jealousy he entered the hotel "with the expectation of finding them both in Leslie's room. . . ." Instead, hearing voices from outside, he stepped onto the balcony. There they sat, Leslie's arm wrapped tightly around the young woman's waist. While deciding what to do next, Killeen spotted an armed George Perine. Knowing him as a friend of Buckskin Frank's, Killeen began backing away as Perine shouted a warning. Leslie pulled his revolver as he rose from his seat. His opening shot was followed by one from Killeen, who jumped Frank and hit him over the head. Killeen then turned toward Perine who fired three rounds. After disarming the younger man, Killeen realized for the first time the seriousness of his wounds.

Despite this testimony, Gray discharged Leslie, claiming he acted in self-defense. Few were pleased. There was, after all, something sordid about the whole business. This was a time when marital infidelity remained a very serious matter. Years before, New York Congressman Daniel Sickles had been acquitted after shooting to death his wife's lover, the son of Francis Scott Key.

Frank Leslie walked away a free man, but he would make his presence known in the years ahead, becoming part of the Tombstone legend. Following Mike Killeen's testimony, George Perine was briefly incarcerated but soon discharged. Within weeks more details of this episode became

---

[49]Lummis, *General Crook and the Apache Wars,* 59-60. For two versions of this man's troubled life, see Martin, *Silver, Sex, and Six Guns,* and Rickards, *Buckskin Frank Leslie.*

public. Questions persisted, especially after August 5 when Leslie married Mrs. Killeen before Judge James Reilly in the parlor of the Cosmopolitan. Many attended, including the Bilickes, lawyer and city councilman Harry B. Jones, saloon proprietor R. F. Hafford, and Tucson banker Charles Hudson, along with many wives and daughters.

Considering the bloody prelude to these nuptials, the whole scene mirrors a strange undercurrent. Yet even John Clum restrained his normal sanctimoniousness: "The *Epitaph* congratulates Mr. Leslie . . . and his most estimable wife upon this happy event, and earnestly wishes them a pleasant voyage over life's troubled ocean."[50] That particular voyage over life's troubled ocean sank under a divorce decree in 1887.

Meanwhile, Tombstone moved ahead, but without the presence of Richard Gird's old friend Thomas Bidwell. Always suffering from poor health, he decided to return to California for an operation on a condition described at one time as a fistula and afterward as stricture of the rectum. Resigning again as justice of the peace, Bidwell had left on June 14, carrying with him a premonition he would not return.

He traveled to Nevada with George Woodbury, a twelve-year-old relative of Gird's wife. At Gold Hill he visited his two children before continuing on to San Francisco. There Dr. Levi Cooper Lane, a professor of surgery at the Medical College of the Pacific, operated only to inform his patient that the procedure had failed. On July 8, 1880, Thomas J. Bidwell died at his room in the Garry boarding house, a block up from Dr. Lane's Mission Street office. He was forty-seven years old.

Both Tombstone papers carried the news of Bidwell's death, mistakenly inferring he had died a wealthy man with two grown children. Actually he had very little money (Bidwell's big dream was running a chicken ranch he hoped Gird would buy between Los Angeles and Santa Monica), and his children were ages five and three. His son became an attorney and died in 1960 at the age of eighty-three; his daughter married and moved east, outliving her younger brother.

Gird considered Bidwell his best friend on earth, often characterizing his feelings as those of a brother—sentiments not shared by his family. Gird's father stated rather bluntly, "Bidwell was a man I did not like, and I cautioned Richard about him." Emma Gird claimed, "While Richard

[50] *Tombstone Epitaph*, August 6, 1880.

seemed so fond of Bidwell the rest of our family . . . disliked him very much. They believed that his designs in seeking Richard's companionship were bad."[51] These feelings increased after Bidwell's death.

Even with no written agreement the closeness of these two suggested a partnership. The Corbins certainly thought so. Philip Corbin recalled that they, Safford, Vosburg, Hamilton Disston, and "I think Mr. Al. Schieffelin" accepted Gird and Bidwell as partners.[52] The elder Corbin was wrong about Albert, but similar suspicions were widely believed around Tombstone. Gird may have treated Bidwell as a brother, but he was never a partner in the Tombstone mines. Worrying about his health, Bidwell trusted Gird to raise and educate his children. After Gird's marriage his wife dismissed that idea. She admittedly had no feeling for those youngsters and resented the intrusion. Besides, Bidwell's sister-in-law would not give them up, having raised them as her own following her sister's death.

After finally selling out at Tombstone, Gird guaranteed their welfare by moving the whole family from Nevada and deeding them half a ranch near Temescal, California. They showed their gratitude by turning around and suing him for half his Tombstone profits on behalf of Bidwell's children. The case was long, messy, and expensive. The court ruled in Gird's favor but its decision was appealed. He agreed to a settlement only to put the whole troublesome matter behind him.

Gird considered the entire affair nothing more than an extortion suit. The case left him deeply disillusioned. Despite his friendship with Bidwell and their many adventures together in California, South America, and Arizona, Dick Gird totally ignored him when outlining his own life for historian Hubert Howe Bancroft in the early 1890s and in writing his 1907 article on the discovery of Tombstone. It was as if, in Richard Gird's mind, Thomas J. Bidwell had never existed.

Back in Tombstone, after Bidwell's death, persistent questions concerning clear title to town lots remained the chief source of trouble. Those first booming years of 1879 and 1880 saw the filing of charges, threats, recriminations, physical violence, loud talk, and drawn weapons. People spoke with raw emotion of forming a vigilance committee to supersede

[51]Deposition of John Gird, New York; and Deposition of Emma Gird, Cedar Lake, New York, *Garner* v *Gird*, 350, 352.

[52]Deposition of Philip Corbin, New Britain, Connecticut, October 31, 1882, *Garner* v *Gird*, 234.

Mayor Randall and Clark, Gray & Company—the common name now used for the townsite organization—to establish some sense of order in the disposition of real estate.

Then, as if Tombstone really needed something more to keep everyone alert, John Clum helped stir up racial unrest. That young journalist often assumed a posture of severe resolve, both in his statements and personal outlook. His prejudices, together with a tendency toward rashness that often bordered on hysteria, as well as his habit of jumping into controversies without thinking them through, are all evident in his handling of the so-called Chinese question.

Although most people in Tombstone were native-born, the ethnic mix included large numbers of western European and Canadian immigrants. Racially, however, the town remained almost exclusively white. Its population on the 1880 census totalled 2,173. By comparison, Tucson, still the largest town in the territory, was about 7,000. For specific individuals the federal census is not always reliable. A few people are listed twice, while others known to be there do not show at all. But these schedules do provide a useful means of measuring any community at that point in time.

Enumerator Philip W. Thurmond, an attorney and former Kentucky legislator who had come into Arizona from Brownwood, Texas, did his best to assure an accurate count: "To avoid disturbing miners sleeping in the day time, Mr. Thurmond has secured a table at Comstock & Brown's [Mount Hood Saloon], where he will be found from 3 o'clock to 5 in the afternoon, and from 8 o'clock to 11 in the evenings. . . ."[53]

Of those finally listed only sixteen were black, with eight of them boarding together with another black family of four. The census also counted forty-four Chinese. Of these, there were nineteen laundry workers and thirteen cooks. Five others worked as laborers and one each gave their occupation as merchant, restaurant keeper, and grocer. Four Chinese women were housekeepers within their own families. Except for one mulatto hotel waiter from Maryland, all the other people shown on the Tombstone census were white. These included over 130 Mexicans who had come north seeking employment, as well as those of Mexican descent born in Arizona, New Mexico, Texas, or California.

Even though the Chinese represented only 2 percent of the population,

[53] *Weekly Nugget,* June 3, 1880.

that fact alone failed to protect them from the twists of irrational excess. In the early summer of 1880, egged on by John Clum and a Contention mill worker named Roger King, many citizens went on an hysterical anti-Chinese campaign.

This was no new phenomenon for southern Arizona. Just the year before Tucsonans complained of their Chinese population: "Last week seven Celestial heroines arrived, which, added to the number already here, make ten in all." The uproar over these ten individuals may seem astonishing today, but it was all taken very seriously at the time. "The Chinese are coming," the press warned, "and with their coming they are bringing their loathsome characteristics." Fearing more Chinese— "Hardly a stage arrives that does not bring one or more Chinamen to our city"[54]—some demanded a special ordinance isolating these newcomers to protect property values.

Now Tombstone seemed prepared to join this noisy upheaval. At a public meeting held on Allen Street in late July, speaker after speaker condemned the Chinese presence. Clum's *Epitaph,* lending its full support, reported, "The petition has been circulated and signed, the circulars printed and distributed, the bell rung and the fire lighted—an omen that the pig-tails were to be fired out." The paper added, "A committee of ten was appointed to convey the sentiments of the meeting to the despised Mongolians. . . ."[55]

Editor Clum also regarded as pertinent similar remarks made by attorney and city councilman Harry B. Jones, adding in summary: "There are but three ways of getting rid of John: First, by modification or repeal of the [Burlingame] treaty; second by violence; and third, by denying them employment and patronage. The first the people of Tombstone can't do; the second is optional with them; and the third presents us much the easiest, quickest and best solution of the case."[56]

The *Epitaph* kept up the pressure. Describing a meeting establishing the Anti-Chinese League "to rid the town of evil," Clum's columns noted: "the property owners of the town were with them in any peaceful efforts to get rid of John Chinaman. The committee appointed at the last meeting to notify the Chinese to leave, reported that they had interviewed the

[54]*Arizona Daily Star,* July 9, 1879.

[55]*Tombstone Epitaph,* July 25, 1880.

[56]Ibid., July 27, 1880.

Mongolians and been informed that they would go when they got ready, or in other words, when the washing gave out."[57]

John Clum was not alone in giving such views public voice. At Tucson his old paper, the *Arizona Citizen,* had earlier printed an advertisement from J. Goldtree & Bro. suggesting: "Avoid Leprosy!/ Patronize White Man's Industry/ Don't Smoke China Cigars/ Avoid all California Trash and Danger by Using Home-Made Brands."[58]

But some in Tombstone felt things had gone too far. Ben Fickas, newly-elected secretary of the League, declined to serve, stating publicly, "Although strongly in favor of restricting Chinese immigration, I am opposed to anything savoring of want of respect to the law of the land."[59] Even Clum backed off after someone attacked a Chinese restaurant owner at Charleston with a club: "It may be that 'the Chinese must go,' but that fact will not justify deeds of bloodshed and murder, and good citizens should take a positive stand against such overt acts."[60]

Aside from mirroring nineteenth-century racial attitudes, none of this made much sense. After all, a handful of Chinese farmers along the San Pedro helped supply Tombstone with much of its fresh produce. But hatred of the Chinese remained common, especially in California. Throughout southern Arizona that summer of 1880, similar sentiment rose and fell with the progress of the Southern Pacific's work crews. At the time enumerator Thurmond counted forty-four Chinese in Tombstone, 843 of their brethren worked near Benson as part of the railroad's grading gangs. Of course this hardly constituted an invasion; all these men were literally passing through.

Many citizens fell victim to the hysteria surrounding this sordid affair, including businesswoman Nellie Cashman. She pointedly advertised that everything sold at her Allen Street store was "all the work of free white labor. NO CHINESE WORK SOLD."[61] On August 8 the *Epitaph* reminded its readers that her store made "a specialty of the goods of the United Workingmen's Association, which employs no Chinamen."

Even with all this raw emotion, not everyone supported the Anti-Chinese League. Indeed, some of the town's chief business leaders and professionals watched with discomfort the rising tide of hatred against this small

---

[57] Ibid., July 29, 1880.

[58] *Weekly Arizona Citizen,* September 12, 1879.

[59] *Tombstone Epitaph,* July 30, 1880.

[60] Ibid.

[61] Ibid., June 26, 1880.

and generally law-abiding segment of the population. Others openly supported the right of the Chinese to live wherever they chose. A few citizens, including the normally somber Earp brothers—who remembered well the stories of the 1871 Los Angeles massacre—went so far as to say they would not stand idle if a mob threatened Tombstone's small Chinese community. Virgil kept U.S. Marshal Crawley Dake fully informed, warning early on of the possibility of riot.

Against such steadily growing opposition, or perhaps from the obvious foolishness of the thing, many began losing the blind fervor needed for a crusade. Clum's grinning rival Artemus Fay summed it up best, writing in early August: "The anti-Chinese movement seems to have entirely disappeared. . . . The legitimate workingmen of the camp had better fight shy of the next self-constituted leader who comes along and endeavors to induce them to make asses of themselves."[62]

As an interesting footnote to this whole affair, John Clum hypocritically changed his mind on the Chinese after they were victimized by his new enemy, the Tombstone Townsite Company. In his December 3 issue the editor proclaimed:

> The Chinese have been informed by Clark, Gray & Co., that rent begins from December 1st. Also several foreigners, who have a very imperfect knowledge of English, have received notice to the same effect. Thus do the schemers seek to prey upon the weak, but they should not forget that the Citizens' League was organized for the specific purpose of protecting those who through any cause whatever are incapable of self-protection.

As the anti-Chinese turmoil faded from public view, real estate troubles continued, with people swarming over lots in their mad rush to get into business. All this running around increased the strain on limited supplies of finished lumber. William Gird, Morse & Company, and other smaller producers in the Huachucas, including the likes of Carr's Mill, Tanner and Hayes, and those at Turnerville in Ramsey Canyon, could not keep up.

But a new firm, and soon to be the largest, L. W. Blinn Lumber Company, began filling the gap. Lewis Blinn came west in 1863. At San Francisco he clerked for Adams, Blinn & Company, a lumber business partially owned by his family. After nine years he went to the state capital where he managed the Sacramento Lumber Company before moving to Arizona. At

[62] *Weekly Nugget*, August 5, 1880.

Tombstone Blinn set up operations in the block south of Toughnut between First and Second streets. There he maintained his office, home, and lumber sheds. He eventually opened outlets along the Southern Pacific from Los Angeles to El Paso.[63]

Blinn, who would make his presence felt in Tombstone, profited from the 1880 building boom. Already San Franciscans Milton E. Joyce & Company had opened a new saloon in the Vizina and Cook Block on the northeast corner of Fifth and Allen. At first the proprietors planned calling their new establishment the Royal or Oriental Palace, but finally settled on Oriental. The thirty-three-year-old Joyce envisioned a palatial scene, and his saloon quickly became an established landmark as well as the site for some of the town's more disquieting episodes.

Customers were treated to impressive surroundings. Light from twenty-eight burners, glittering through crystal chandeliers, bounced off a polished bar along the east wall. Painted white with gold trim and elegant carving, it was originally made for the Baldwin Hotel in San Francisco. After Mr. Baldwin rejected it as too small, Joyce & Company stepped in, and along with two handsome sideboards, shipped the pieces to Tombstone. The divided back portion of the Oriental, floors covered with Brussels carpet, served as a club and billiard room. The house even supplied stationery for its patrons.

As townspeople marveled at the splendor of the Oriental, construction began on an elegant new hotel across from the Cosmopolitan. It was here that Charlie Calhoun built the first structure on what became the townsite. By 1880 Comstock and Brown's Mount Hood Saloon occupied that ground, described as "one of the pioneer houses of this town, and a favorite resort, where can be seen groups of men assembled reaching into the hundreds."[64] But Comstock and Brown dreamed of bigger things. In pouring over plans for the Grand Hotel they set a new standard for Tombstone.

The rather impressive results opened with much fanfare on the evening of September 9. To the locals it proved inspiring. Tombstone's physical appearance often seems drab from surviving photographs. Yet many enterprising businessmen painted their buildings an imaginative array of colors,

[63]*Langley's San Francisco Directory*, 1860-1872; and Guinn, *Historical and Biographical Record of Los Angeles*, 534-537.

[64]*Weekly Nugget*, June 10, 1880.

with yellow, red, and Prussian blue particular favorites. Even though much of the town still looked rough from the outside, interiors of many businesses and private homes boasted all the luxury and comforts modern 1880 had to offer.

Citizens crowded the Grand Hotel's opening as if tourists on pilgrimage. What they saw pleased the mind as well as the eye. Surely any town with such an elegant symbol was one worth coming to. Passing through one of several arched entrances, all eyes were drawn to the heavily carpeted staircase with its walnut balustrade. On the second floor they visited the so-called bridal chamber, taking up half the building's front with its own private parlor. Brussels carpet covered the floor. Costly oils hung on the walls. Walnut and silk furnishings dominated. The only blank space awaited a piano still in transit. Along the main corridor were sixteen smaller bedrooms filled with rich carpets and walnut furnishings. Thick mattresses caught the eye of many an envious miner.

Downstairs crowds glanced into the offices and reading cove before stopping to gaze at the dining room with its three chandeliers and walnut tables covered with cut glass and fine china. Kitchen service was dominated by a twelve-foot Montagin range with patented boiler offering hot and cold water. The well-stocked barroom, leased to separate managers, and soon graced by a top-of-the-line Monarch billiard table, filled the east half of the main lobby. The office occupied a space to the left of the staircase, isolated from the reservation desk by a French plate-glass mirror. To either side were engravings of English turf scenes.

The Grand Hotel's elegance and charm helped inflate Tombstone's self-confidence that summer of 1880; its residents conveniently sidestepping questions surrounding the townsite patent, Frank Leslie's escapades, the anti-Chinese fiasco, and other misgivings. The growing population dreamed of their town becoming a new county seat. Even as Tombstone searched for a civilized identity the sounds of construction at least pointed toward material progress. In time violence and other disasters overshadowed common expectation, but few worried now about the future. Amidst all the self-congratulations, Pima County Sheriff Charles A. Shibell had already appointed a new deputy for the Tombstone District—a fellow named Wyatt Earp.

Chapter 7

The Earp Brothers

Although in residence only twenty-eight months, the Earp brothers have become Tombstone's focal point. Thanks to an ascendancy of folklore over fact, events are now chronicled as having taken place before or after their famous gunfight with the Clantons and McLaurys. That clouded story, together with Doc Holliday's checkered career, has been told many times. Yet few recognize the simple but tragic error of time and place lying at the core of this legend. Instead, a series of deadly encounters has evolved into something sadly simplistic. An easily seduced sense of history is often used by modern aficionados to satisfy some personal need for heroic imagery. Dedicated disciples willingly embrace all the melodrama while ignoring the truth.

What the Earps actually did fails to compare with the contributions made by local entrepreneurs or that army of nameless men who worked the mines. By contrast, the brothers and their enemies are remembered today only for specific acts of violence. Rising from some deep-rooted psychological urge to glorify such behavior, popular culture has reinvented the Earps, particularly Wyatt, and pawned them off to that rather pathetic world populated by shallow, one-dimensional folk heroes.

It seems strange, considering the interest in Wyatt Earp's life over the last sixty years or so, but there is still no reliable biography of the man. Truths of one sort or another can be found, but published sources all fail to provide a satisfactory explanation of this man's long and often troubled life; a life filled with bizarre contradictions of character and purpose. Despite all the lies and legends swirling beyond the bounds of reality, Wyatt Earp and his brothers were hardly as evil as their detractors suggest, nor the dedicated guardians of goodness their unashamedly ardent admirers wish us to believe. As with everyone else, these men must be judged within the limited confines of their own time, place, and background. Even acknowledging those incidents of violence and adventure, these men, for the most part, led surprisingly ordinary lives.

Of course the question remains whether anyone, even a peripheral character like Wyatt Earp, can really be understood as a whole. Or is it simply a case of defining the possibilities after studying surviving fragments of contradictory evidence? Mark Twain understood this dilemma: "What a wee little part of a person's life are his acts and his words. His real life is led in his head, and is known to none but himself. . . . Biographies are but the clothes and buttons of the man—the biography of the man himself cannot be written."[1]

Although outwardly cold and taciturn, the Earp brothers were hardly seen as unique. They gloried in their triumphs and cursed their defeats. Itinerant gamblers and erstwhile lawmen, the brothers obsessively guarded their private lives. Viewed as clannish by much of the world around them, this closeness of purpose at times faltered under the pressure of events. Those bitter experiences would haunt the family for decades. Woven into all this turmoil, their women suffered much in Tombstone. Never accepted into the town's leading social circles, the four Earp wives remained isolated—often at the insistence of their husbands—in a community none of them remembered with fondness.

Five of the brothers eventually settled in the booming Arizona silver camp. Optimistically calculating all the possibilities Tombstone could offer, they began arriving on December 1, 1879. They came with a vague plan of entering the transportation business. For years Wyatt dreamed of living life as a wealthy capitalist. During his days in the Kansas cattle centers of Wichita and Dodge City, he watched with envy the respect shown visiting Texas cattle barons. Toying with that image, Wyatt Earp, too, could see himself someday as an important cattleman. It made little difference that he knew nothing whatever about the inner workings of the livestock empire he saw himself heading.

The mining tycoons that often strolled Tombstone's streets prompted a similar vision. Wyatt Earp knew of the power and influence wielded by western bonanza kings. Blinded by misguided dreams of fame and fortune, he dabbled in mining property for the rest of his life. Earp's failure came from grasping more at the image of a successful entrepreneur

---

[1]Clemens, *Mark Twain's Autobiography,* 1:2.

than in trying to understand the reality of such undertakings; nor did he show any inclination to study and learn the necessary details of the various enterprises he hoped to dominate.

When the most influential investors visited Tombstone, coming as many of them did in private railroad cars to Benson and later to Contention City, checking into the finest reserved suites at the Grand Hotel or Cosmopolitan, men such as Wyatt Earp felt intimidated by these elegant displays of wealth and power. Secretly seeing himself in that world, Wyatt unknowingly shared this common fantasy with hordes of disillusioned contemporaries. But as a popular character type, Wyatt Earp has become representative of something far more complex, something buried deep within the dream corner of the American psyche.

All his life Wyatt Earp longed for the respect shown the empire builders. Yet for him it would not happen. Instead, Earp remained a rather nondescript figure traveling the shadowy boundaries of the underworld. The limit of his abilities and the reality of his life forever blocked passage into that gilded world he hoped to enter. Still, he was no naïve simpleton. One Arizona historian, who interviewed Earp toward the end of his controversial life, concluded quite correctly that Wyatt was "a very suave and crafty dissimulator."[2] Only after his death would Wyatt Earp become a respected, larger-than-life heroic symbol. But even that would not last, and today this mythical view of the man appears tarnished to objective researchers not blinded by hero worship.

Wyatt resigned as assistant marshal of Dodge City in September 1879, and together with his second wife, Mattie, traveled to his brother's home at Prescott, Arizona. In New Mexico they picked up Doc Holliday, then breaking the monotony of his small dental practice with visits to the gaming tables. Doc's common-law wife, Mary Katherine, (a refugee of Tom Sherman's Dodge City saloon and dance hall[3]) came with him. They all rode in Wyatt's wagon. "It was during our trip from Las Vegas to Prescott that Doc and Wyatt became such good friends, which meant the end of the happiness we had enjoyed," Kate confessed. "I loved Doc and thought the world of him, and he was always kind to me until he got mixed up with the Earps."[4]

[2]Lockwood, *Pioneer Days in Arizona*, 283.

[3]1875 Kansas State Census, Dodge City, Ford County, Sheet 5, Line 3.

[4]Memoirs of Mary Katherine [Holliday] Cummings.

Others, too, remembered Doc Holliday's initial friendliness. "He was quite a noted character in those days," recalled a later governor of New Mexico. "I met him quite frequently and found him to be a very likable person."[5] Even Virgil Earp admitted, long after Holliday finally joined them in Tombstone: "There was something very peculiar about Doc. He was gentlemanly, a good dentist, a friendly man, and yet outside of us boys I don't think he had a friend in the Territory."[6]

Despite his own wife's foreboding, Virgil chafed with rumors about Tombstone. An older brother, James, arrived with his family. Shelving plans to rendezvous at Leadville, Colorado, the men now looked southward. Virgil had orchestrated the Earps' latest adventure. A younger sibling, Morgan, would soon follow from California. At Prescott the family carefully selected belongings for their trip. Included was Allie's sewing machine, which Virgil agreed to take only after confronting his wife's stubbornness. They passed out of the mountains to Wickenburg and then pushed on to Phoenix over the same route taken nearly three years before by Gird and the Schieffelins. Before long they reached Tucson. There, on November 27, 1879, U.S. Marshal Crawley P. Dake appointed Virgil a deputy for southern Arizona. The marshal needed no prodding, familiar as he was with Earp's reputation for steadfastness.

Moving south along the San Pedro beyond Ohnesorgen's, all three Earp wagons pulled to the edge of the road to allow passage of an incoming stage. As the driver cursed and snapped his whip over the heads of his team, Virgil explained to his wife that the U.S. Mail enjoyed the right of way. The stage driver then made a serious mistake. Racing by, he deliberately pulled close enough to graze one of Virgil's horses. Seeing his animal's blood, Earp started in pursuit, followed closely by his brothers who had no idea what was wrong. At a stop five miles ahead Virgil confronted the driver and beat him nearly senseless as Wyatt and James watched in silence. The dazed man finally mumbled an apology through loosened teeth, surrounded by stunned bystanders wondering who these people were. After the stage reached Tombstone the story made the rounds, convincing many that these somber newcomers were not to be trifled with.

---

[5]Otero, *My Life on the Frontier*, 216, 218.

[6]*San Francisco Examiner*, May 27, 1882.

Once in Tombstone Wyatt dismissed the family's plan of starting a stage line. While looking over the town and making themselves agreeable to local political and business leaders, the brothers began visiting saloons and testing the gaming tables. As Allie recalled with mounting disgust:

> the men couldn't find anything to do somehow. I could see right off none of them had come to Tombstone to do any prospectin'. . . . Wyatt had no intention of workin'. You couldn't hardly get him to split any firewood, he was that careful of them long slender hands of his. He was always kneadin' his knuckles and shufflin' cards to keep his hands in shape for gamblin'. But mostly he was out sizin' up the town as he said.[7]

To help meet expenses and keep food on the table, the Earp wives took in laundry and did sewing for a penny a yard. As for accommodations, Allie remembered: "We happened on a one-room adobe on Allen Street that some Mexicans had just left. It didn't even have a floor—just hard packed dirt, but it cost forty dollars a month. We fixed up the roof, drove the wagons up on each side, and took the wagon sheets off the bows to stretch out for more room. We cooked in the fireplace and used boxes for chairs."[8]

Wyatt may have been uninterested in physical work, but he gave his occupation as "farmer" on the 1880 census. Virgil did the same rather than describe himself as a federal officer. In a blush of honesty James admitted being a saloonkeeper—actually running a small sampling room at 434 Allen Street. James, his wife, and teenage stepdaughter were listed twice on successive days. On the first enumeration, however, all seven of the clan were shown living together.[9] Two others would join their siblings in Tombstone. Morgan arrived within weeks, followed later by his wife, Louisa. Eventually the youngest of the brothers, twenty-five-year-old Warren Earp, made an unheralded appearance. But during the June census those three were still living in southern California with Nicholas Earp and his wife.[10]

During their tours around town the brothers made the acquaintance

---

[7]Waters, *The Earp Brothers of Tombstone,* 90-91.

[8]Ibid., 90.

[9]1880 United States Census, Tombstone, Pima County, Arizona, E.D. 2, Sheet 7, Lines 40-46; and Sheet 13, Lines 14-16.

[10]1880 United States Census, Temescal, San Bernardino County, California, E.D. 63a, Sheet 56, Lines 11-16.

of Wells Fargo agent Marshall Williams, an affable twenty-eight-year-old New Yorker who would play an important role in the drama about to unfold.

Williams had come to Arizona by way of California and Utah. Earlier he worked for David Neahr, a mining man from Picacho, California, then doubling as a freight forwarder for the Southern Pacific as it pushed rails toward Yuma. Although Neahr chased bigger dreams involving mining and the railroad, Williams, a man of limited vision, severed the connection to branch out on his own. At Tucson he opened a grocery on Meyer Street south of Dick Brown's saloon. But Tombstone's glitter sucked in the young man, together with his wife, three-year-old daughter, and infant son (Marshall Jr. would die at age seventeen months). Besides working as agent for H. C. Walker & Company and Wells Fargo, Williams opened a cigar, tobacco, and stationery store, then became correspondent for the *Boston Economist.*

Impressed by the powerfully built Virgil Earp, especially after the new deputy U.S. marshal helped arrest a counterfeiter of Mexican and American silver coins, Williams deliberately gave all the brothers a closer look. Taken in by their seemingly calm determination, and hearing from Wyatt an account of his exploits as a Kansas lawman, Williams finally offered Earp a position as shotgun messenger. This not only allowed Wyatt a few extra dollars, but it gave him an opportunity to study the surrounding country and meet area residents. Maneuvering himself into a position, however minor, with the prestigious firm Wells, Fargo & Company also provided a degree of respectability—an important consideration to a man like Wyatt Earp.

On the night on July 21, 1880, thieves stole six government mules from Camp John A. Rucker in the Chiricahua Mountains. Its commanding officer, an English-born 1st lieutenant (also officially designated a brevet captain) named Joseph H. Hurst, set out in pursuit of the purloined animals accompanied by four soldiers. They followed a fairly clear trail toward Tombstone. There Hurst prevailed upon Virgil Earp to act in his capacity as deputy U.S. marshal, thus adding the weight of that office to the chase. This was important since the military had no law enforcement authority over civilians despite the nature of the offense. Virgil agreed and asked his brothers Wyatt and Morgan, as well as Marshall Williams, to ride with him.

At Charleston Wyatt claimed he met a man named Dave Estes (probably Daniel E. Estes, then a miner from Camp Rucker), who said he saw the mules at McLaury's ranch, north of the Charleston narrows on the Babocomari. There, Earp's informant insisted, the government markings were being altered; the normal "U S" left shoulder brand changed, unimaginatively, to "D S" (Hurst later described them as reading "D 8"). Before any arrests could be made McLaury's friend, Frank Patterson, took Hurst aside for a private conversation. Afterward the officer told Virgil he had reached an understanding that would retrieve the government's property, contingent upon the deputy U.S. marshal and his party returning to Tombstone. Earp reluctantly withdrew. Hurst and his soldiers also left empty-handed. With no power of arrest the military had been outmaneuvered.

Recriminations soon ricocheted through Tombstone newspaper columns. Lieutenant Hurst fired the opening salvo in the *Epitaph* on July 30, offering a $25 reward for the arrest and conviction of three men he tentatively identified as Pony Diehl, A. T. Hansbrough (Augustus S. Hansbrough), and Mac Demasters (Sherman McMasters). Hurst also offered a like amount for each of the six mules recovered. He then charged Frank Patterson, Frank McLaury, and Jim Johnson with assisting the thieves in hiding the animals and changing the brands.

Frank McLaury replied with some heat in the August 5 issue of the *Weekly Nugget*. He said the lieutenant quietly spoke of stealing the mules himself, a charge no one took seriously. McLaury claimed innocence in the whole affair, calling Hurst "a coward, a vagabond, a rascal and a malicious liar." Frank ended his defense with spirit: "My name is well known in Arizona, and thank God this is the first time in my life that the name of dishonesty was ever attached to me. . . . I am willing to let the people of Arizona decide who is right."

Since nothing could be legally proved one way or the other, the matter quietly dropped from public attention. Until the stolen mules episode, however, the Earps and McLaurys had gotten along rather well. But this brief encounter north of Charleston created a dangerous realignment of relationships that ended in an unfortunate and explosive climax fifteen months later.

Aside from sizing up Tombstone, guarding Wells Fargo shipments, and helping his brother when called upon, Wyatt Earp found time to regularly visit the county seat at Tucson. He not only toured the gambling joints but took time to expand his list of acquaintances. These included the county sheriff, Charles A. Shibell. As he had done earlier with Marshall Williams, Wyatt now made certain the sheriff appreciated his law enforcement experiences at Wichita and Dodge City. Shibell then occasionally used Wyatt as a special deputy. His performance was such that on July 27, 1880, Shibell hired Earp as one of his regular deputies at Tombstone.[11]

John Clum reacted warmly to the news: "The appointment of Wyatt Earp . . . is an eminently proper one, and we, in common with the citizens generally, congratulate the latter on his election. Wyatt has filled various positions in which bravery and determination were requisites, and in every instance proved himself the right man in the right place."[12] Keeping everything in the family, Morgan Earp took over Wyatt's old job with Wells Fargo.

Despite his appointment and future political plans, Wyatt had not yet registered to vote. Virgil had signed up on January 20, but for some reason Wyatt, Morgan, and James waited until September 27 to add their names to county rolls.[13] Becoming a deputy sheriff was important to Wyatt Earp. The appointment not only provided the authority he craved, it also placed him in a position to witness the political workings of the county from the inside. Meanwhile, he hoped to convince fellow citizens of his rather formidable talents as a peace officer.

Even a cursory examination of the record proves Wyatt was an excellent choice. When tracking an offender no distance became an obstacle, no man's reputation deterred Shibell's new deputy. Wyatt Earp turned in a remarkable performance, so much so that his efforts easily overshadowed all the other officers working in and around Tombstone at that time.

But this devotion to duty remains one of the most glaring contradictions of Wyatt Earp's personality. On the one hand he could be a steadfast officer—as he proved in Dodge City and again working for Shibell

[11] W. S. Earp, Deputy Sheriff Appointment, July 27, 1880; sworn before Justice of the Peace Mike Gray and filed July 29, 1880, Pima County Recorder's Office.

[12] *Tombstone Epitaph*, July 29, 1880.

[13] Pima County Great Register, No. 2495, V. W. Earp, Precinct No. 17, January 20, 1880; No. 3194, Wyatt Earp; No. 3258, Morgan Earp; and No. 3267, J. C. Earp, Precinct No. 17, September 27, 1880.

at Tombstone—while on the other he never passed up a chance to fleece some poor sucker in a card game. Wyatt Earp earned a rather unsavory reputation among police in various cities as a card hustler and confidence man. Between stints as Dodge City's assistant marshal, he and Dave Mather—the ubiquitous "Mysterious Dave"—were kicked out of Mobeetie, Texas, for running the gold brick scam.[14]

Charlie Shibell made his decision to add another deputy none too soon. The district wallowed in fresh episodes of senseless violence. Both justice courts were up and ready for business. Mike Gray stood as the old hand adjudicating trouble while James Reilly, an eccentric lawyer and one-time journalist, occupied the second bench. To fill Judge Bidwell's sudden vacancy, the board of supervisors at first favored Andrew J. Felter. Reconsidering, they appointed the politically well-connected Reilly instead.[15] He was a man with wide experience who would leave his mark on Tombstone's volatile history.

Born in northern Ireland, James Reilly came to America as a teenager in 1849. He joined the army soon after stepping off the ship. During a frontier career spanning a decade, he earned the rank of sergeant-major. This experience may have contributed to the bluntness of character so evident in Tombstone. After discharge he became a freighter, first in Texas and then Arizona. While farming and running a grist mill in Mexico, he killed a man. Jailed and then detained in what amounted to house arrest, Reilly did not return to Arizona until late 1866. He found work as a teamster, mine laborer, prospector, and steamboat wood chopper before organizing a small freighting firm between Colorado River mining camps. Later he became a military contractor, hotel proprietor, and developed small mercantile interests.

Reilly studied law and was elected Yuma County district attorney in 1876. Only then did he get himself admitted to the bar. Two years later he started the *Yuma Expositor.* Soon afterward he moved it to Phoenix. There his iconoclastic political views cost him the county printing contract. Selling out, the ex-editor packed up his three-hundred-volume law library and headed south, reaching Tombstone as a forty-nine-year-old

---

[14]McIntire, *Early Days in Texas,* 131-132. For an explanation of how this incredulous scam worked, see Quinn, *Gambling and Gambling Devices,* 248-257.

[15]"Board of Supervisors," *Arizona Daily Star,* July 9, 1880.

bachelor on May 12, 1880. Residents soon found Reilly a man unlikely to compromise. Eight years before he had argued politics with Tom Bidwell and claimed, "I think . . . we did not speak or have a conversation afterward until in Tombstone."[16] Now he sat on the bench ready to dispense frontier justice to an apprehensive clientele. And business was heating up.

In the early evening of July 22 Tom Waters lost his life over his choice of apparel. That morning he bought a black and blue plaid shirt. Touring saloons following good luck at the gaming tables, Waters endured some good-natured ribbing. But as his alcohol level rose he began taking offense. In the back room of Tom Corrigan's Alhambra he declared he would fight the next challenger. At the front door he met E. L. Bradshaw, an older and smaller man. The two were friends, having prospected together in earlier days. Unaware of the threat, Bradshaw merely mentioned Tom's new shirt, whereupon Waters knocked him cold with a fist to the eye. Crossing Allen Street, he repeated the performance on a young man named Ed Ferris.

Regaining consciousness, Bradshaw washed the blood from his face. At his cabin he bandaged his left eye and armed himself. Returning to town he calmly took a seat in front of Vogan and Flynn's saloon and waited. Watching as Waters finally left the Alhambra, he hurriedly crossed over, asking his friend, "Why did you do that?" Hearing Waters reply, "Because you made fun of me," Bradshaw pulled his revolver and started shooting. Waters turned to flee but fell suddenly to the sidewalk, a bullet hole under his left arm, two in his back, and another at the top of his head, the slug angling down toward the neck.

Acting Coroner James Reilly held an inquest. Mike Gray followed up with a preliminary hearing, after which, denying a defense motion for bail, Bradshaw—a panel member during the Killeen inquest—was transferred to Tucson to await action of the grand jury. Wyatt Earp eventually served thirty-two subpoenas in the case. Of this episode George Parsons recorded, "Too much loose pistol practice."[17] Despite Waters's disagreeable behavior, his funeral procession, Clum explained, "was quite a large one and embraced many of our most prominent citizens."[18]

---

[16]Testimony of James Reilly, *Garner v Gird,* 308; McClintock, *Arizona,* 2:504; Lutrell, *Newspapers and Periodicals of Arizona,* 95; and *Portrait and Biographical Record of Arizona,* 155-156.

[17]Parsons, *Private Journal,* July 24, 1880, 142.

[18]*Tombstone Epitaph,* July 24, 1880.

Around daybreak on July 30 Allen Street again echoed with hostile gunfire. This time Roger King, a leading figure in the anti-Chinese hysteria, shot and killed his one-time cabinmate, a tin-roofer named Thomas Wilson. Frank Leslie testified that a bleeding King stumbled into his bar at the Cosmopolitan demanding cartridges, claiming Wilson had beaten and tried to rob him. King found his prey outside the Headquarters Saloon below Fourth Street, and the two men opened fire. Wilson ran through the building with King in hot pursuit. Witnesses heard more shots from the rear of the saloon. There Wilson lay dead with a bullet through his heart. King calmly walked up Allen Street and surrendered to Constable Daniel Miller. A coroner's jury heard testimony, and a warrant was issued by Mike Gray charging King with murder.

All this gunplay prompted George Parsons to complain: "Our town is getting a pretty hard reputation. Men killed every few days—besides numerous pullings and firings of pistols and fist fights. One man in town now with a $1200 reward for him. Others known to be convicts."[19]

Before arrangements were completed to escort King to Tucson, John J. Pace, a thirty-eight-year-old miner from Virginia, handed Deputy Babcock a message from Charlie Shibell ordering him to surrender custody. Luckily Wyatt Earp chose to intervene before the older man naively complied. Simply put, the new deputy questioned the order's validity on the strength of the sheriff's name being misspelled. Earp telegraphed Tucson and learned the message was indeed a forgery. Shibell ordered Earp to arrest Pace. Wyatt did so after turning King over to the embarrassed deputy for removal to Tucson. Later Babcock tried to salvage his reputation by claiming he had detected the forgery himself and only gave the impression of going along in order to discover if others were involved. No one believed him. Congratulations came quickly from John Clum: "The frustration of the plan is entirely due to the judgement and good sense displayed by officer Earp."[20]

Soon afterward Burleigh Mining Company official Ward Priest and a female companion drove past a group of men having fun at the expense of a drunk. Mistaking the remarks as directed at him, Priest angrily returned to Allen Street and assaulted one of the group, a young auctioneer named James Henley. The two had had an earlier argument. A

[19]Parsons, *Private Journal,* July 30, 1880, 147.

[20]*Tombstone Epitaph,* August 3, 1880.

warrant was given Wyatt Earp, who arrested Priest and hauled him before Judge Reilly.

Earp enjoyed a busy week, having already arrested a Mexican for brutally beating a woman. This case would be dismissed because the victim proved too drunk at the time of the assault to identify her attacker. Later Justice Gray handed Wyatt a warrant charging Thomas A. Bordan (Bortan), a twenty-one-year-old clerk at Camp Rucker, with attempted murder. Earp returned the accused to Tombstone. Again Gray dismissed the charges after a short hearing. Wyatt then joined the futile search for a disgruntled ex-employee from the Contention who had stabbed night foreman William Shaw in the back after being fired. Superintendent Josiah White offered a $100 reward—without success—for the man's apprehension.

Miners had more to fear than arbitrary dismissal. On August 11, 1880, they crowded a meeting at Ritchie's Hall, above Brown's Restaurant on Fifth Street, to discuss working conditions and decide whether or not to organize a union. Owners and representatives of the major producers, including Richard Gird, Josiah White, E. B. Gage, Thomas Farish, Ward Priest, Amos Stowe of the Empire, A. H. Emanuel of the Vizina Consolidated, and five others, angrily responded by issuing a manifesto against union activity: "Having established wages in this camp upon a scale more liberal than exists in any mining district elsewhere, we claim the right to employ such men as we see fit, without dictation from any outside source whatever; and to this end we have given instructions to our foremen . . . to give employment to no one identified with the so-called 'Miners Union.' If the mines can not be worked without such assistance we will close them down."[21] More would be heard on that subject, but the harsh tone succeeded in intimidating workers for nearly four years.

In a move seen by many as long overdue, the Tombstone Common Council finally agreed to regulate the carrying of concealed weapons by issuing Ordinance No. 9 on August 12, 1880: "it shall be unlawful for any person not an officer of the law to have or carry in the Village of Tombstone any fire-arm, knife or other dangerous weapons concealed about his person, without a written permit from the Mayor. . . ."[22] Violators faced a fine as high as $50 and/or thirty days in jail.

[21] *Weekly Nugget,* August 12, 1880; and *Tombstone Epitaph,* August 12, 1880.

[22] *Tombstone Epitaph,* August 14, 1880.

All this came too late to be of any help to Mike Killeen. Ever since his encounter with Buckskin Frank on the balcony of the Cosmopolitan Hotel, residents stood divided over the guilt of Leslie's friend George Perine. William T. Lowery, one of the dead man's former employers, claimed to have uncovered new evidence. On the morning of August 14 he swore out a complaint. The court handed the warrant to Wyatt Earp for service. Fearing Perine might escape after warnings from friends, the new deputy organized a small posse to assure the man's capture. Morgan Earp arrested Perine between Richmond and Old Tombstone and turned him over to Deputy Sheriff Babcock.

The next day Wyatt learned of five horses stolen from Robert Mason, a Contention City merchant and hotelkeeper. He asked his brothers Virgil and Morgan to take up the chase. They tracked the animals to Charleston, only to discover the thieves held a commanding lead heading south toward Sonora, and thus gave up pursuit. While there, however, the brothers heard that Deputy Sheriff Scow from Camp Grant had arrested a man thought to have stolen some horses and mules from that place three months before. In a quirk of justice, a Mexican had just stolen a mule from the Camp Grant offender. Virgil and Morgan tracked the thief. "He showed fight," the press reported, "but gave up when a six-shooter was run under his nose by Morgan Earp."[23] The brothers turned their frightened captive over to the deputy at Charleston before returning to Tombstone.

Meanwhile, George Perine was scheduled to appear before Judge Reilly. The examination lasted ten days and produced some interesting testimony, delighting an ever-curious public. As it turned out, Frank Leslie had been expecting some kind of trouble that night. Twice he and Mike Killeen drank together at Lowery and Archer's saloon, east of Tasker and Pridham's and a stone's throw from where the dance party was held. Exactly what they said is unclear. In all likelihood a threat of some sort was made. Afterward Leslie armed himself with two revolvers, confiding to Perine, "George, I fear somebody is going to make a break to-night."[24]

Townspeople understood that Frank Leslie's interest in another man's wife had precipitated this tragedy. Killeen's own statements, those given to Mike Gray and then to others attending him at his deathbed, impli-

[23] Ibid., August 17, 1880.

[24] Testimony of N. F. Leslie, *Tombstone Epitaph*, August 24, 1880.

cated Perine. The dying man's only regret was that he would never get revenge. Dr. Henry M. Matthews testified that from the angle of Killeen's chest wound the shot could not have been fired by Leslie, who admitted falling to his knees during the scuffle.

All this new testimony changed few minds, but a fresh controversy had already erupted. City Councilman Harry B. Jones, one of the attorneys, had exchanged words on the street with Judge Reilly the day the hearing opened. Entering Reilly's court, Jones listened silently as the judge demanded an apology, explaining he could not remain on the premises without one. Jones refused and was ordered out.

The next morning he reappeared and the argument continued. Reilly directed the officers to remove him. None complied until Jones pulled a revolver and began reciting his constitutional rights. Reilly jumped the angry attorney and the two exchanged blows. Wyatt Earp stepped between them, placing both men under arrest. Jones was taken before Judge Gray and charged with assault. Released on his own recognizance, he boldly reentered Reilly's court, only to be fined $25 and ordered to spend a day in the county jail.

Earp took Jones into custody a second time but allowed the attorney to await the coach while he returned to the courtroom. Still enraged, Reilly demanded that the deputy appear the next morning to show cause why he should not be held in contempt for not removing Jones when first ordered. Wyatt explained that he was already under instructions to deliver Jones to Tucson and that he could not be in two places at once. Earp again reminded the judge to consider himself under arrest as soon as the day's proceedings ended. Reilly fumed. After court Wyatt escorted the red-faced jurist to Judge Gray's office. He, too, was released on his own recognizance. But Reilly never forgave Shibell's new deputy for his involvement in this embarrassing incident.

Wyatt was not the only officer to run afoul of Justice Reilly. Constable Miller picked up policeman James Bennett on a complaint of false arrest filed by Charlie Calhoun and Ed Peacock. They had been hauled in for disorderly conduct while under the influence. Bennett released them the next morning after they paid him $10, explaining the money was a forfeiture for their appearance. Calhoun now claimed an injustice because they had not been allowed to appear before a magistrate. Reilly ordered the officer held for the grand jury.

At Tucson Judge John S. Wood dismissed the Bennett case as there was no proof of malice. He then cited the careless manner and defective condition of records from justice courts, resulting in dismissals and undue expense to the county. Reilly, applying a heavy dose of predictable sarcasm, took it all personally and responded with a lengthy defense of himself "and all other equally ignorant, careless and incompetent Justices. . . ."[25]

Now, hearing the contempt charge against Harry B. Jones, the Tucson judge faced another example from Reilly's court that convinced him he had been right all along. Citizens found this latest episode amusing, even for Tombstone. "What a complication of affairs," one editor chuckled. "He . . . came up on a writ of habeas corpus before Probate Judge Woods [*sic*]. . . . The order of commitment proving defective, and the punishment imposed for the contempt greater than a Justice can inflict, the prisoner was at once discharged from custody on motion of the District Attorney."[26]

The two Earp brothers (Virgil accompanied his brother taking the unrepentant attorney to the county seat), returned to Tombstone late in the evening of August 20. Trying to assess politics while at Tucson, they now prophesied one segment of the upcoming county elections: "the Sheriff's fight is generally conceded to [J. H.] Hewitt, as a dark horse, Shibell being considered a dead duck, from the fact that he is believed to be and always has been a Republican. That he, in all probability, will not go before the Democratic Convention, but run as an independent candidate."[27] Only time would judge the brothers' skill as political prognosticators.

Of more immediate concern was the response to James Reilly's judicial outburst. Over a hundred citizens signed a petition urging the judge's resignation, characterizing his conduct as disgraceful and incompetent. Included were major political, business, and professional figures such as Mayor Randall, Artemus Fay, Dr. Henry M. Matthews, banker Heyman Solomon, and mining men Ward Priest and Edward Field. Less substantial citizens were represented by the likes of Peter Spencer, one of the original group arrested in the Van Houton murder case. Both

[25] *Tombstone Epitaph*, September 5, 1880.

[26] *Arizona Daily Star*, August 19, 1880. For more on Tucson's reaction, see *Arizona Daily Star*, August 18, 1880; and *Daily Arizona Citizen*, August 18, 1880.

[27] *Tombstone Epitaph*, August 21, 1880.

Wyatt and Morgan Earp signed. Virgil's name is missing but he was preparing to travel to San Bernardino to visit his family and rejoin his wife, who had left three weeks before.

Considering Reilly's stubborn personality, his reaction came as no surprise. He simply ignored everyone and refused to resign. Sending his response to John Clum for publication, the beleaguered jurist attacked his accusers as having "acted with less dignity, deliberation, justice and consideration towards your fellowman than I have done." He then singled out thirty of the signatures as belonging "to a class whose good opinion I never coveted."[28]

Reilly turned his back on the affair and went about the business of presiding over the examination of George Perine. After hearing the testimony of more than twenty witnesses and enduring five hours of closing arguments, Reilly decided the evidence justified sending the case to the grand jury. With Perine unable to post the $5,000 bond, Wyatt Earp transferred him to Tucson on the afternoon of August 27. In October the grand jury refused to indict, and the case was dropped.

While the Perine hearing was winding down, Dr. Henry Hatch and a man named Welch argued over the ownership of the lot on the southwest corner of Fourth and Fremont. Welch lived there in a small cabin he had built. Hatch, claiming never to have relinquished ownership, showed up one morning and began sawing down several posts Welch had placed around the property. Hearing of the attack on what he considered his home, Welch confronted the ill-mannered intruder. Dr. Hatch pulled his revolver.

This rather volatile medical man had reason to believe weapons offered a workable solution to his problem. Only four months before he had resorted to six-gun justice backing Mike Gray off some Fremont Street real estate: "Words ran high, pistols were drawn and an excited scene followed. . . ." Gray's son Richard "received a cut in the face from the pistol, in the hand of the Doctor, but not a serious one."[29] Perhaps mindful of that incident, Welch now retreated and swore out a warrant which Wyatt Earp served that same afternoon.

Confronted by Shibell's cold and taciturn deputy, Dr. Hatch, a forty-

---

[28] Ibid., August 27, 1880.

[29] *Weekly Nugget*, May 6, 1880.

year-old Englishman, meekly handed over his revolver and submitted to arrest without saying a word. Judge Reilly found him guilty of exhibiting a deadly weapon in a rude, angry, and threatening manner and fined him $100, much to the delight of Mike Gray. The upper portion of the lot in question eventually became the property of cattleman Henry C. Hooker. Dr. Hatch quietly settled for half the adjacent one.[30] In time Hooker leased the corner to Thomas Moses for the Capitol Saloon.

Wyatt Earp kept busy. Early in the evening of August 24, at a place near Watervale called Mineral Springs, teamster George McKinney shot and killed Captain Henry Malcolm, a fifty-two-year-old barman, part-time carpenter, and all-around colorful character.

The chief witness was Henry Forest, himself just released from jail for fast driving. He claimed the victim had come to the aid of a man named Mason. McKinney wanted to retrieve a pistol, but the intoxicated Mason refused its return. Captain Malcolm interrupted the brawl by smashing an unloaded shotgun over McKinney, forcing him off the downed man. Ordered home, McKinney reluctantly agreed on condition he got back his .38 Smith & Wesson. Mason tossed the revolver from his cabin into the dirt. Then, for whatever reason, the various antagonists agreed to have a drink together. Moments later shots rang out. Malcolm, hit four times, called for his partner and older brother Samuel as he staggered and fell. Mason, although a major participant in the original quarrel, escaped unhurt.

Wyatt and Morgan Earp rushed to the scene but returned empty-handed. Around noon the next day Wyatt learned McKinney was at his father's house near an arroyo leading from Watervale to the San Pedro. The older man told the deputy his son was taking a bath about a quarter mile away. Earp intercepted him on the trail and made the arrest. McKinney claimed Malcolm had advanced upon him in a threatening manner and that he only fired in self-defense. The prisoner insisted he would have surrendered the day before but feared mob retaliation. After all, Captain Malcolm, a Mexican War veteran and later an officer in the Mexican army during the Maximilian interlude, was a respected citizen, as his well-attended funeral suggested.

---

[30]Assessment Roll of the City of Tombstone, H. C. Hooker, N 90 ft., Lot 10, and W 1/2 Lot 9, Block 17, Paid December 3, 1881, 35; and Henry Hatch, E 1/2 Lot 9, Block 17, Paid January 24, 1882, 13.

Wyatt delivered his prisoner to answer the murder charge. Judge Gray ordered McKinney held without bail and issued Earp subpoenas to assure the appearance of witnesses. While serving these papers the deputy sheriff happened upon John M. Glenn, an assault and battery suspect being removed from Contention City owing to the illness of the local magistrate. Wyatt willingly took custody and handed Glenn over to Mike Gray, who dismissed the charges: "The bulk of the evidence showed that no weapon was used in making the attack, that in fact it was nothing more than a drunken row at most."[31]

Some objected to Gray's sense of justice in deciding such cases. One Sunday morning assault and battery incident in mid-September, ending with a pistol shot fired into the front wall of the Alhambra, drew only a $10 punishment. After Wyatt Earp arrested James Henry for beating his wife, the judge fined the offender $7.50 (a favorite sum in Gray's court) and bound him over to keep the peace for $200 more, prompting Clum to suggest, "Mrs. Gray ought to sue for a divorce from the Judge for letting the wretch off so easily."[32]

Judge Gray finally opened the McKinney examination on the morning of September 3. As expected, witnesses contradicted one another. In the end, however, Morgan Earp, acting as a special deputy, took the prisoner to Tucson. There, following a two-day trial in mid-October, a jury acquitted the defendant after deliberating only twenty minutes.

Wyatt took time from his busy schedule of card games and serving as Shibell's deputy to join the town's first fire department, soon known officially as Tombstone Engine Co. No. 1. Earp hoped to impress potential voters with his sense of civic responsibility. Joining was also a good idea socially, as it provided another way to develop and strengthen friendships. The fire company carried with it elements of a fraternal organization. Wyatt saw it all as one more step toward fulfilling his long-range political goals. Having an organized department also made practical sense in a community with so much wooden construction. The town had already witnessed a number of small fires. Luckily all were easily contained.

Not wishing to tempt fate, a small group gathered in early September

---

[31] *Tombstone Epitaph,* August 29, 1880.

[32] Ibid., October 5, 1880.

The town's first firehouse opened in the summer of 1881. The date on the sign simply acknowledges the fire company's early organizational efforts. Wyatt Earp, the group's secretary, is standing at the far right, second row.
Courtesy, Arizona Historical Society.

at Kelly's Wine House—then located in the Heitzleman Building near the Grand Hotel—to organize the engine company. The assembly elected Harry B. Jones, the lawyer and city councilman, president and Wyatt Earp secretary. Committees were formed to solicit subscriptions for equipment and to draft a constitution and bylaws. At the next meeting, held five days later at Danner and Owens's Hall, liquor dealer Leslie F. Blackburn became foreman. He brought with him experience as a fire department volunteer at both Virginia City and San Francisco. Other members included saloonman Milton Joyce, City Marshal Fred White, Wells Fargo agent Marshall Williams, newspaper publisher Artemus Fay, and newcomer John H. Behan.

That second meeting had its lighter side: "an interloper loaded to the water line with 'Oh, be Joyful' put in an appearance, and after repeated warnings to keep quiet which he paid no heed to, was unceremoniously pitched through a window by Marshal White. . . ."[33] The assembly broke up with all those in attendance rather pleased with themselves.

Wyatt kept busy hauling in social misfits and making arrests in less violent cases, together with working numerous civil filings, attachment suits and the like, all the while wondering if his fellow citizens bothered to take notice. Earp constantly calculated every move by how it benefited himself and his family.

All these efforts to build a constituency were seriously threatened on October 10 by Wyatt's twenty-nine-year-old friend Dr. John Henry Holliday. That Sunday evening the Georgia-born, Pennsylvania-trained dentist quarreled with Johnny Tyler at the Oriental. Fearing trouble between these two hot-headed gamblers, mutual friends separated and disarmed both. Proprietor Milton Joyce ordered Tyler out.

Turning on the intoxicated dentist, Joyce berated him for starting a row. Doc continued arguing until the exasperated saloonman quite literally tossed him into the street. Holliday, a small man physically but unpredictable when aroused, stomped back through a side door demanding his revolver. Ignored, he rushed to his boarding house for another one, angrily opening fire upon his return. Joyce got off one shot but missed. Closing on Holliday, he beat the drunken man senseless with a six-shooter. Only then did Joyce discover he had been wounded in the right hand. His partner E. P. Parker, standing behind the bar when the gunfire exploded, was shot through the big toe of the left foot.

City Marshal White and Officer Bennett rushed in to disarm the parties. They escorted the wounded proprietor from the premises after helping Holliday to a chair, "it being generally thought, from his bloody appearance, that he was severely, if not fatally, hurt. Such, however, proved not to be the case, and he was arrested by Deputy Marshal Earp."[34] Joyce swore out a complaint. Holliday furnished the $200 bond demanded by Judge Reilly. All three injured parties took to their beds. For a time friends feared that Joyce might lose his hand, but the rugged saloonman from San Francisco recovered. Holliday also eventually survived that night's brush with the law.

---

[33]Ibid., September 15, 1880.

[34]Ibid, October 12, 1880.

Doc Holliday had not come to Tombstone with Wyatt Earp in December 1879. Instead he stayed in Prescott pushing his luck at the gaming tables. For a time he shared living quarters with carpenter Richard Elliott and John J. Gosper, later Arizona's acting governor. Doc finally linked up with Wyatt at Tombstone during the late summer of 1880. He registered to vote but turned quickly to gambling and the bottle, making as many enemies as friends.[35] Suffering from consumption, Holliday's mood seldom seemed affable. Kate's absence did not help. The two had quarreled early on at Prescott and she refused to accompany him. Kate seldom visited Tombstone, and then only to try her best "to break up the partnership between Wyatt and Doc."[36] Leaving Prescott, she stayed about a week at Tip Top before going on to Globe and opening a boarding house for miners.

Wyatt, who acknowledged the gambling dentist had once saved his life in Dodge City, had little time to worry about Doc's latest dilemma, despite their friendship. On the same night as Holliday's less-than-bravura performance at the Oriental, Jerry Barton killed E. C. Merrill at Charleston. Deputy Sheriff Milton McDowell called for Earp's assistance in transporting his prisoners to Tombstone. Wyatt and one of his brothers responded, hauling in Frank Ray, Jim Clark, and Al Wicher.

Those three had been carousing drunkenly with Merrill at the time of the tragedy, firing their revolvers while ignoring McDowell's commands to stop. At Barton's saloon they demanded gambling money but were ignored. Merrill, at whose Grand Central Store Oliver Boyer killed Martin Sweeney in 1878, later returned, pulled a gun, and threatened Barton. The twenty-seven-year-old barman drew his own pistol and instantly killed Merrill with three well-placed shots. Justice of the Peace Burnett set bail at $2,500. Barton's friends Frank Stilwell and John Campbell furnished the funds. Burnett fined Merrill's three companions $150 or an equal dollar-a-day jail rate. Unable to pay, they accompanied the two Earps to Tombstone, there to await transportation to the county lockup.

Wyatt had briefly turned away from law enforcement to help his

---

[35]1880 United States Census, Prescott, Yavapai County, Arizona, E.D. 26, Sheet 4, Lines 22-24; and Pima County Great Register, No. 3357, J. H. Holliday, Precinct No. 17, September 27, 1880.

[36]Memoirs of Mary Katherine [Holliday] Cummings.

brothers build houses for the family on property they owned at the southwest corner of First and Fremont. Below these dwellings they erected stables for their horses, including racing stock they hoped would earn extra money on Tombstone's half-mile track near Watervale.

Duty soon called Wyatt from his carpentry chores to serve a grand larceny warrant issued from James Burnett's justice court against Peter Spencer. (Clum's newspaper reported this arrest taking place on October 23, but Wyatt's county expense account gives the date as October 8.[37]) At Tombstone Spencer had purchased two stolen Mexican mules. He resold the animals, but the original owner recognized his property. Tracing the transactions, he swore out the complaint against Spencer, who eventually secured his freedom, but not without some bitterness against Wyatt. The paths of these two would cross again in the months to come.

Earp then came to the aid of City Marshal Fred White. Just past midnight on October 28, 1880, responding to a group of revellers discharging their revolvers into the night sky, White tried disarming William "Curly Bill" Brocius. Brocius was in town from the San Simon Valley. He had joined a number of other saloon-touring free spirits—the whole group mistakenly identified as "Texas Cowboys." At Tom Corrigan's Alhambra someone suggested they go up the street. Near the corner of Sixth shots were fired. Mine laborer Andrew McCauley turned to Charleston resident James Johnson, then working part time as a miner, and said, "Let's get out of this." As they ran to the south side of Allen Street, McCauley heard Brocius call out to the others, "This won't do boys." Curly Bill then joined McCauley and Johnson seeking shelter behind a small cabin set back from the street nearly halfway to Toughnut.

Wyatt Earp stepped from the Bank Exchange Saloon after all the commotion disrupted his card game. From the corner of Fifth and Allen he heard shots and clearly saw pistol flashes a block away. Unarmed, the deputy sheriff asked his brother for a weapon. Morgan pulled back his coat, showing that he, too, carried no gun, suggesting instead that he ask Fred Dodge. The Wells Fargo informant quickly offered Wyatt his revolver.

---

[37] *Tombstone Epitaph*, October 24, 1880; and Wyatt S. Earp, Deputy Sheriff Expenses, October 8 through November 9, 1880, Miscellaneous Records, Pima County Recorder's Office.

Approaching Brocius and the marshal (McCauley and Johnson had by then backed away another ten or fifteen feet), Earp heard White's stern voice: "I am an officer, give me your pistol." Brocius slowly pulled his revolver from its holster. Just as Wyatt grabbed Curly Bill from behind, White reached for the gun, demanding, "Now, you God damn son-of-a-bitch, give up that pistol." As he spoke White tried jerking it from Curly Bill's grasp. A shot rang out, the bullet striking the surprised officer in the left groin, penetrating the lower intestine. Earp pistol-whipped Brocius to the ground.

Dodge and Morgan Earp rushed over. The cabin, near the later site of the Bird Cage Theater, was one they shared before Morgan's wife, Louisa, arrived. The Wells Fargo man recalled Wyatt telling him to snuff out the smoldering fire burning White's clothes, caused by the closeness of the shot and the black powder then used in cartridges. Wyatt arrested Brocius and together with his brothers Virgil and Morgan rounded up a half-dozen others. Andrew McCauley escaped arrest and Wyatt Earp released another offender, Jerry Atkinson, after he furnished a cash bond. Some criticized Earp for setting the man free, but Wyatt defended himself by showing a copy of the judge's release order.[38]

As for the others, Gray fined Andrew Ames, a miner from Virginia, $40 for carrying a concealed weapon and discharging it on the street. Edward Collins, James Johnson (accused earlier in the army mule fiasco; he would die seven months later at Galeyville after accidentally shooting himself), and a man named Richard Loyd all paid $10 on weapons charges. Gray dismissed the case against Frank Patterson, a perennial troublemaker, after it was proved that in this instance he had tried controlling the situation. Later that same day, October 28, on motion of Sylvester B. Comstock, the town council appointed Virgil Earp assistant marshal at a salary of $100 a month.[39]

Tucson attorney John Haynes represented Curly Bill before Judge Gray. Reacting to all the hard feelings building in town, they waived examination. Gray ordered the case transferred to Tucson. Rumors of a vigilance committee helped speed up the process of removing the prisoner. Wyatt, Virgil, and Morgan Earp, together with George Collins, took their man in tow while others helped guard the buggy all the way

[38]*Tombstone Epitaph,* October 29, 1880.

[39]Minute Book A, Common Council, Village of Tombstone, October 28, 1880, 9-10.

to Benson. On the train Brocius asked Wyatt to recommend a good attorney. Earp suggested Hereford and Zabriskie. Curly Bill declined, confessing James Zabriskie had once prosecuted him in Texas. He then reportedly admitted killing a stage driver in El Paso County.

Wyatt Earp's expense account shows he claimed $2 for arresting Brocius, seventy cents serving a subpoena, $30 for "taking prisoner to jail 75 miles," plus another $20 covering four guards one way and $10 for one guard's round trip. Two more dollars paid for meals, and $18.92 was charged for everyone's train ride.[40] With Brocius tucked safely away in a jail cell, residents awaited reports on Marshal White's deteriorating condition. By mid-morning, October 30, doctors openly confessed he had no chance to survive. White gave an official statement on those events surrounding the shooting, and then, in the presence of his father and friends Ward Priest and Henry Fry, died within the hour. He was just thirty-one years old.

Curly Bill survived the law's relentless urge to punish, primarily due to Fred White's deathbed statement that the shooting was accidental. Wyatt Earp's testimony supported that conclusion. During a hearing before Justice Joseph Neugass at Tucson in late December, gunsmith Jacob Gruber demonstrated that the .45 Colt revolver Brocius carried was mechanically damaged. It could be fired from the broken half-cock spur if the hammer was mistakenly lowered into that position. Others testified that the defendant tried restraining his intoxicated and overzealous companions after the night's merriment got out of hand. Carefully reviewing the evidence, Neugass ordered the prisoner discharged.[41]

Brocius learned nothing from this brush with the law. Within the month he and a companion terrorized Contention City, firing their weapons at passersby and robbing one business of $50. Warrants were issued but Curly Bill held off the local deputy with obscene threats and a Henry rifle. They continued their drunken revelry at Watervale, challenging county officers to come and take them. It all lead one disgusted citizen to suggest: "The terror these men have caused the traveling public, as well as the residents along the San Pedro, is having a serious influ-

[40]Wyatt S. Earp, Deputy Sheriff Expenses, October 8 through November 9, 1880, Miscellaneous Records, Pima County Recorder's Office.

[41]*Daily Arizona Citizen*, December 27, 1880. This issue contains the testimony of Wyatt S. Earp, James R. Johnson, Andrew McCauley, Jacob Gruber, and others.

ence, and this scab on the body politic, needs a fearless operation to remove it. It is no trouble to find the rascals, for they are not hiding, and they defy the law."[42]

Marshal White's funeral had taken place under the auspices of the fire company at 2:30 on the afternoon of October 31. Mourners crowded the building—a new two-story adobe called Gird's Hall on the northwest corner of Fourth and Fremont—long before services began. Both the common council and White's comrades at the fire company issued resolutions of respect. Methodist preacher J. P. McIntyre presided over the service, with miner Reuben F. Coleman manning the organ. After singing one final hymn the crowd formed a procession to the cemetery.

Everyone stood by the grave in respectful silence. John Clum grumbled over McIntyre's remarks: "To say that the reverend gentleman trenched upon the bounds of common sense is but to echo the sentiment of the vast congregation. . . . The reverend knows just as much about the Great Unknown as any living creature who has never been there, and his captious flings at the fireman's resolutions were as injudicious as they were ill-timed." The combative editor, however, ended his account soberly: "The cortege following the murdered Marshal to the grave was the largest ever seen in our embryo city. It embraced all classes and conditions of society, from the millionaire to the mudsill and numbered fully 1,000 persons."[43]

While friends buried Fred White, news reached Tombstone that Jerome B. "Jerry" Atkinson, the man Wyatt had released from custody on Judge Gray's order, was found murdered at Groton Springs, some thirty miles distant on the Southern Pacific line. On the night of White's shooting he had been discussing a possible cattle deal with the San Simon crowd, but few took him seriously as a participant in the tragedy. Atkinson carried about $100 at the time of his death. Evidence at the scene suggested he had resisted his attackers.

Following the sad drama of Marshal White's funeral, residents were in no mood to sympathize with lawbreakers. They applauded Wyatt's

[42] *Arizona Weekly Star,* January 27, 1881.

[43] *Tombstone Epitaph,* November 2, 1880. For additional references to the whole episode surrounding the death of Marshal White, see *Tombstone Epitaph,* October 28, 29, 31, 1880; *Daily Nugget,* October 30, 31, November 2, 1880; Minute Book A, Common Council, Village of Tombstone, October 30, 1880, 11; and Lake, ed., *Under Cover for Wells Fargo,* 10, 235-236, 241-242.

arrest of James "Red Mike" Langdon in front of the Bank Exchange Saloon. Langdon was wanted in Virginia City for murdering a man the year before with a Bowie knife. Unfortunately, authorities in Nevada, apparently wanting nothing more to do with the unruly fugitive, now refused to issue an extradition warrant. In Tombstone the prisoner was eventually set free.

As the town debated issues of law and order, Virgil Earp decided to run for city marshal in the special election hastily scheduled for November 12. He abandoned plans, announced the day before White was shot, to challenge Isaac Roberts and Owen Gibney as constable for Precinct No. 17. Incumbent Daniel Miller had skipped out on a prospecting trip to New Mexico. The council then appointed six extra policemen, including Roberts, Bob Hatch, and Ben Sippy to serve temporarily under Virgil and keep the peace during the regular general election of November 2.[44] Earp had no part in this selection process—he was simply given the list and ordered to hand out badges.

Pleasing many citizens, the hot-tempered James Reilly had decided not to seek reelection as justice of the peace. Instead three new justices would be chosen. None should have taken office until January, but Andrew J. Felter was elevated to the bench two weeks after the election because of Reilly's then-unexpected resignation. But no county race was more hotly contested than that of sheriff, a bitter fight pitting Democratic incumbent Charlie Shibell against his Republican challenger Bob Paul.

Charles A. Shibell, born in 1841 at St. Louis and educated at Davenport, Iowa, arrived early in Arizona. Giving up his job as a Sacramento store clerk, Shibell saw Tucson for the first time in 1862 as a teamster with General Carleton's California Column. After discharge, he settled in and devoted the next dozen years to various pursuits including mining, ranching, freighting, and farming an area north of the Mexican border. He also served as treasurer of both the Tucson and Citizens Building and Loan Associations. By 1870 Shibell became interested in politics and was named deputy collector of internal revenue. Nearly five years later he was appointed a deputy sheriff. As a Democrat Shibell was elected Pima County sheriff in 1876. His record was exemplary. In

[44]Minute Book A, Common Council, Village of Tombstone, November 1, 1880, 12.

1880 he was running for a third term but faced a determined Bob Paul.[45]

Robert H. Paul, a man with long experience as a peace officer, was born in Massachusetts in 1830, the son of a Canadian father and Portuguese mother. As did so many New Englanders before him, he began his working life as a cabin boy on a whaling ship, and enjoyed the adventure of being briefly marooned on Tahiti. Excited by rumors of gold, he left Honolulu and sailed into San Francisco Bay in early 1849. After mining five years in the Sierra foothills, Bob Paul turned to the law. At Calaveras County he first became a constable of Campo Seco township, then a deputy and later undersheriff, before finally being elected sheriff in 1859. He would lose his race for a third term.

By 1874, after again trying his luck as a mining man, Paul accepted a position with Wells Fargo, this time working as a special detective and express messenger out of Tulare County. As such he came to Arizona four years later, accompanied by his wife, three children, and mother-in-law. He quickly established a reputation as a resolute officer, so much so that Crawley Dake would appoint him a deputy U.S. marshal. Tucson Republicans saw in this quiet Californian a man capable of winning the sheriff's office. As their candidate Bob Paul faced Charlie Shibell on November 2, 1880.[46]

After a hot campaign Shibell appeared the winner in a close race; only 42 votes separated the two candidates. That is, until Republicans began questioning the returns, especially those from Precinct No. 27. There Bob Paul lost by a staggering margin of 103 to 1. The balloting took place at Joe Hill's house at San Simon Cienega, north of Apache Pass, with Ike Clanton appointed inspector and John Ringo and A. H. Thompson serving as poll judges. Deep-rooted distrust surrounded these choices.[47] Supervisors' clerk and former Pima County sheriff

[45]Farish, *History of Arizona,* 1:127-132; McClintock, *Arizona,* 3:888-891; Adams, *History of Arizona,* 3:537-538; and Mike Anderson, "Posses and Politics in Pima County," 253-282.

[46]Bancroft, *History of Arizona and New Mexico,* 620, n7; Pima County Great Register, No. 2283, R. H. Paul, Precinct No. 1, December 31, 1879; No. 6968, Robert Havlin Paul, Precinct No. 1, November 1, 1880; 1880 United States Census, Tucson, Pima County, Arizona, E.D. 39, Sheet 40, Lines 28-33; *Tombstone Epitaph,* September 24, 1880; *Weekly Arizona Citizen,* September 25, 1880; and Ball, *United States Marshals of New Mexico and Arizona Territories,* 128, 166.

[47]Pima County Board of Supervisors Minute Book, 1:424, 426-427; *Arizona Daily Star,* October 14, 1880; and *Weekly Arizona Citizen,* October 16, 1880.

William Sanders Oury originally planned to locate the balloting at Phin Clanton's ranch straddling the Arizona-New Mexico border. But since no one could determine on which side of the line his house stood, the move was made to San Simon.

The Clantons had successfully cultivated strong political ties with many Democrats and some Republicans. The same could not be said for Ringo. Although he had hoped to be named San Simon's delegate to the county Democratic convention in August, he did not register to vote until late October and then requested its cancellation.[48] Ten years earlier, living with his mother at San Jose, California, he worked as a farmer.[49] Ringo later gained a small reputation in Texas as a troublemaker before moving to Arizona. In December 1879 he shot Louis Hancock in the neck at Safford after the unfortunate man declined to have a whiskey with him. Hancock preferred beer.[50] Ringo delayed the case, claiming in a letter to Shibell, "I got shot through the foot and it is impossible for me to travel for awhile."[51]

Ringo's parents had been married at his mother's home in Clay County Missouri in 1848, following Martin Ringo's service as a private with Colonel Alexander Doniphan's Missouri Mounted Infantry during the Mexican War. John Ringo, however, was born in 1850 in his father's home state of Indiana at a place then called Green's Fork. They soon returned to Missouri where John spent his boyhood. The family traveled across the plains in 1864 and settled in California, but without Martin, who accidentally shot and killed himself on the trail near Glenrock, Wyoming. Young John suffered a painful mishap of his own when one of the wagons rolled over his foot. By his late teenage years Ringo was little more than a farm laborer with a serious drinking problem. Later his misdeeds so disgusted his family he found himself ostracized. Only within the narrow confines of legend does his life have meaning, and his legend is mostly fantasy.[52]

---

[48]Pima County Great Register, No. 5372, John Ringo, Precinct No. 27, October 21, 1880.

[49]Colahan and Pomery, *San Jose City Directory and Business Guide*, 115.

[50]*Arizona Daily Star*, December 14, 1879; and Territory of Arizona vs. John Ringo, In Justice Court, Edward Tuttle, J. P., Precinct No. 9, Pima County, December 9-10, 1879, 46-47.

[51]John Ringo, San Simon Valley, N. M., to Charles Shibell, Tucson, March 3, 1880. Courtesy of Robert N. Mullin.

[52]Mexican War Military Record, Martin Ringo, National Archives and Records Service; Woodson, *History of Clay County*, 107; and *Journal of Mrs. Mary Ringo*, June 7, July 30, 1864.

Neither Charlie Shibell nor Bob Paul cared much about John Ringo. Instead, Paul concerned himself with his pending suit alleging election fraud after both sides formally requested the poll lists. A trial was finally scheduled for mid-January 1881. Shibell had already been sworn in for another four-year term. The hearing produced some interesting testimony.

Irregularities had clearly taken place, not only at San Simon, but at Tres Alamos and Tombstone as well. Shibell supporter Leslie F. Blackburn, a saloonman and liquor dealer from San Francisco, claimed nearly 30 Tombstone votes shifted illegally to Paul out of a combined total of 819. Others acknowledged errors but disputed the numbers. Evidence concerning San Simon seemed the most damaging. Witnesses claimed nowhere near 104 votes had been cast. The place itself was little more than a faint smudge on the map. It became a shipping point for the California Mining District, on the northeast side of the Chiricahua Mountains some twenty miles south, but even eight weeks after the election it could boast only the freighting depot, a single store, one restaurant, a saloon, and some twenty inhabitants scattered about in small adobes and crude shacks.

To deceive curious onlookers, the ballot box was stuffed with Shibell's votes before the polls opened. William Sanders Oury testified that he had taken "the certificates of registration to San Simon myself, and gave them to Isaac Clanton." Oury later visited Clanton at his father's ranch south of Charleston. Asked why, he answered, "That's my business." When pressed, he responded angrily, "I shall not answer," but quickly added, "I thought his character had been attacked, and I wished to have him come in and clear himself. . . . my reason for going for Clanton was that my friendship for him was so strong; might have had other objects in view."[53] Concerning what these "objects" might be, Oury unfortunately refused to elaborate.

The whole proceeding lasted several days, and in the end Judge C. G. W. French disallowed the San Simon returns, giving victory to Bob Paul. Shibell's forces appealed. The question languished with the territorial supreme court for nearly three months. Charlie Shibell continued to act as Pima County sheriff until the high court ruled against him.

[53] *Weekly Arizona Citizen*, January 30, 1881.

Before any of this could be decided, Virgil Earp was caught up in the special city marshal's race. It started out as a three-man contest with James Flynn, the former saloon partner of Jim Vogan, the first to step forward. Earp quickly followed. Benjamin Sippy, a miner from Pennsylvania accused with seventeen others for the past two months of squatting on property claimed by the Vizina Consolidated Mining Company, was the last to announce.

Virgil campaigned hard and kept arresting troublemakers. The day after the general election he picked up Owen Gibney, one of his former rivals for constable, for carrying a concealed weapon and firing it on the street. Judge Gray fined the chagrined defendant $7.20 on each count. Earlier, Gibney had withdrawn from the constable's race and was replaced by Hugh Haggerty.

On the eve of the election James Flynn dropped out, despite widespread support, joining instead with T. E. Fitzpatrick as co-proprietor of the Cosmopolitan Saloon—soon renamed the Stock Exchange. With Flynn's unexpected departure everyone predicted a tight race between Virgil Earp and Ben Sippy. A week before the balloting, newspaperman Artemus Fay described Earp "as a fearless and efficient officer [who] will serve the city faithfully if elected."[54] Three days later he mentioned Sippy's strong campaign, noting, "The Marshalship for the next two months will be a profitable office, as he will collect the city taxes."

Voters crowded the polls on November 12. Unfortunately for the Earp brothers, Virgil lost this bid for public office by a vote of 259 to Ben Sippy's 311.[55] Earp resigned as assistant marshal two days later, collecting his final $54.83 from the city.[56] Clum remarked, "Mr. Sippy will make an efficient Marshal, and should receive the support and assistance of all good citizens."[57] Virgil fell back on his position as a deputy U.S. marshal.

Four days after the election Marshal Sippy found himself facing a new ordinance governing the police force. He now needed the council's approval for any appointments he made. They also retained the right to discharge individual officers. To reduce repeated complaints and protect

[54]*Daily Nugget,* November 4, 1880.

[55]Minute Book A, Common Council, Village of Tombstone, November 13, 1880, 15.

[56]Ibid., November 15, 1880, 18.

[57]*Tombstone Epitaph,* November 13, 1880.

the new department's image, the ordinance decreed: "No policeman when on duty shall enter any public drinking house, gambling house, house of ill-fame or place of public amusement, unless in the discharge of his official duty. . . ."[58] In addition, all constables, sheriffs, or deputy sheriffs residing within Tombstone's corporate limits were authorized to make arrests for local violations.

Three days before Virgil's defeat, Wyatt, who had supported fellow Republican Bob Paul, resigned as Shibell's deputy.[59] His last official act was to again arrest attorney Harry B. Jones, this time for failing to respond to a summons. Whether or not Shibell forced his ungrateful deputy to resign is unclear. Yet something of the way the community viewed Wyatt's service can be seen in the November 12 issue of the *Daily Nugget*: "Wyatt Earp's resignation as deputy sheriff was heard of by his many friends with regret. During the time he has held the office he has been active and prompt in the discharge of all duties and every citizen had the consciousness that his life and property were as well protected as they could be by any single officer."

Accepting Wyatt's resignation on November 11, Shibell replaced Earp with John Harris Behan, a longtime Arizona political insider. Behan opened his office with Harry B. Jones before moving into the back of an Allen Street cigar store.[60] Artemus Fay said of the new deputy: "Intelligence, experience and courage are in him united and the residents of this section are to be congratulated that they will have so efficient an officer."[61] Time would tell. Two weeks later Shibell repaid Leslie Blackburn's support by also naming him as a deputy.[62] Blackburn, who in 1868 had killed a man in self-defense at Virginia City, Nevada,[63] disappointed many, appearing more interested in working as a bill collector for commissions than as a peace officer. Even so, within four months Crawley Dake appointed him a deputy U.S. marshal.[64]

---

[58]Ordinance . . . Establishing Village Police, *Daily Nugget*, November 17, 1880; and Minute Book A, Common Council, Village of Tombstone, November 16, 1880, 21.

[59]Wyatt S. Earp, Resignation as Deputy Sheriff, November 9, 1880, Pima County Recorder's Office.

[60]John H. Behan, Deputy Sheriff Appointment, November 10, 1880, Pima County Recorder's Office; and *Daily Nugget*, December 2, 1880.

[61]*Daily Nugget*, November 12, 1880.

[62]Leslie F. Blackburn, Deputy Sheriff Appointment, November 24, 1880, Pima County Recorder's Office.

[63]*Daily Territorial Enterprise*, April 17, 18, 1868.

[64]*Arizona Weekly Star*, March 24, 1881; and *Weekly Arizona Citizen*, March 27, 1881.

Wyatt's resignation worked unexpectedly against him. Surrendering his badge of respectability, Earp slipped back into that shadowy world of the saloon and gambling hall. With this change of emphasis from peace officer to saloon habitué, many began looking at Wyatt with curious suspicion. Citizens were no longer so sure about this somber man as they had been during those forceful weeks he served as Charlie Shibell's deputy.

Wyatt's decision may have left him without any official position, but his strategy seemed logical. It was no mystery that a new county, embracing the Tombstone Mining District, was being considered. By supporting Bob Paul, Wyatt Earp hoped to prove his loyalty to Republicans, thus advancing his scheme of influencing those in power to help make him sheriff. With a new county Wyatt would have lost his position anyway, since Shibell's jurisdiction would no longer include Tombstone.

Some disapproved of the brothers' pragmatic maneuvering, one dissenter noting: "The 'Earp' family flopped for 'Paul' on election day, thinking he was a dead winner, and went against 'Little Charlie,' but they failed to connect. Their ingratitude to one who had always been their friend has been marked by his many friends in Tombstone, and retribution politically has already reached one."[65] The Earps slipped back into political limbo, but their day would come.

---

[65]"Hawkeye" to the editor, November 12, 1880; *Arizona Daily Star*, November 14, 1880.

Chapter 8

New County and Old Troubles

Despite political setbacks, Virgil and Wyatt Earp stayed close to the action. Since arriving in 1879 the brothers had dabbled in marginal mining properties, water leases, and real estate. Back in Dodge City one report spoke in exaggerated terms of a $30,000 sale: "The boys should have had a quarter of a million for this valuable property, but not having capital with which to work it, they let it slide. . . ."[1] Locations such as the Long Branch and Dodge reflected Wyatt's Kansas years. The Mattie Blaylock he named for his wife. Wyatt, Virgil, and James participated together in most of these. With other claims they brought in investors such as C. G. Bilicke or Tucson retailer Albert Steinfeld. More often the outside partners were R. J. Winders or A. S. Neff.

Robert Winders, a Pennsylvanian twenty-seven years Wyatt's senior, had come to Tombstone with his family by way of Texas in 1879. He kept chasing mining property even after opening a keno game with Charlie Smith at the back of Danner and Owens's saloon. Winders traveled widely throughout southeastern Arizona and northern Mexico searching for mineral sites. Early on he became chief mining advisor to the Earp brothers. Their friendship endured, the bond itself not broken until Winders died on a trip to San Antonio a decade later.

Andrew S. Neff, a thirty-six-year-old teamster and small-time merchant, got involved with the Earps in a number of claims. None brought him the financial security he longed for. Neff had served in an Iowa infantry regiment during the Civil War. Afterward he wandered, unable to settle in one place for long. Marrying at Fort Smith, Arkansas, he lived in Nebraska, Los Angeles, Phoenix, and Tip Top before coming to Tombstone. Even his place of birth is not consistent: at one time or another he claimed France, Ohio, and Illinois. Neff kept searching for

[1] *Ford County Globe*, March 23, 30, 1880.

opportunities to help support his wife, two daughters, and infant son. His Tombstone years remained ones of struggle. He ended up running a second-hand store at Raton City, New Mexico.

In early November 1880 Wyatt Earp and Andrew Neff sold their interest in the Comstock Mine to A. H. Emanuel at an undervalued price. Neff and three of the Earps then bonded the Grasshopper over to him for an equally low sum. Reports three months before falsely suggested they had transferred that property to San Francisco mining speculator Robert F. Pixley. Emanuel, older and wiser than Wyatt, knew more about mines than Winders, Neff, and all the Earps put together.

Alfred H. Emanuel was by then superintendent of the Vizina Consolidated. Born in Philadelphia, he traveled to San Francisco by way of Panama in 1850. A year later he met Mike Gray for the first time, the two becoming life-long friends. Emanuel worked as an apprentice for commission merchants Bryant and Paxton before joining shipping agent Thomas Cundell. A decade later he changed professional emphasis and moved to Nevada. In Arizona he drew upon those new experiences with mining operations, especially on the Comstock Lode, where he worked two years as foreman of the famous Gould & Currey and nine months with Isaac James on the Yellow Jacket at Gold Hill.

Emanuel had started a livery business at Virginia City in 1864 with Charles H. Light, an earlier lessee of the Exchange Billiard Saloon at the International Hotel. The two eventually became Nevada freighters into Signal for the McCracken works. They shifted to Tombstone by January 1880 and began hauling ore for the Contention and Grand Central, before Emanuel took a position with James Vizina and made even more money at the expense of Andrew Neff and the Earps.

In fairness, the brothers did more than simply play at being mine owners. Four days after resigning as a deputy sheriff, Wyatt joined Winders in carrying out the survey for the 1st Northern Extension of the Mountain Maid—owned jointly with Virgil and James—covering a portion of the townsite later described as Earp's Addition. It must have been something to see. Winders, aided by Thomas Kelly, worked as a chain carrier with Wyatt Earp standing by as flagman. Deputy U.S. Mineral Surveyor Henry G. Howe attested to the work after George

Meley and J. H. Holliday witnessed the $500 improvement affidavit.[2] Considering their later elevation as frontier icons, it is amusing to picture Wyatt Earp and Doc Holliday playing such mundane roles in a mining survey.

More in keeping with his family's future image, Virgil Earp entered a rifle tournament in late November at a temporary shooting gallery set up behind Jim Vogan's saloon. Organizers flooded the town with small green handbills to assure a packed house. Virgil won $5 at fourth place, with a sixty-four out of a possible score of seventy-five. Wesley Fuller took home $15 for second place. Johnny Behan finished out of the money at number eight. The $20 first prize went to Henry A. Plate, a representative of his family's San Francisco firm specializing in firearms, ammunition, and other sporting paraphernalia.

As the Earps shifted their efforts from law enforcement to budding capitalism, John H. Behan opened his career as Shibell's newest deputy by arresting James Ryan, detained after trying to burglarize a miner's cabin. Behan enjoyed a mildly successful stint as a Pima County officer. At one point he chased an offender to within six miles of Fort Bowie to make an arrest. In fairness, however, his record hardly compares with that of Wyatt Earp. But then neither did any of Shibell's other appointments to the district.

After just three weeks in office Johnny Behan overturned his sulky while speeding. Little damage was done except to the new officer's pride. Speeding cases, averaging fines of $5, filled justice dockets toward year's end. Shortly after Behan's mishap a two year old was run over at the upper end of Fremont. The child escaped without serious injury. A couple days later John Clum wrote: "Fremont street seems to have been selected as a speeding ground for our fast trotters. Children should be kept out of the street, as it often happens that a valuable buggy is ruined from running over a child—and buggies cost money."[3]

At least children thus injured enjoyed the best of 1880 frontier medical science. Although outnumbered by lawyers, more than a half-dozen

[2]Survey Field Notes of the 1st Northern Extension of the Mountain Maid Mining Claim, General Land Office No. 6716, Mineral Certificate No. 78, Lot No. 62, November 10-16, 1880; and Records of Mines, Pima County Recorder's Office, Book J, 432-433.

[3]*Tombstone Epitaph,* December 14, 1880.

doctors worked in Tombstone that fall. Earlier practitioners included a woman, Mrs. Marion Webb, M.D., who opened an office near Sixth and Fremont in late April. She stayed through mid-September before moving on. That void was quickly filled by a rather colorful character in the Tombstone drama, Dr. George Emory Goodfellow.

Born in 1851 at Downieville, California, Goodfellow studied engineering at Berkeley before entering Annapolis in 1872. Tossed from the Naval Academy after only six months for striking a black man, he turned to medicine. Graduating from the University of Wooster Medical Department at Cleveland in 1876, Goodfellow began practice as a two-year contract surgeon at Fort Whipple. He then opened an office in Prescott. Discouraged with that decision, he again linked up with the army, this time as an assistant surgeon at Fort Lowell near Tucson. It proved a short tenure, the contract annulled at his own request.

Goodfellow planned to stay in Arizona despite his deep prejudice against Mexicans. After several exploratory visits to Tombstone he decided to settle there permanently. He arrived September 20. The new start did not go well after discovering his belongings rifled and his surgical instruments stolen by a government teamster. A speedy investigation led to the offender's arrest and the property's return. Resupplied, Goodfellow found space on the top floor of the new J. V. Vickers building on Fremont Street, three lots west of Fifth. (Later Dr. T. W. Seawell moved his practice there, openly advertising "Venerial [*sic*] diseases a specialty," thus highlighting another problem plaguing Tombstone's frisky population.) By mid-November, with help from Nellie Cashman and other citizens, the energetic Dr. Goodfellow opened the town's first hospital in Mrs. M. L. Woods's former restaurant at New Boston.[4]

The doctor became close friends with his office landlord. Both were nearly the same age and driven by visions of success, although in different fields. As a young man John Vickers left his home in Chester County, Pennsylvania, for a business career in New York City. After eight years he traveled to Tombstone, "when only those of mettle and courage succeeded," to dabble in real estate and the cattle business. In time he became president of the sprawling Chiricahua Cattle Company and

[4]Quebbeman, *Medicine in Territorial Arizona,* 18, n113; *Polk's Medical Register and Directory,* 314, 160; Pima County Great Register, No. 4220, G. E. Goodfellow, Precinct No. 17, October 11, 1880; *Tombstone Epitaph,* September 21, October 7, November 16, 1880; and *Daily Nugget,* September 9, 23, October 7, 10, November 12, 21, 1880.

served in a variety of political positions. He died in Los Angeles in 1912 as a sixty-two-year-old millionaire.[5] His Fremont Street medical tenant admired his tenacity.

Goodfellow would make his own mark on Tombstone, in time becoming a strong supporter of the Earp brothers. A skilled surgeon and proficient boxer, Goodfellow occasionally paraded his volatile personality. Nor do events suggest he mellowed much with age. Some years later he stabbed *Epitaph* paper carrier Frank V. White with a double-edged weapon resembling an Italian poniard for "annoying him and on numerous occasions [brushing] against him in a manner calculated to be insulting."[6]

Much was heard from the good doctor during Tombstone's earlier troubles. In the process he became Arizona's leading expert on gunshot wounds, even publishing papers on that subject in medical journals. In one he referred to the many Colt and Winchester firearms as simply "toys with which our festive or obstreperous citizens delight themselves."[7] At the same time he studied botany and mastered half-a-dozen languages.

Men like Goodfellow saw a future in Tombstone. Town lot titles might be questioned by the cautious, but caution was in short supply. Visions of wealth and success clouded rational thinking. The sights and sounds of heavy construction pacified most doubters.

William Gird had built a two-story adobe on the northwest corner of Fourth and Fremont (Marshal White's funeral was held there), displacing the old post office put up by Al Schieffelin and T. J. Bidwell. Postmaster John Clum temporarily moved one lot west, until Gird decided to expand to a sixty-foot front. With that decision the post office shifted diagonally across the intersection into one of Appolinar Bauer's remodeled buildings. Clum seemed satisfied with his new surroundings as it allowed many more boxes, helping him keep abreast of the growing population. He had already decided to rebuild his Fremont Street newspaper office, located two lots west of Gird's, into a two-story adobe. The new *Epitaph* opened in December, allowing the boastful editor's remark:

---

[5]Spaulding, *History of Los Angeles*, 3:523-524; *Trow's New York City Directory, 1872-1880*; Wagoner, *Arizona Territory*, 520; *Los Angeles City Directory, 1895–1912*; and The Estate of John V. Vickers, First and Final Account, June 19, 1914, Superior Court of the County of Los Angeles.

[6]*Tombstone Epitaph*, August 10, 1889.

[7]Goodfellow, "Cases of Gunshot Wound of the Abdomen," 214.

"Visitors are invited to call and examine for themselves the results of a business only seven months established."[8]

Meanwhile, after much bureaucratic wrangling in Washington, the government had finally issued the Tombstone townsite patent on September 22, 1880.[9] The instrument was mailed to James S. Clark, not to Mayor Randall. Suddenly, because of disagreements within the Interior Department, it was ordered returned. Again addressed to Clark, the document arrived a second time on November 5. All this heightened concern since the original application to the Florence Land Office was in the name of the mayor and common council, not Clark, Gray & Company.

That same afternoon a group from the newly formed and generally bipartisan Citizens' League called upon the mayor to question him about the growing controversy. Randall willingly confessed to not having the papers. James Clark had already sent them to Tucson for filing. The mayor also admitted not reading the patent but instead had listened as it was read to him. At a later meeting attended by George Parsons, Randall "gave his solemn promise that he would not deed until the patent was in his hands and not then until [a] meeting of the Council was called when he would be guided by their wishes."[10]

The Citizens' League then met in the parlor of the San Jose House to discuss strategy. They were stunned to learn from Councilman Harry B. Jones that Randall intended to deed nearly the entire surface of the townsite to Clark, Gray & Company, "irrespective of occupants or improvements, or the rights of individuals or the community," Clum quickly noted.[11] Parsons claimed the mayor told Jones, "There's more sugar in the other crowd." Randall was to get $2,500 and Jones $500. Parsons predicted Jones's "complicity will operate against his acting for the people now."[12]

In this instance Alder Randall proved a man of his word. On November 8, without any reference to the city council, he issued a transfer deed giving James S. Clark a one-fifth undivided share, Maurice E. Clark two

---

[8]*Tombstone Epitaph,* December 11, 1880.

[9]Certificate No. 177, Miscellaneous Records, General Land Office, 5:154-156; and *Tombstone Townsite Title Abstract,* 12.

[10]Parsons, *Private Journal,* November 7, 1880, 179.

[11]*Tombstone Epitaph,* November 7, 1880.

[12]Parsons, *Private Journal,* November 7, 1880, 180.

undivided one-fifth interests, and John D. Rouse and Mike Gray each a one-fifth undivided share. At the request of James Clark this controversial deed was filed and recorded the following day.[13]

All this was done despite provisions specifying the patent was to be held "in trust for the several use and benefit of the inhabitants of the town. . . ."[14] Outraged citizens crowded public meetings and lighted bonfires amid cries of fraud, a mood reminiscent of the summer's anti-Chinese upheaval. Nor did news of a nearly empty treasury calm anyone. No fees from property transactions had ever been collected. Despite all the activity in this bustling mining camp, city coffers contained only a few hundred dollars. Estimates of lost revenues greatly exceeded that amount. Clum reminded his readers, "the annual tax upon the citizens must be correspondingly increased."[15]

The government's deputy surveyor of mineral lands, Michael Kelleher, immediately began questioning the patent's boundaries: "the town can be located in either of two places. . . . If there was a recognized city government here, with town records accessible to the public, we might get a sight of the patent itself and find it all right; but from the copy published it is all wrong and crooked."[16] As it turned out there were errors in the original document, forcing a supplemental patent to be issued in 1883 "for the purpose of correcting an error as to the boundary line . . . which is still intact and uncancelled."[17]

John Clum jumped on the issue. His anti-Chinese crusade may have fizzled from confusion and embarrassment, but with the townsite patent fiasco he knew he was on to something. It is now almost impossible to explain his motives. He may have been genuinely concerned for his community's well being, or he might simply have manipulated public fears for personal political gain. Clum launched bitter attacks through his newspaper, helping create a hysterical atmosphere that lessened any chance of compromise. The townsite proprietors, ruthless men and survivors all, held firmly to the position that they had started the town and

[13]Transfer Deed, November 8, 1880, *Tombstone Townsite Title Abstract*, 15.

[14]Certificate No. 177, Miscellaneous Records, General Land Office, 5:154; and *Tombstone Epitaph*, November 11, 1880.

[15]*Tombstone Epitaph*, November 10, 1880.

[16]Ibid., November 12, 1880.

[17]Supplemental Patent, Miscellaneous Records, General Land Office, 5:497-499; and *Tombstone Townsite Title Abstract*, 13.

expected to enjoy a profit. Clum's constant complaining stiffened their resolve, setting the stage upon which none of the players planned giving an inch.

Newcomers tried ignoring these issues, mostly without success. Hotel registers bear witness to a steady flow, marking arrivals from Oregon, Nevada, California, Boston, Philadelphia, New Orleans, and countless places in between. Many drifted in from other camps already in decline, including Virginia City and Dick Gird's old Signal townsite. Clark, Gray & Company induced many to try Tombstone by circulating a catalogue highlighting the town's investment possibilities. To any and all self-deceiving believers, Clum gave warning: "Our advice to strangers is to read and retain the catalogue, and also to retain their money until the matter is adjudicated in the courts."[18]

No one seemed immune from this growing controversy. After Marshal Sippy arrested Red Mike for disorderly conduct, Judge Gray dismissed all charges on the defendant's solemn pledge to behave himself. Ignoring his promise, the unwanted Nevada fugitive and a companion promptly jumped one of Gray's lots near Sixth and Fremont. The disillusioned justice "had a little interview with him and they concluded to vacate."[19]

Perhaps as a welcomed respite, not every disagreement involved disputes over real estate. Around the time of Red Mike's latest ruckus, Edward Benson, co-proprietor of the O.K. Corral, argued with a man named Kellogg over a woman at Emma Burns's sporting house. Kellogg dared the quarrelsome southerner to arm himself, which he did, stomping back carrying a double-barreled shotgun. Perennial troublemaker Charlie Calhoun shoved a six-shooter out the door, ordering Benson's departure or he would blow the top of his head off. The grim-faced liveryman fired both barrels and made a run for it, only to be captured outside town by Marshal Sippy and Officer Bennett. George Parsons remarked, "Benson . . . shot Calhoun but didn't hurt him much."[20] Dragged before Judge Gray, the repentant offender paid a $20 fine for resisting the officers, $10 for firing a weapon on the street, and was discharged.

[18] *Tombstone Epitaph,* November 28, 1880.

[19] *Daily Nugget,* November 19, 1880.

[20] Parsons, *Private Journal,* November 13, 1880, 182.

Three days later it appeared there might be more gunplay. Workmen arriving for the early shift at the Gilded Age found their entry blocked by one of James Clark's men standing guard with a shotgun. The mine's owner, Edward Field, still engaged in his long-running feud with Clark, Gray & Company and distrustful of Tombstone justice, rushed to Tucson to protect his rights in the county courts. At Citizens' League meetings he had repeatedly advocated his well-known position regarding overlapping mining claims and townsite titles. But now, reported the *Nugget,* "During his temporary absence a cabin has been built over his shaft by the opposing party, and thus the matter stands."[21]

Within two weeks building contractor John Hanlon scuffled with C. G. Bilicke. The hotel proprietor had tried sinking a shaft in Hanlon's lot on the east side of Fifth north of Toughnut to run a tunnel into the Mountain Maid Mine. Bilicke had Dick Gird's permission to do so as the property sat atop the Good Enough claim, still part of the Tombstone Mill and Mining Company. Al Schieffelin settled the dispute by compromise, suggesting Bilicke work some distance away from the irritated Irishman's office.

Town lot troubles stayed in the public eye through the contentious columns of Tombstone's newspapers. Regarding Clark, Gray & Company, the *Nugget*'s tone appeared less strident than that embraced by Clum's *Epitaph.* Originally the two editors battled across traditional political lines. Before long they began exchanging journalistic pot shots of a more personal nature, such as Fay's less-than-flattering observation: "Neighbor Clum knows of lots of women in Arizona who raise cotton every time they draw their breath. Is it possible that it was while in pursuit of such information as this that he got bald-headed?"[22]

Clum responded in kind, but over the town lot issue he raised the heat, characterizing the *Nugget* as the Townsite Company's organ. In turn, Fay described his rival as "The daily 'slop bucket' down the street. . . ."[23] Growing tired of exchanging these endless insults, Fay reopened negotiations to sell out to Tucson journalist Harry M. Woods, his printer Elanson S. Penwell, and a man named Mitchell. Of the expected new regime

[21]*Daily Nugget,* November 19, 1880.

[22]*Weekly Nugget,* August 5, 1880.

[23]*Daily Nugget,* November 21, 1880.

Clum remarked, "As they are gentlemen of experience . . . we look for a long and useful life of the *Nugget.* Success to the enterprise."[24]

Within two weeks the *Nugget* congratulated Clum on his new baby daughter, while offering kind words regarding his wife's deteriorating condition. Mrs. Clum died late in the afternoon of December 18, leaving behind her husband, daughter, young son, mother, and a saddened community. Of Mary Clum, George Parsons wrote: "Half of the present female population or more could be better spared than she."[25] The editor shook off the effects of grief and buried himself in politics.

Disagreements over property rights kept multiplying. New York lawyer Thomas Wallace and Irish butcher Thomas Ward both claimed a lot near the Vizina and Cook Block. Ward filed suit, holding a deed from Clark, Gray & Company showing a $500 sale price. By late November, perhaps impatient with the pace of justice, he and two friends raided the disputed ground in Wallace's absence "and began to demolish the habitation . . . and to fire his goods and chattels into the public highway, the same being Allen street."[26] An enraged Wallace swore out a complaint after Ward ignored his protests. Marshal Sippy arrested the disgruntled butcher and his companions, as attorney Wallace tried repairing the damage under the gaze of an amused crowd of onlookers. Clum sided with Wallace, but Ward ended up with the property.[27]

An even more dramatic confrontation, arising from a long-standing feud, erupted ten days later. Back in September the always-volatile James Reilly had built himself an office near the southwest corner of Fifth and Fremont. Unfortunately he picked a spot claimed by James S. Clark of the Townsite Company. When Clark returned from Tucson a few days later, "He smiled grimly while looking at Judge Reilly's new office on his pet lot."[28] Not wishing an open clash with the unpredictable Irishman, Clark chose to bide his time. With Reilly now safely out of office at the justice courts, the company decided the time had come.

Early on Saturday morning December 4, 1880, James Clark and Mike Gray accompanied a dozen hired hands and tried to physically carry Reilly's small building into Fremont Street. A crowd quickly gathered

[24]*Tombstone Epitaph,* December 1, 1880.

[25]Parsons, *Private Journal,* December 18, 1880, 190.

[26]*Tombstone Epitaph,* November 25, 1880.

[27]Assessment Roll of the City of Tombstone, T. B. Ward, Lot 19, Block 19, filed January 31, 1882, 32.

[28]*Tombstone Epitaph,* October 2, 1880.

shouting insults. Reilly's caretaker ran for Ben Sippy. Whether intimidated by the marshal or the angry spectators who could just as easily turn into a lawless mob, Gray's men refused his order to proceed. One called out that he expected his $10 anyway. To deafening applause Sippy demanded, "This thing has got to stop right here." Someone loudly suggested they shove the house back onto the lot. As volunteers stepped forward, Clark and Gray hurriedly braced the building with two poles.

The townsite proprietors loudly ordered everyone off their property. Gray's son pulled a revolver, but his father dissuaded him from doing anything foolish, causing Clum to smirk in a tone of subdued disappointment, "the old man was right, for once: a single shot from that direction would have settled the townsite question very effectually." A woman pushed through the crowd and confronted Clark, asking if he intended moving her house. In reply he said, "he would if it were not an adobe." To that a "bystander remarked to the Louisiana land-grabber that an absent man, a defenseless woman, or a Chinaman were their only victims, and asked him why he did not tackle some one who was able to defend himself. He did not answer."[29]

Amid rumors of Clark and Gray planning to move the corner house to block Reilly's dislodged building, George Parsons and other armed volunteers spent the night guarding the property. The diarist feared someone might try blowing up the place, but the only episode worth mentioning proved less dramatic: "A couple of drunks during the night were fooling around, trying to find their way home probably as one said something about the house being in the way." The next morning curious bystanders and league members "made a rush and in a very few minutes had the house back where it belonged. . . . That's over and the Citizen's [*sic*] won."[30]

Marshal Sippy's resolve during the Fremont Street standoff endeared him to many, while the whole affair helped deepen the resentment against the townsite organizers. After Reilly returned from Tucson he began building a stronger fence, then filed suit against Clark, Gray & Company alleging $300 damage to "his property." Reilly eventually won title to the disputed lot.[31]

[29]Ibid., December 5, 1880.

[30]Parsons, *Private Journal,* December 4, 5, 1880, 187.

[31]Assessment Roll of the City of Tombstone, James Reilly, N 100 ft. Lot 9, Block 18, filed January 13, 1882, 24.

After nearly a month subscribing funds for legal expenses, the Citizens' League, in the name of John Clum, J. V. Vickers, and Appolinar Bauer, appeared before the territorial supreme court asking an injunction against Clark, Gray & Company. On December 2 Associate Justice Charles Silent issued a temporary restraining order against the Townsite Company, the Village of Tombstone, and Mayor Alder M. Randall. Deputy Sheriff Behan served the papers six days later. This battle was far from won, but Clum happily reported that the news fell "like a wet blanket upon the ambitious firm of Gray & Clark."[32]

John Clum hoped to build on the public's resentment against the Townsite Company—resentment he had so thoughtfully encouraged—and win the mayor's chair. Staff members at the *Nugget* laughed off the suggestion. Clum, assuming a self-righteous posture, ignored them. But others opposed to Randall's regime felt the fiery editor too controversial, preferring instead a more levelheaded administrator. On December 27 they chose as their candidate Robert Eccleston, a Gold Rush '49er and brother-in-law to banker Charles Hudson. Eccleston was no stranger to Arizona, having served as clerk of the Papago Indian Agency before joining the Tucson firm of A. D. Otis & Company, dealers in general hardware and building supplies. He moved to Tombstone in February to open a branch store on Fremont Street west of the Melrose Restaurant.

Now he thought he would be running for mayor. During a later meeting at William Gird's lumberyard the Citizens' League reconsidered. Randall, who had left for Los Angeles on vacation, did not even dream of running for a second term. So, perhaps they did need a proven foe of Clark, Gray & Company to challenge Democrat Mark Shaffer, a respected businessman and president of the Miners' Hospital Association. "Eccleston was not strong enough to run," Parsons confided to his journal.[33] As a newly appointed executive of the league he felt obliged to explain to his friend this change in strategy. Parsons, who had just shared dinner with the Eccleston family, was clearly embarrassed at the prospect but forged ahead. Clum, already chosen chairman of the league, leaped at this opening to grab the nomination.

---

[32]*Tombstone Epitaph*, December 9, 1880.

[33]Parsons, *Private Journal*, December 28, 1880, 193.

All those months championing property rights of ordinary citizens worked in Clum's favor. On election day, January 4, 1881, he crushed Shaffer, whom Parsons confessed "made a poor exhibition of himself inside the polling enclosure in a desperate effort to retrieve the day."[34] The final count was 532 to 165. Other results elevated George Pridham, Godfrey Tribolet, Julius Kelly, and H. S. Gray to the city council. Benjamin Sippy, helped by his Fremont Street performance at Reilly's office, retained his post as town marshal. This time he defeated Howard K. Lee, 556 to 125.[35] Virgil Earp had declined to run, although Wyatt was at one time rumored to be considering a challenge.[36]

Even Clum's critics accepted his victory. At Tucson one old adversary admitted:

> The election of John P. Clum to the office of Mayor of Tombstone indicates the pulse of the people on the townsite question. Mr. Clum has been one of the foremost in standing up for what he believed to be the people's rights. His election is a strong endorsement of his conviction. The people of Tombstone have made a good choice. Their Mayor has every qualification to fill the trust.[37]

Clum's election brought with it feelings that perhaps, at last, everything would be set right. The disappointment came later.

Responding to this rise in municipal optimism, however misguided, residents often documented letters home with newspaper clippings and photographs. C. S. Fly supplied most of these latter mementos at prices ranging from twenty-five cents to $3. For this, and other circumstances unconnected with the camera, he is now linked to the history and legend of Tombstone.

Fly's father, Boon, a carpenter and farmer from North Carolina, had settled in Andrew County, Missouri, with his English-born wife, Mary. There Camillus Sidney Fly was born in 1849. Intrigued with the possibilities offered by the Gold Rush, the family traveled by ox team across the plains to California. They settled in Santa Rosa, where the elder Fly helped build the town's first hotel. By 1851 they moved east into the

[34]Ibid., January 4, 1881, 196.

[35]*Arizona Weekly Star,* January 6, 1881; and Minute Book A, Common Council, Village of Tombstone, January 11, 1881, 31.

[36]*Daily Nugget,* December 3, 1880.

[37]*Arizona Weekly Star,* January 6, 1881.

Napa Valley, Boon having purchased about a thousand acres from early California traders Thomas O. Larkin and Jacob Leese. The family devoted its energies to farming, planting orchards, and tending livestock. In that tranquil setting Boon and Mary Fly raised their six sons and three daughters.

Camillus married a divorcée, Mary E. "Molly" Goodrich, at Napa in late September 1879. She was a quiet but strong-willed woman who had traveled overland with her family from Illinois in the spring of 1853. Her father had come back for them after struggling two years in the gold fields. They settled first at Snow Point in Nevada County, followed by moves to Solano, Napa, and Contra Costa counties. Molly showed an early interest in photography and developed considerable skill, although in a more primitive place than Napa. Following their marriage, both she and C. S. were praised by the local editor, who remarked, "they will engage in the photographic business in some place not definitely fixed upon at present."[38] The couple arrived in the San Pedro Valley in December 1879 and within a month set up a temporary gallery at Charleston.[39]

Hoping for greater prosperity, they moved to Tombstone. In early April, Fly's younger brother Webster checked into Charlie Brown's Mohave Hotel. Reunited, the trio worked first with Norwegian photographer Carsten A. Holstead. He and Camillus soon planned another gallery for Harshaw Camp. But Fly, standing over six feet tall and known to his friends as "Buck," decided to stay in Tombstone.[40] He enjoyed its wide-open atmosphere and earned a good living making portraits, documenting developments in the district, and speculating in mining property.

The family purchased the lot at 312 Fremont Street and built themselves a small boarding house, with a photo gallery detached to the rear across a narrow stoop. Their neighbor to the west was pioneer lumber-

[38]*Napa Daily Register,* September 30, 1879; Marriage Record, Camillus Sidney Fly to Mary Edith Goodrich, September 29, 1879, Napa County Recorder's Office, Napa, California, Book B, 240; and *History of Contra Costa County,* 601-602.

[39]*Weekly Arizona Citizen,* January 17, 1880.

[40]1880 United States Census, Tombstone, Pima County, Arizona, E.D. 2, Sheet 15, Lines 7-10; *Weekly Nugget,* April 8, 1880; and Cochise County Great Register, No. 432, Webster Washington Fly, Tombstone, November 26, 1881, No. 472, Camillus Sidney Fly, Tombstone, December 8, 1881. For a brief background sketch on this family, see: *History of Solano and Napa Counties,* 847-848.

man and ex-mayor William A. Harwood. He occupied a modest frame dwelling near the corner of Third on ground purchased from Al Schieffelin and T. J. Bidwell. In time that rather unimpressive space between Harwood and Fly's became the single most famous piece of Tombstone real estate.[41]

As Camillus Fly toured the district taking photographs, conversation turned from John Clum's election victory to early maneuvers in the legislature promising a new county with Tombstone as county seat. Yet before any hard news on that subject arrived from Prescott, Virgil Earp once again stepped into the limelight. The deputy U.S. marshal held no grudge against Ben Sippy, although he may have chuckled after the new police chief admitted losing his ivory-handled revolver. The embarrassed officer was forced to advertise a $10 reward for its return. Soon afterward Virgil rushed to Sippy's aid during the so-called Johnny-Behind-the-Deuce affair.

That small-time gambler, a former porter at Tucson's Palace Hotel whose real name was Michael Rourke, shot and killed W. P. Schneider at Charleston on January 14, 1881, amid a conflicting array of circumstances. John Clum, always accepting melodrama as a fact of life, characterized the incident as a "Brutal Murder of an Upright Citizen . . . by a Desperado."[42] The eighteen-year-old Rourke took exception, claiming that Schneider, chief engineer of the Corbin works, had threatened him with a knife. "I was excited, of course," he later admitted,

> for the man was twice as big as I was. I'm not certain whether the knife was open or not, for his hand was sideways to me, and I could only see the end of the handle. He continued to crowd against me, and I pulled my gun and shot him. I was so excited when I shot him that I dropped my revolver and ran a little ways. I always tried to keep out of such trouble, and when I saw the blood—well, I didn't hardly know what I was doing, I guess.[43]

Constable George McKelvey, who witnessed the shooting, arrested Rourke. Fearing trouble from townspeople and angry mill workers spilling across the river after learning of Schneider's death, the lawman

---

[41]Assessment Roll of the City of Tombstone, C. S. Fly, Lot 3, Block 17, paid December 2, 1881, 9; and W. A. Harwood, Lots 1 & 2, Block 17, paid December 19, 1881, 13. Harwood bought these two lots for $1,000. *Weekly Nugget*, March 25, 1880.

[42]*Tombstone Epitaph* (weekly), January 17, 1881.

[43]*Weekly Arizona Citizen*, January 15, 1881.

asked two citizens to help guard his prisoner. McKelvey then ordered Constable Bell to get a wagon to transport Rourke to Tombstone. Angry bystanders saddled up and followed. Slowing down to give the horses a rest, McKelvey watched his pursuers gain ground. Nearing Tombstone, he spotted Ben Sippy and warned him of the danger. Riding double, the marshal carried Rourke into town. The terrified boy jumped off Sippy's horse in front of Jim Vogan's Allen Street resort—the sampling room, gambling joint, and tenpins alley next to the Bank Exchange—and began frantically begging protection. He also reportedly blurted out a confession "that he had killed his man."

As word of Schneider's death reached Tombstone a fresh mob began taunting the officers. Against these threats Marshal Sippy stood determined to transfer his prisoner to Tucson. Confused legend names Wyatt Earp as holding off the mob single-handedly with a double-barreled shotgun after Sippy and Deputy Sheriff Behan refused to help. It is clear from all contemporary accounts, however, that Sippy took charge immediately and was then assisted by Behan, Virgil Earp, Wesley Fuller (described at one point as a constable), and "a strong posse well armed." A desperate situation, to be sure, but certainly not one defused by a single brave man.

Placing their prisoner in a wagon, Clum reported, the officers moved "down the street, closely followed by the throng, a halt was made, and rifles leveled on the advancing citizens, several of whom were armed with rifles and shotguns."[44] Rourke sat between Sippy and Behan, with Fuller, Virgil Earp, and George McKelvey riding in back. Ten to a dozen heavily armed men escorted them as far as the Dennis ranch, some fifteen miles beyond Tombstone. From there the lawmen and their prisoner continued on alone. Despite rumors that members of the mob had taken the cutoff between Tombstone and Pantano, they locked their prisoner in the county jail without further incident. Public opinion strongly supported one editor's description that Johnny-Behind-the-Deuce "had about as narrow an escape from death as often falls to the lot of ordinary mortals."[45] In mid-April he escaped from custody and never did stand trial.

[44] *Tombstone Epitaph* (weekly), January 17, 1881.

[45] *Weekly Arizona Citizen*, January 15, 1881.

With Schneider's death behind them, district residents again shifted their attention to Prescott and the rumors of a new county. Chief backers of the plan included mining company superintendent and Charleston merchant William Kidder Meade, a later U.S. marshal and a Democratic member of the legislature since 1879, as well as Harry M. Woods and Benjamin A. Fickas.

Woods, a thirty-two-year-old native of Alabama, had traveled to California with his family as a child. He moved to Arizona in 1878 and became associated with the *Arizona Star* at Tucson, then worked as a reporter for the *Daily Record* before going to Tombstone. At the time of the county bill's passage his old paper described him: "By trade he is a printer, by profession an editor; in the first he ranks as one of the very best in the Territory, in the latter he is above average, and has been as extensively copied as any writer in the Territory."[46] After giving an occasional helping hand to Artemus Fay, he and his partners began the long process of buying out the *Nugget*. During legislative debates on the county bill, Woods served as a newly elected member of the territorial House of Representatives.

Born in Indiana, Benjamin Fickas helped edit a newspaper in San Diego for a couple of years in the early 1870s. He came to Arizona for the first time in 1876. Soon he was back on the coast, only to return two years later. In the fall of 1879 Fickas served as chief clerk of the Tenth Legislative Assembly, before moving to Tombstone as an erstwhile mining man, bookkeeper, and mayoral hopeful. After becoming a school trustee he associated briefly with the anti-Chinese crusade. During the controversial elections of 1880, the thirty-two-year-old Fickas won a seat in the Legislative Council. There he joined William Kidder Meade in championing the county bill.

The question of a new county had been raised before. Politicians and businessmen at Tucson hoped to delay the process as long as possible, although Charlie Shibell did sign a supporting petition. Most disagreed with the sheriff and resisted the idea of losing those sizeable tax revenues generated by the district. Still, as the area grew in importance, change became inevitable. Proposed boundaries stirred up fresh disputes, Pima County authorities wanting the San Pedro River recognized

[46]*Arizona Weekly Star*, February 10, 1881.

as the natural dividing line. That idea raised howls of protest in Tombstone and was quietly dropped.

Debate then erupted over what name to choose. Harmless suggestions included La Plata and San Pedro. Some felt it inappropriate to honor an individual, so complaints echoed angrily along capital hallways over the possibility of naming the new county Cochise (actually the first spelling was "Cachise," which more accurately reflected the Apache pronunciation). After supporters, chiefly Harry Woods, placed that name before the legislature, Captain Nathan Sharp, a cantankerous sixty-one-year-old farmer from Hayden's Ferry, voted no because "he had suffered too much from the depredations and murderous attacks of that bloodthirsty savage to immortalize his name by naming after him a county, that in the future would probably become the wealthiest in the Territory."[47] In spite of Sharp's deeply personal objections the name passed. By late May the spelling "Cochise" became generally accepted.

Debate on the bill creating Cochise County opened on the morning of January 17. By early afternoon, following a particularly noisy session, the measure finally passed the lower house 15 votes to 9. All five Pima County representatives from Tucson closed ranks and voted no. Four members from Tombstone, plus J. K. Rodgers of Smithville (from soon-to-be-created Graham County) voted in favor. Tombstone legislator John McCafferty originally voted no, but after being pressured behind closed doors changed his mind, siding with Rodgers, Thomas Dunbar, E. H. "Fatty" Smith, Moore K. Lurty, and Harry M. Woods.

In the upper house the next day the council referred the matter to committee. Led by the adroit political maneuvering of William Meade and Benjamin Fickas, the measure passed and was sent to the governor's office for signature on January 19.[48] Frémont delayed action, prompting the observation: "If he has had the bill ten days, exclusive of Sundays, and has not signed it then it has become a law nevertheless, as the Statutes of the United States is peremptory on this subject."[49]

The legislature then passed a forty-one-page act, approved on Febru-

---

[47]Ibid.

[48]*Session Laws of the Eleventh Legislative Assembly.* Also see *Arizona Weekly Star,* January 20, 27, February 3, 10, 17, 1881; and *Weekly Arizona Citizen,* January 22, 30, February 6, 13, 20, 1881.

[49]*Weekly Arizona Star,* February 3, 1881. This referred to an act of Congress, passed during the summer of 1876, "relating to the approval of bills in the Territory of Arizona."

ary 21, incorporating the city of Tombstone[50]—two years later they also "Re-Incorporated" Tucson.[51] Of course, the board of supervisors had already tackled that question for the booming silver camp back in 1879, but with a new county the town would no longer lie within their jurisdiction. Authorities at the capital hoped to lessen well-known controversies, especially those surrounding Tombstone's highly volatile real estate market.

Concerned more with the present than the future, an immediate scramble took place for prestigious and lucrative county offices. The power to fill these slots rested with the Republican governor, but Frémont deferred to legislative compromise regarding most nominations. What resulted for Cochise County was an acceptable mix. Republicans Lyttleton Price and John O. Dunbar became district attorney and county treasurer, respectively. On the three-man board of supervisors Milton E. Joyce and Joseph Dyer were Democrats, with Joseph Tasker the sole Republican. That party also won the coroner's office after the appointment of Dr. Henry M. Matthews.

Both Wyatt Earp and Shibell's deputy Johnny Behan coveted the sheriff's office. Not only would this position bring with it considerable authority but also a comfortable income. Sheriffs in Arizona, to help finance their office expenses and payrolls, were authorized a percentage of the taxes they collected. In wealthy areas, such as one including the Tombstone Mining District, the possibilities seemed limitless. As a Republican, and thus a member of the governor's political party, Wyatt falsely believed he enjoyed an inside chance at the appointment. In truth, Earp's naïve misunderstanding of how Arizona politics really worked doomed all hope. He simply fell victim to an entrenched old-boy network.

John Harris Behan knew all the right people at the capital. Indeed, as a Democrat he had served with many of them in the legislature, first in 1873 from Prescott and again six years later from Signal. He had proven himself over many years in Arizona. By contrast Wyatt Earp remained, at least politically, a relatively unknown newcomer. Behan was an old-

[50]*Acts and Resolutions, Eleventh Legislative Assembly*, 37-78.

[51]"An Act to Re-Incorporate the City of Tucson," March 7, 1883. Hardy, comp., *Charter and Ordinances of the City of Tucson*, 1-36.

timer by Arizona standards, having lived in the territory since 1863, coming from his Westport, Missouri, home by way of San Francisco. Settled in Tucson as an eighteen-year-old day laborer, he was later hired by the government to help supply local military encampments. Before long he moved to Prescott and ran a small freighting business hauling goods to nearby mining camps. He entered politics, eventually becoming recorder and then sheriff of Yavapai County. As sheriff he was once rather rudely assaulted by a group of irate Chinese laundrymen.

All the excitement surrounding Tombstone prompted Behan's move south. He had scouted the area as early as 1879 but did not locate there permanently until late summer 1880, leaving behind a profitable saloon business at Tip Top. Behan registered to vote at Tucson on September 13. Two days later he and his son reached Tombstone and checked into the Grand Hotel. Young Albert was simply visiting his divorced father. Within a month, and after joining the volunteer fire department, Behan leased the bar at the Grand Hotel, causing the *Epitaph*'s remark: "Johnny has a host of friends all over the Territory, and we predict for him success in his new field." Using these contacts, Behan easily secured the appointment as Cochise County's first sheriff.[52]

The Democratic Party dated its origins in the San Pedro Valley from a meeting three years before at the ranch of William L. Hayes, a young man from Missouri with his eyes on the future. A southern conservative tradition has marked its appeal in the area ever since. Hayes's later lumber partner Francis Tanner and Ed Schieffelin's friend Charles Bullard—elected a delegate to the County Democratic Committee—helped with the initial organization. Quite naturally Cochise County Democrats joined with those at Prescott favoring Behan's appointment strictly along party lines.

Wyatt Earp's Republican credentials meant little to Governor Frémont. Democrats dominated Arizona politics. Although Frémont had been appointed by the president, and thus did not need the approval of local voters for his actions, he still preferred harmony over discord and almost always went along with the Democratic majority over such things

[52]McClintock, *Arizona,* 3:627-628; Wagoner, *Arizona Territory,* 511, 514; *Federal Census—Territory of New Mexico and Territory of Arizona, Excerpts . . . 1864,* 62; *Weekly Arizona Citizen,* March 7, 1879; 1880 United States Census, Tip Top, Yavapai County, Arizona, E.D. 22, Sheet 1, Line 31; Pima County Great Register, No. 2668, John H. Behan, Precinct No. 1, September 13, 1880; and *Tombstone Epitaph,* September 15, October 24, 1880.

as selecting officials for some new county. The governor, who cared little for Arizona, cared even less about the relative merits of Earp or Behan and simply went along with the Prescott consensus. None of this mattered much to Wyatt Earp. He felt betrayed.

After all, even though convinced he would be named sheriff, Johnny Behan admitted he had approached Wyatt "and told him that I knew I would get the appointment . . . and that I would like to have him in the office with me." Responding to this unexpected generosity, Earp thanked him for his kindness but insisted, Behan recalled, "that if he got the office, he had his brothers to provide for and could not return the compliment. . . . I said I asked nothing if he got it. . . ."[53]

When pressed to explain not bringing Earp into his office after becoming sheriff, Behan said he learned of Wyatt warning Ike Clanton about a pending subpoena over the San Simon election fraud. Intending to serve the document, Behan freely admitted he had no idea where the Clanton ranch was located. Virgil Earp gave him directions.

Johnny made an evening ride to Charleston with fellow deputy Leslie Blackburn and a young man named Lawrence Geary. Two riders, whom Behan assumed to be Virgil and Doc Holliday, passed them on the road. Afterward, seeing Wyatt and the combative dentist at Charleston, the deputy confessed his mistake. Earp explained to Behan that he was there to recover a horse stolen from him some time before. Sherman McMasters claimed to have seen the animal at Charleston in the possession of young Billy Clanton. Wyatt was waiting for papers to be forwarded from A. O. Wallace's justice court at Tombstone so he could legally take possession.

The deputy thought no more about it until Ike confessed that Earp had told him about the impending subpoena. Clanton armed his crowd to resist service. It was this, Behan insisted, that dissuaded him from offering Wyatt a position with the new Cochise County sheriff's office. The truth to all this is impossible to prove now, but it should be understood that everyone made their statements some months later under rather trying circumstances.

But in that spring of 1881 Behan confidently occupied the sheriff's

[53]Deposition of John H. Behan, Territory of Arizona vs. Morgan Earp, et al, Defendants, Document No. 94, In Justice Court, Township No. 1, Cochise County, A. T. (Hayhurst typescript), 36.

chair. He set up his first office at the Dunbar brothers' corral on Fifth Street above Fremont, prompting Clum's humorous observation: "and if Johnny Behan were not very good-natured he would get mad when the boys sing out 'First stall to the right.'"[54]

Wyatt Earp had clearly lost that round of political infighting, but he calculated to await his turn, patiently prove himself further to the citizens of Cochise County, and run for the office. The plan called for what today would be characterized as a carefully crafted public relations ploy. He would show himself as an agent of law and order. Immediately clouding that strategy, however, two of Earp's Dodge City pals, dandies Bat Masterson and Luke Short, drifted in to survey Tombstone's gaming tables. Masterson returned to Kansas within eight weeks, but not before witnessing the shoot-out between his two friends Short and Charlie Storms.

Luke Short, whose physical appearance did justice to his name, dealt faro at the Oriental for Lou Rickabaugh. Charlie Storms had arrived only days before from El Paso. He was a well-known sporting man, drawn west by the California Gold Rush, who gambled widely in mining camps such as Pioche, Virginia City, Deadwood, and Leadville. On February 25, 1881, he argued with the twenty-six-year-old Kansas faro dealer over the turn of a card. As Storms had been drinking, Masterson separated the two. Returning to the Oriental corner early that afternoon, the nearly sixty-year-old Storms demanded satisfaction. Luke Short, who never drank while gambling, calmly obliged. He "dodged behind the corner post and fired at Storms," reported the press. "In all three shots were fired, one hitting Storms and killing him. Our informant thought that Storms did not fire, but had his revolver in his hand. . . . public opinion seemed to be that the shooting was justifiable."[55]

George Parsons added some rather colorful details:

> today the monotony was broken by the shooting of Chas. Storms. . . . Shots—the first two were so deliberate I didn't think anything much was out of the way, but at next shot I seized hat and ran out into the street just in time to see Storms die—shot through the heart. Both gamblers. . . . Trouble brewing during night and morning and S[torms] was probable aggressor though very drunk. He was game to the last and after being shot through the heart by a desperate effort steadying revolver with both hands fired—four

[54] *Tombstone Epitaph*, April 3, 1881.

[55] *Arizona Weekly Star*, March 3, 1881.

> shots in all I believe. Doc. Goodfellow bro't bullet into my room and showed it to me. 45 calibre and slightly flattened. Also showed a bloody handkerchief, part of which was carried into wound by pistol. Short, very unconcerned after shooting—probably a case of kill or be killed.[56]

Short's response to the aggressive behavior of Charlie Storms became the talk of the town, easily blending into conversation about another killing along the river. Cattleman John H. Slaughter had taken revenge on a destitute black man he had befriended, who in return assaulted his wife and stole a rifle and two blankets from his house near Hereford.

Learning of the attack from a messenger sent to Charleston, Slaughter rushed home. Sorting out details, he and two companions took up the trail the next morning. Slaughter was a small man physically but tough as nails. Two years after his death it was noted in the Arizona Supreme Court, after a description of the rugged little Texan threatening to kill a man, that that fact alone was "sufficient to make anyone turn pale who knew John Slaughter."[57]

The daring cattleman now found his prey

> in the brush about two miles from the house, where he had evidently camped for the night, and where he still lay apparently asleep. The negro sprang up on the near approach of the party and grasped the rifle which he had stolen the day before, but on Mr. Slaughter firing a bullet over his head and telling him he must die, he weakened. Four bullets from the rifles of the party sent him to account to a higher tribunal for his crimes. . . .[58]

Both Slaughter and Short escaped punishment for resorting to personal violence. John Slaughter would one day be elected sheriff of Cochise County. Luke Short, seeing things somewhat differently, realized his days in Tombstone were over. He soon followed Bat Masterson back to Dodge City. Wyatt Earp was not sorry to see them go. As a known friend of both it must have reminded many of Doc Holliday's earlier escapade at the Oriental. In Wyatt's mind such relationships could easily discourage voter trust. Yet he still had trouble separating himself from their world. Essentially that milieu of the gambler and the gun were Earp's as well.

---

[56]Parsons, *Private Journal,* February 25, 1881, 209.

[57]In the Supreme Court of the State of Arizona: Manuel Garcia and Jose Perez, Appellants vs. State of Arizona, Appellee; brief on Behalf of Appellants, January 1924, 10.

[58]*Tombstone Epitaph,* as quoted in *Weekly Arizona Citizen,* February 13, 1881.

With Wyatt's resignation as Shibell's deputy and Virgil's failure against Ben Sippy, the Earp brothers lost much of their official clout. Indeed, with Johnny Behan installed as sheriff, Virgil's commission as a deputy U.S. marshal remained the family's only badge of authority, although they still maintained close ties with Wells Fargo. True, the Earps kept busy gambling, investing in mines and real estate, and cultivating friendships for the future, but with the exception of Virgil coming to the aid of Marshal Sippy during the Johnny-Behind-the-Deuce affair, events seemed to have passed them by.

Then, just weeks after Wyatt's disappointment over losing out to Behan, the Earps were once again chasing lawbreakers. A band of bumbling road agents tried to rob the night coach between Tombstone and Benson.

Chapter 9

Road Agents and the Earps

In early March 1881, after several disagreements with his partners at the Tombstone Mill and Mining Company, Richard Gird decided to sell out. Some estimates exaggerated the price at $1.2 million. The actual figure was half that, and Gird paid two-thirds to the Schieffelins. Nothing was held back, Albert remembered: "I sold 100,000 shares of Mr. Gird's stock that was standing in his name, and I understand that Mr. E. A. Corbin sold [the remaining] 25,000. . . ."[1] Five weeks later the company went ahead and paid another dividend. Gird bought himself a ranch in southern California. He still held interests at Tombstone with stock in the Huachuca Water Company, as well as some scattered real estate deeds and mining claims. Normally his selling out would have dominated all local gossip, but everyone's attention soon shifted elsewhere.

On the evening of March 15 Kinnear and Company's coach—the one named Grand Central—pulled out from Tombstone with half-a-dozen passengers on its way to Benson. On top were driver Eli P. "Bud" Philpot and Robert H. Paul, riding shotgun to protect Wells Fargo's treasure while awaiting the court's decision on the contested Pima County sheriff's race. At Watervale they took on two more passengers, a miner named Riley and Peter Roerig, a French Canadian who had given up his job at the Tranquility to examine mining sites in Montana. Riley found a seat inside while his friend rode on top behind Philpot and Paul. From Watervale they traveled to Contention City, stopping to change horses and take delivery of six silver bars from the Contention Mill.

Passengers climbed back aboard and the stage pulled out, rolling past the cemetery towards Drew's Station, a ranch and rest stop less than two miles beyond. Some two hundred yards below that point, as the coach slowed to negotiate a small incline from one of the many washes cutting the route, armed men stepped onto the road. Their leader shouted, "Hold!" Sensing a robbery, Bob Paul yelled back, "I hold for no man!" as he unleashed a

[1]Deposition of Al. E. Schieffelin, San Francisco, July 1882, *Garner* v *Gird*, 339.

double-barreled shotgun blast into the shadowy figures at roadside. The would-be robbers fired several rounds in return. Bud Philpot tumbled forward from his seat. Another bullet mortally wounded Peter Roerig.

The terrified horses pulled the stage clear of the incline and past Drew's Station at full gallop. Men there, alerted by the sounds of gunfire and the stage roaring past, rushed to the scene in time to see the attackers disappear into the darkness toward the Dragoon Mountains. They found Philpot's body lying in the road. Someone raced to Contention City to alert Arthur C. Cowen, local postmaster and Wells Fargo's resident agent. He rode for Tombstone to spread the news; Contention still lacked direct telegraph service. A later search of the area revealed thirteen empty cartridge cases, two loaded cartridges, and three masks made of sea-grass rope untwisted to form crude beards and wigs.

Aboard the moving coach Bob Paul retrieved the fallen reins with some difficulty and rolled on into Benson. It was nearing eleven o'clock. He telegraphed Marshall Williams at Tombstone, who received the message shortly before Cowen rode in on his sweat-stained horse from Contention City.

As the news spread pandemonium broke out. An excited Dr. Goodfellow burst in on George Parsons, demanding to borrow a rifle. Parsons, his chess game interrupted, agreed only after receiving assurances from his headstrong friend that no one was about to be shot. Grabbing a revolver, he followed Goodfellow. "Men and horses were flying about in different directions," the diarist recorded. "A large posse started in pursuit." As the officers set out, Parsons and others blinded by the excitement began shadowing "several desperate characters in town, one known as an ex stage robber,"[2] after their own plans to cut off the outlaws at San Simon fell victim to repeated delays and lack of information.

Drs. Goodfellow and Matthews conducted postmortems at Benson. The twenty-eight-year-old Philpot, who proudly carried a gold watch from Wells Fargo, was an experienced stage driver from the coast and well liked around Tombstone. His death brought with it demands for justice or revenge. He left behind a young wife, an eight-year-old son, and a sister in California.[3] His body was returned there on March 21. Mourners crowded

[2]Parsons, *Private Journal,* March 16, 1881, 214-215.

[3]1880 United States Census, Calistoga, Napa County, California, E.D. 77, Sheet 16, Lines, 1-4. On this census E. P. Philpot gave his birthplace as Ireland. Registering to vote in Pima County on December 28, 1880 (Great Register No. 7239, Precinct No. 17), he claimed American birth. The *Weekly Arizona Citizen* reported on March 20, 1881: "He was born in an emigrant wagon on the Platte river. . . ." That version also mistakenly gave him four children.

the funeral at Calistoga. Back in Tombstone plans were soon underway for a benefit at the Turn Verein Hall, featuring singers, dancers, musical numbers, dramatic readings, stage sketches, and other examples of local talent. On behalf of his company J. D. Kinnear organized a separate fund for the dead man's family. Together these efforts raised more than $500.

The Philpot and Roerig murders outraged ordinary citizens, but some hoped "the whole affair will probably end in measures which will effectually abate the cow-boy nuisance, as the ruffians are stealing all of the stock in the valley and riding into and capturing small towns, and generally exercising highway terrorism which has been permitted too long. Only last night fourteen head of fine horses were run off from the vicinity of the station where the attack on the coach occurred."[4]

Sheriff Behan organized a posse. Since the stage carried several mail sacks, Virgil Earp also swung into action as a federal officer, calling upon his brothers Wyatt and Morgan for assistance, along with Bat Masterson—a passenger on the stage according to Bob Paul—together with "several other brave, determined men."[5] Wells Fargo agent Marshall Williams rode with the Earps, representing his company's interest in the affair.

They followed an ill-defined trail toward the Dragoon Mountains. Wood choppers confirmed the passage of at least three men. Some distance beyond the wood camp the trail suddenly cut back toward the San Pedro, crossing the river below Tres Alamos. The fugitives did everything possible to throw off the chase. They rode single file, along rocky ledges whenever possible, and used the river itself as a roadbed at several points. At a small ranch run by blacksmith William Wheaton, the pursuers found a horse played out from hard riding. There, along the river, they also lost the trail.

Continuing down the San Pedro toward the tiny river settlement of Redington, just beyond the Cochise County line some thirty-five miles north of Benson, they captured an associate of the bandits named Luther King. He was caught loitering at Redfield's ranch. As they rode up the officers spotted King milking a cow, a Winchester rifle nearby and two revolvers and a cartridge belt strapped to his waist. Confronted by this somber show of force, the young man quietly surrendered. Under a rather

[4] *Arizona Daily Star*, March 24, 1881.

[5] *Tombstone Epitaph*, March 16, 1881; and V. W. Earp to C. P. Dake, telegram, March 21, 1881, as quoted in Ball, *United States Marshals of New Mexico and Arizona*, 118. For Bob Paul's reference to Masterson as a passenger on the Benson stage that night, see *Arizona Republican*, June 26, 1892.

tense interrogation he confessed that the horse found at Wheaton's was indeed his. As it turned out the Redfields had supplied the others with fresh mounts.

When the sheriff and deputy U.S. marshal rode onto their property in 1881, the reputation of the Redfield family remained ambiguous, despite occasional thrusts at respectability. Leonard and his younger brother Henry, a harnessmaker, ran two adjoining ranch sites, farming and raising some cattle. A year and a half before, the Pima County Board of Supervisors had appointed Leonard justice of the peace for the lower San Pedro district. At one point he was even considered a possible candidate for the legislature. During the controversial November elections the elder Redfield served as a polling inspector for Precinct No. 13 at Nicholas Soza's house.

Leaving their home in Tulare County, California, the family had lived in the San Pedro Valley since 1876. An angry mob would lynch Leonard at Florence in 1883 for his participation in the Riverside stage holdup between that place and Globe, resulting in the death of Wells Fargo messenger Johnny Collins. Henry moved to Benson and became a respected citizen.[6]

Still wondering if the Redfield brothers were in any way involved in the holdup attempt, Virgil Earp turned his single captive over to Sheriff Behan for transport back to the county lockup. Virgil did so only with the understanding that King was a federal prisoner.

The Earp brothers and Bob Paul led the posse after the other suspected outlaws, identified now as Bill Leonard, Harry Head, and Jim Crane. Behan and Marshall Williams returned to Tombstone with the talkative Luther King. The Earps and their companions followed the faint trail from Redfield's toward Tucson. The tracks then abruptly doubled back to Tres Alamos. The outlaws now seemed headed for their New Mexico sanctuary. The officers followed but realized they were losing ground on mounts jaded from the chase. After several days Wyatt and some of the others returned to Tombstone. Virgil, Morgan Earp, and Bob Paul relentlessly continued their pursuit.

At Tombstone Wyatt conferred with Marshall Williams and Wells

---

[6]For more on this family, see *Arizona Weekly Star,* July 10, 1879; 1880 United States Census, San Pedro River, Pima County, Arizona, E.D. 7, Sheet 25, Lines 13-15; *Memorial and Biographical History of the Counties of Fresno, Tulare, and Kern,* 647-648; McClintock, *Arizona,* 3:248-249; *Portrait and Biographical Record of Arizona,* 754; and Miller, *Arizona: The Last Frontier,* 207-218.

Fargo's chief of detectives, James B. Hume, who had arrived from San Francisco at his company's insistence. Resupplied with horses and provisions, Johnny Behan again took up the chase accompanied by Buckskin Frank Leslie, newcomer William M. "Billy" Breakenridge, and Ed Gorman. On March 24 they rejoined the Earps and Bob Paul at Helm's ranch in Cochise Canyon on the east side of the Dragoons. There the outlaws had stolen fresh horses, including a racer owned by Tombstone saloonman Bob Archer. Even with the services of an experienced Indian scout like Frank Leslie, chances for success seemed to be slipping away.

Then, to the intense embarrassment of the sheriff's office, Luther King escaped from custody by simply helping himself to a revolver and walking out the back door. Newly appointed undersheriff Harry M. Woods missed the maneuver as he watched attorney Harry B. Jones prepare a bill of sale for a horse King was selling liveryman John O. Dunbar. This trio claimed their backs were turned for only a moment before they glanced up and noticed the prisoner gone. "His confederates on the outside probably had for him a horse in readiness and he decamped into the darkness, in what direction no one knows," confessed the *Nugget*.[7] The *Epitaph* added, "King did use the boys at the Sheriff's office rather rough, that's a fact. He never said a word about going—never whispered 'tra-la-la.'"[8]

At Tucson the press joked: "And now Tombstone is 'kicking' about the insecurity of its jail. Five prisoners have escaped from the jail and officers within the past few months."[9] George Parsons was not amused: "King the stage robber escaped tonight early from H. Woods who had been previously notified of an attempt at release to be made. Some of our officials should be hanged. They're a bad lot."[10]

Days before the affair near Drew's Station, Virgil and Morgan Earp had chased two of the earlier escapees through the Sulphur Springs Valley, a Frenchman named Paullin and a horse thief from Silver City, New Mexico, known as Red. Both eluded capture and fled into Sonora. There Paullin outsmarted Deputy U.S. Marshal A. F. Burke, awaiting papers from Governor Frémont at Prescott, by writing "a long and very friendly letter to Deputy United States Marshal, Virgil W. Earp . . . stating that he was very sorry he had been obliged to leave Tombstone without bidding

[7] *Tombstone Daily Nugget*, March 29, 1881, as quoted in *Arizona Weekly Star*, March 31, 1881.

[8] *Tombstone Epitaph*, March 30, 1881.

[9] *Arizona Weekly Star*, April 7, 1881.

[10] Parsons, *Private Journal*, March 28, 1881, 217.

the Earp boys good-bye, and asking Virgil to forward his clothes, books, papers, etc., to Hermosillo." Reading the letter himself, the deputy naïvely assumed "Virg would hardly thank him for bringing back to captivity such a dear friend . . . . So he let Paullin meander peacefully away. . . ."[11]

Meanwhile, the posse following the Benson stage outlaws rode through the Sulphur Springs Valley. At Galeyville, a three-month-old mining camp on the northeast side of the Chiricahuas, they learned the fugitives were headed for a ranch near Cloverdale Springs, New Mexico. Following directions to that place, they ran out of food and water after crossing the border into what was then still part of Grant County. Reaching their goal, the posse found the place deserted. Virgil's horse stood near exhaustion. Considering their options it was decided that Behan, Leslie, Gorman, and Breakenridge should retreat to Barlow and Price's ranch for more provisions, while Virgil, Morgan Earp, and Bob Paul walked their horses and continued the pursuit.

After once again joining forces, the posse fell back for fresh mounts at Galeyville, fast becoming a favorite hangout for Cochise County malcontents. Refused aid, Paul rode slowly for Tucson while the others returned empty-handed to Tombstone on April 1. They received a warm welcome. At one point the men had gone sixty hours without food and thirty-six without water. Behan repaid Breakenridge by appointing him a deputy sheriff. Clum boasted their feat "will pass into our frontier annals,"[12] but the Earps were clearly disappointed, especially after learning of Luther King's mysterious escape.

To make matters worse, their younger brother Warren had been arrested and fined $25 for firing a pistol within the city limits. Then, to their growing dismay, whispered suspicion added Doc Holliday's name to the list of suspects in the Philpot and Roerig murders. Those rumors simmered for several weeks as common street gossip before boiling over into serious court action. Concerning lawlessness in southeastern Arizona, however, there was more to excite public imagination than merely suggesting suspects or pondering the circumstances surrounding one bungled stage holdup.

A handful of unsuccessful stage robbers posed no more than a passing threat. Others, members of a loose confederation, represented a more serious challenge to authorities on both sides of the border. Tradition holds

[11] *Tombstone Epitaph*, April 8, 1881.

[12] Ibid., April 2, 1881.

that atop this criminal pyramid stood the Clanton family. Although involved in questionable activities, the Clantons were hardly seen as social outcasts. They counted among their friends many important businessmen and well-placed political figures of both parties in and around Tombstone and throughout a large portion of southern Arizona. Even the Earp brothers, early on, enjoyed some rather friendly contacts with family patriarch Newman Haynes Clanton.

Newman Clanton was born in Davidson County, Tennessee, in 1816. Raised on stories of that county's most illustrious citizen, early American military and political hero Andrew Jackson, Clanton also dreamed of excitement and adventure. In early 1840 he married Mariah Kelso at her family's home in Callaway County, Missouri.[13] Newman Clanton and his teenage bride then settled on a local farm and started a family. After the birth of three sons, including Joseph Isaac "Ike" Clanton in 1847, they moved for a time to Illinois, where a number of other Tennessee Clantons had settled. A daughter was born there in 1852, after which they returned to Missouri.

One of Mariah Clanton's brothers, Isaac Kelso, came back from California in 1856. Although murdered for his money shortly thereafter,[14] stories told of the Golden State lingered in his brother-in-law's mind. Before trying California, however, N. H. Clanton and his growing family settled in Texas to farm and raise a few scattered head of cattle, first in Dallas County and then farther to the southwest in Hamilton County.[15] In Texas other children were born, including William H. "Billy" Clanton in 1862, bringing the total to seven.

Both Newman and his eldest son John W., then working as a freighter, heeded the call of the Confederacy during the Civil War. Neither man's record proved exceptional. Newman, who stood six-foot-one, enlisted as a private in a Home Guards outfit. Judged overage at forty-six, he saw no active infantry duty. John Wesley Clanton's service was marred by charges of desertion. Yet at the time of his death in 1916, family tradition claimed J. W. "was a Confederate scout during the Civil War."[16]

---

[13]Marriage Record, Newman H. Clanton to Mariah P. Kelso, January 5, 1840, Callaway County Recorder of Deeds, Fulton, Missouri, Marriage Book B, 40.

[14]*History of Callaway County*, 922.

[15]1860 United States Census, Precinct No. 6, Dallas County, Texas, Sheet 352, Lines 1-10.

[16]Santa Rosa, (Calif.), *Press Democrat*, April 20, 1916; and Death Certificate No. 123, John Wesley Clanton, April 19, 1916, Sonoma County Recorder's Office, Sonoma, California.

After Appomattox Newman Haynes Clanton finally made the move to California. The family took the southern route and by early September arrived at Fort Bowie. Crossing through Apache Pass, the Clantons first saw the area of southeastern Arizona with which their name would long be associated. Mariah died during the journey. Clanton and his now-motherless children continued on, hoping to find a better life on the celebrated Golden Coast.

In California the family split up, with John Wesley Clanton marrying teenager Nancy B. Kelsey and settling on her parent's Inyo County farm. His younger brother Phineas lived with them. Newman Clanton established himself in what was then still part of Santa Barbara County to begin raising corn, barley, and beans. On the 1870 census he listed the value of his farm at $1,600 and that of his personal property at $750.[17]

At first things went well enough, but trouble, brought on by themselves and especially Ike's poor judgment, soon threatened the family's fragile situation. Their main problem rested with the ownership of the nearly fifty-thousand acre Rancho La Colonia, below Hueneme Point and south of present-day Oxnard. Squatters, including the Clantons, claimed the land as part of the public domain. The recognized owners thought otherwise, and they were powerful fellows indeed—including Thomas A. Scott, a former assistant secretary of war and president of the Union Pacific Railroad.

The subsequent events, although bloodless, became known as the Hueneme War.[18] During the course of this rather intense standoff, Ike Clanton, then employed as a day laborer, made the mistake of threatening the life of surveyor Johnny Stow. The next morning Scott's agent Thomas R. Bard, later a major figure in the Union Oil Company and a United States senator from California, armed himself with a Winchester rifle and called Ike from his shack. Bard gave the young hothead a quick lesson on the facts of frontier life, warning him rather forcefully not to repeat his mistake.[19]

That incident, and other problems related to opening a store at the site of Bard's proposed supply wharf, soon convinced Newman H. Clanton

---

[17]1870 United States Census, Township No. 1, Santa Barbara County, California, Sheet 4, Lines 38-40; and Sheet 5, Lines 1-2.

[18]*History of Santa Barbara County,* 380-385; and *Memorial and Biographical History of the Counties of Santa Barbara, San Luis Obispo and Ventura,* 203-204.

[19]Hutchinson, *Oil, Land and Politics,* 1:5, 176, 179; and numerous conversations between Hutchinson and the author.

that he was about to be overwhelmed by circumstances. It was time to retreat. Never again would he allow himself to be caught so far outside the prevailing political and economic scheme of things. These were lessons Ike never learned.

By 1873 the Clantons returned to that openly wild area of southeastern Arizona they had first seen eight years before. They started out at Globe City but by early August moved into the Gila River Valley near old Camp Goodwin, a post abandoned by the army in early 1871. They chose a good location, farming 120 acres "not very far from Camp Apache and Bowie, and within reasonable distance of the important mining camp of Clifton. All these places must have much grain, vegetables and all sorts of farm and dairy products."[20]

Within a year the place was called Clantonville. Momentum grew to have Newman appointed justice of the peace in 1874. He eventually branched out from farming and for a time ran the Clanton House Hotel, a small, ramshackle affair near Fort Thomas on the Gila River. Not wishing to repeat his California mistakes, Clanton became friends with a number of well-placed individuals. These included cattleman Henry C. Hooker and San Carlos Indian Agent John P. Clum. Newman Clanton clearly saw the economic potential of supplying contract beef to government reservations.

As early as 1874 John Wesley Clanton, who along with his family had joined Newman at Camp Goodwin, began looking into the possibilities of stock raising in the Sulphur Springs Valley. Yet two years later all the Clantons—except Phin, who resided at Globe, and Billy, who attended a Catholic school in Tucson—still lived on the Gila River site. Later they briefly settled at Pueblo Viejo, a small Mexican community southeast of Safford.

Not until 1878 did they begin moving into the San Pedro Valley. There they established a cattle and dairy ranch less than four miles below the Gird and Corbin mill sites. In time they developed other locations, including Phin's ranch on the Arizona–New Mexico border. Ike briefly ran a restaurant at Millville before Newman became interested in town lots at Charleston and in some rather disappointing mining claims in the Huachuca Mountains. Even so, the family hardly realized they had a rendezvous with history and would soon become part of America's frontier folklore.

---

[20] *Weekly Arizona Citizen*, December 13, 1873.

To support his family's growing business interests, Newman Haynes Clanton became associated with a loosely organized band of smugglers and cattle thieves. Profits were high and the risks low in this sparsely settled corner of the territory. William Morris Stewart, a San Francisco lawyer and former United States senator from Nevada, recalled a Charleston resident describing their ranch as Fort Clanton.[21]

Clanton's new allies were known as "Cowboys"—a term then in general use by northern Arizona Republicans against their southern Democratic counterparts, especially those with ties to Texas or Lincoln County, New Mexico. In southeastern Arizona the term was further refined to identify marauding outlaw bands. As attorney Stewart described them: "The cowboys around Tombstone frequently engaged in bloody conflicts. They were divided into clans. They 'collected' stock in Texas, New Mexico, and old Mexico, herded their booty in Arizona, and sold it in California or around the mining camps. These cow-boys, of course, were the renegades. There were lots of other cow-boys who did a legitimate stock business, but they were all fighters."[22]

Many of Clanton's colleagues, at least by local standards, stood out as respectable citizens. After all, most Arizona cattlemen shared with him similar activities and profits. In secret federal reports the names of such esteemed cattlemen as Henry C. Hooker and Walter Vail appear in reference to the illegal cattle trade. Few were above suspicion. Several prominent merchants and various professionals in both Tombstone and Tucson were mentioned with interesting regularity during discussions of Mexican smuggling.

Clanton's motives can be seen more accurately in terms of economic survival rather than a simple criminal impulse. In today's terms he would be seen as a middleman or broker. None of the Clantons ever faced indictment for their activities—not that such practices went unnoticed or unchallenged by authorities. Officers in and around Tombstone, Charleston, Contention City, and other smaller communities tried within the limits of their various jurisdictions to stem the growing tide of banditry operating on both sides of the border. For a time Wyatt Earp became marginally involved in these efforts as a deputy sheriff, as did Virgil in his capacity as deputy U.S. marshal.

Undeterred by reports of marauding banditti, people kept flooding

[21]Brown, ed., *Reminiscences of Senator William M. Stewart,* 269. [22]Ibid., 266.

into Tombstone. A number of important Nevadans now abandoned Virginia City for Arizona's newest silver bonanza, all hoping to repeat earlier successes. These included Frederick A. Tritle, who had helped organize the Belcher Mining Company on the Comstock, got involved in politics, and worked nearly a decade as a stockbroker at Virginia City. In late 1880 Tritle brought his wife and five children to Arizona. Within months he and William B. Murray opened an office on Allen Street as stock and mining brokers. They prospered, linking clients by telegraph with the nation's major stock exchanges. Later Tritle was appointed Arizona's sixth territorial governor, picked to preside over a legislature derisively known as the "Thieving Thirteenth."

John S. Young, another of the Nevada crowd and a former Virginia City politician, pursued a more practical course. He opened an eatery in an adobe near Seventh and Safford before relocating to more spacious accommodations on Allen Street. His Brooklyn Restaurant stood on the earlier site of M. Calisher & Company's first mercantile.

Most of these Storey County expatriates congregated and reminisced at Thomas Moses's Capitol Saloon. Although his was one of the smallest watering holes in Tombstone (measuring a mere thirty by thirty feet), Moses became one of the town's more popular barmen. He had already seen much of the West and had done everything from freighting, working as an army sutler, and running a Seattle hotel to serving as a sheriff in Colorado. At Virginia City he was twice elected a justice of the peace. With the Comstock's decline Moses moved to Tombstone and opened his saloon with Canadian Andrew J. Mehan. This arrangement lasted only a short while, although the two continued other business dealings.

Moses stayed in Tombstone less than two years before moving on, first to Colorado, then to California, and finally Washington. Sailing to Seattle, he worked as a clerk and lived with his son-in-law, John E. Clancy. That family ran a string of saloons and gambling houses in the Northwest. Avoiding Alaska's harsh winters at the turn of the century, Wyatt Earp ran a faro table in the rear of Clancy's Seattle franchise.

Tombstone's reaction to the Benson stage murders stood in sharp contrast to that shown earlier killings. It seemed to mirror a degree of maturity for a place most observers still considered a rough community. But the slow process of change had begun, despite embarrassing setbacks. Such was demonstrated by an item published just days after the attack on Kinnear's coach: "What action if any did the Coroner take looking to an investigation of the human arm which was found on the street the other day? It surely demanded some attention."[23]

Then, hotel keeper Carl Bilicke quarreled with carpenter Charles Wilkes near the shaft of the Mountain Maid Mine. Enraged, some witnesses said, Wilkes reached for his revolver. Crossing the street at that moment to defend his father, young Albert Bilicke pulled his own weapon and fired. He hit Wilkes behind the left ear, the bullet coming out near the nose. Dr. McSwegan did not think the wound mortal but felt his patient might lose the hearing in that ear. Marshal Sippy arrested Albert and picked up a discarded revolver, dropped Wilkes's friends insisted, by the elder Bilicke. Judge Wallace set bail at $500. Helped by his family's prominent position in the community, Albert avoided any serious legal entanglements.

Such setbacks to civility could not overshadow signs of material progress. That same month workmen began installing the district's first telephone system, connecting the Grand Central and Contention mines with their mills on the San Pedro. Although many did not see it as beneficial, there were now forty-two attorneys "in this unhappy camp," with more of their number on the way. Some found solace in the town's sixty saloons; at least they outnumbered the lawyers.

Undersheriff Harry M. Woods endured the criticism for Luther King's escape, but it did not dampen his business instincts. By April his on-and-off-again negotiations with Artemus Fay finally ended with his purchase of the *Nugget.* John Clum seemed pleased, at least for the moment: "it has passed into the hands of gentlemen qualified to make it more of a credit to the camp . . . it will hereafter be under the control editorially and otherwise of Harry Woods himself." Richard Rule, a staunch Democratic journalist and behind-the-scenes political manipulator, "will have charge of the local department."[24] Rumor suggested that Hugo Richards, a major player in the Democratic Party at Prescott and friend of Johnny Behan, was a silent partner.

Clum may have been right about Harry Woods. Within weeks the new editor criticized himself in his own newspaper. In response to an inaccurate Associated Press dispatch claiming the first Cochise County grand jury had considered indicting Sheriff Behan over King's escape, Woods

[23] *Tombstone Epitaph,* March 22, 1881.

[24] Ibid., April 1, 1881.

wrote, "We do not propose to discuss the matter . . . further than to say that if any blame should attach to any one in regard to the matter, it should rest upon the shoulders of the Under Sheriff. . . ."[25]

Leaving the paper, Artemus Fay worked to develop the Nugget Mine near the old Brunckow before moving on to Dos Cabezas. There he was appointed justice of the peace. In early 1882, in partnership with Cochise County Treasurer John O. Dunbar, he started the short-lived *Dos Cabezas Gold Note.* After his wife's death, Fay finally moved north to publish another newspaper at Flagstaff. But Tombstone had lost an experienced journalist.

On May 11, hoping to find an undiscovered niche in an already crowded market, Thomas Gardiner, local agent for the Associated Press and Wells Spicer's old boss on the *Arizona Quarterly Illustrated,* began publishing the *Tombstone Union.* It lasted less than a month, Gardiner claiming, "The reason . . . is the shrinking of business here, caused by conflicting titles and rings sucking the life's blood of the community."[26] While there was much truth to all that, Tombstone was simply too small to support three daily newspapers. Without the ardent political alignment represented by the Democratic *Nugget* and the Republican *Epitaph,* it is doubtful the town could have supported two. Toward the end of the year Tom Gardiner went on to become a co-founder of the *Los Angeles Times.*

Artemus Fay was gone, but his old rival John Clum, who had just bought out Thomas Sorin's one-third interest in the *Epitaph,* stumbled over a hornet's nest of his own making after Sorin left to become financial agent for the Mesa Consolidated Mining Company. Newcomers mingling with old timers all hoped the town's troubles with local government had ended. People wanted confidence in Clum's promised reforms. The mayor and council then sorely tested the public's resolve by debating a controversial new ordinance covering business fees.

Following Tombstone's legislative incorporation, the city government reissued its ordinances. No. 4 caused the trouble. Licenses were required for "any vocation, trade, calling, business or employment. . . ."[27] These ranged from $200 a year for banking houses to $60 for laundries. For more unsavory outlets, such as brothels selling alcohol, the yearly fee reached $240. A single billiard table cost $24 a year. Only the profes-

[25] *Tombstone Daily Nugget,* June 8, 1881.

[26] *Tombstone Epitaph,* June 12, 1881, quoting a telegram sent the Associated Press in San Francisco on June 8.

[27] Ordinance No. 4, *Tombstone Epitaph,* April 13, 1881; and Minute Book A, Common Council, Village of Tombstone, April 6, 1881, 56.

sions—doctors, lawyers, and, interestingly, newspapers—were exempt. Stirring more resentment, assessed quarterly fees against hotels and other lodging houses were to be based on monthly receipts. Owners cringed at the thought of their profits becoming public knowledge.

Because of earlier town lot scandals, the sorry state of municipal finances was generally known but misunderstood. Voters had believed Clum's assurances to change all that. They assumed fees from property sales would correct any imbalance. Thanks to Alder Randall's chicanery, however, the city was forced to fill its coffers with a general tax, while Clark, Gray & Company carried on as usual, albeit with a lower profile.

But the fees now demanded struck many as arbitrary. Fifty firms and individuals signed a hastily drawn petition of protest. These included merchandiser David Calisher, saloonmen Milton Joyce and Wehrfritz and Tribolet, as well as new restaurateur John Young. Even the notorious Buckskin Frank Leslie's name appeared. Clum countered with the claim, likely true, that "most of this loud talk was done by parties who pay little or no tax whatever." More importantly, he pointed out what many had forgotten: "The city of Tombstone at present does not own a single foot of ground within its limits; hence the only revenue must be from taxation."[28]

Frankly, many did not give a damn. Tombstone always was a self-centered sort of place. As a mining town its very existence depended less on social commitments than on a steady ore supply. No one realistically considered a long-term future. Even the great Comstock was in decline, and Tombstone was no Virginia City. Citizens tended to think of themselves first and worry about their community second, if at all. As one visitor observed that spring of 1881: "In no mining section with which I am acquainted does there exist a less public spirited class of merchants, take them as a whole, than, I regret to say, may be found in the city of dolorous name. . . . Too much selfishness and no public spirit will in time kill any place."[29]

John Clum, who tended to be thin-skinned when attacked, took all the criticism personally, even if he did try absolving the mayor's office by placing blame squarely on the shoulders of the city council: "permit me to state that the language of your petition is too unjust and discourteous to

[28] *Tombstone Epitaph* (weekly), April 11, 1881.

[29] *Weekly Arizona Citizen*, May 15, 1881.

deserve anything but contempt in its treatment at my hands. . . . those communications which unjustly attempt to reflect upon the dignity and honor of our councilmen will be summarily relegated to that boundless oblivion which their nature may deserve."[30]

Many found such remarks arrogant. Facing revolt, Clum tried defending himself, backpedaling with a lengthy but reasonable explanation comparing Tombstone's fees with examples from Virginia City. But for John Clum the damage was done, forever compromising his position as a man of the people. While no one longed for Alder Randall's return, Clum's tarnished reputation as a reformer caused him discomfort in the months to come. Fleeing criticism, he left town with his children and mother-in-law on the evening of April 19. In New York he hoped to find relief negotiating the purchase of some long-sought fire-fighting equipment. George Pridham served as acting mayor during Clum's absence, but the council eventually retreated and did not impose the new fees.

Saloonmen Wehrfritz and Tribolet may have grumbled over Ordinance No. 4, but they had lost none of their optimism for making money in Tombstone. That spring they added a second story to their popular resort, providing more than twenty new offices at the center of the main business district. Looking for more space, Dr. Goodfellow eventually moved in, as did county coroner Henry M. Matthews, several lawyers, and Deputy U.S. Marshal Virgil Earp. Across Fifth Street Vizina and Cook—following their neighbor's example—also expanded upward, providing a large gaming room above the Oriental for Lou Rickabaugh, a gambler and self-styled speculator friendly with the Earps.

Similar signs of progress began popping up along Fremont, slowly becoming a center for government offices. County Recorder A. T. Jones abandoned his cubbyhole on Fourth Street for Bauer's spacious new adobe (originally intended as a brewery) east of the O.K. Corral's rear entrance and R. E. Farrington & Company's Papago Cash Store. The staff hauled in records and furniture as two puppies scurried obliviously underfoot. Named Contention and Toughnut, these undisciplined mutts enjoyed a free run of the place, delighting many patrons while annoying others. In the basement Clement Eschman, a druggist by training who had come to Tombstone from Nevada, opened a saloon and German delicatessen called "The Grotto" with his partner William Alderson. Suitable for fam-

[30] *Tombstone Epitaph*, April 13, 1881.

Looking north over properties of the Tombstone Mill and Mining Co. in 1881. Many landmarks are visible, including Schieffelin Hall, the new Wehrfritz Building, Tasker & Pridham, the Gird Block, Russ House, and the Grand Hotel. To the far left are buildings of the Vizina Consolidated Mining Co.

Courtesy, Arizona Historical Society.

ilies, they also claimed the coolest spot with the coldest beer in town. The *Nugget* moved in next door.

After a shaky start, hampered by lot jumping and lawsuits, the Mining Exchange began building its two-story headquarters across from the *Nugget* between the *Epitaph* and Gird Block. Citizens willingly embraced these false symbols of permanence. Their pride reached its peak, however, over Schieffelin Hall, a huge adobe edifice taking shape on the northeast corner of Fourth and Fremont, planned as a setting for offices, public meetings, and theatrical performances.

All this construction served as an antidote for various municipal disappointments. Public gloom crystallized after R. J. Winders filed a felony complaint charging Alder Randall with malfeasance while sitting in the mayor's chair. After listening to several days of testimony, Justice of the Peace Wells Spicer handed down his decision on June 2. He ably outlined Randall's role in deeding the townsite to Clark, Gray & Company, James Clark's questionable involvement with the townsite patent, and their blatant disregard for the wishes of the community.

It was all true of course, but Arizona had no law covering such circumstances (at that time the core of the entire civil and criminal codes for the territory filled a single 692-page volume). So long as bribery was not proved citizens could only punish elected officials through the ballot box. Once out of office there was no remedy. Spicer quoted one statute section detailing punishment for fraudulent conveyances,[31] but this, he reasoned, applied only to personal, not official, transgressions. With the law seemingly silent, Spicer grumbled with frustration as he dismissed the case.

Wyatt Earp viewed all such legal imperfections with callous indifference. His interests lay elsewhere. Despite recent setbacks, Earp still dreamed of taking over the sheriff's office. Content to await the 1882 county-wide elections, he hoped to impress voters by conspicuously parading his abilities. How better than to apprehend some outlaws? Perhaps he could still get his hands on those responsible for attacking the Benson stage. Wells Fargo offered $3,600 for Leonard, Head, and Crane, but Earp's eyes remained focused on his cut of Cochise County's sizable tax revenues after becoming sheriff.

To carry out his scheme he needed confederates with ties to the outlaws. Wyatt approached Ike Clanton, Frank McLaury, and Joe Hill as the

[31]Hoyt, *Compiled Laws of the Territory of Arizona,* (367) Sec. 131, 93-94.

most likely candidates. The four met in a small yard behind the Oriental. There Earp explained that if successful he would grab the glory, they could split the reward, and he promised to keep the arrangement confidential. Clanton found the plan agreeable. He was having his own troubles with Bill Leonard over a ranch site in New Mexico and wished his rival out of the way. Fearing the three fugitives might make a fight of it, Ike wanted assurance that Wells Fargo intended to pay dead or alive. Earp approached Marshall Williams for confirmation. In turn, the agent telegraphed his San Francisco superiors and received a coded reply from General Superintendent John J. Valentine agreeing the company would pay either way.

Wyatt passed on the information while huddled with Ike and Joe Hill in front of a small cigar store near the Alhambra. Clanton then demanded more proof than Earp's word. According to Wyatt, Williams could not find the original message amid the piles of paper cluttering his desk, but hurried over to Western Union's telegraph office behind Heitzelman's jewelry store. Earp borrowed the copy, with no explanation to Williams after he again decoded the message, to convince his distrustful co-conspirators. Afterward Wyatt filled in Frank McLaury on the details. All three Cowboys seemed pleased. Joe Hill started at once for New Mexico, hoping to entice Leonard, Head, and Crane back into Cochise County with plans to rob a paymaster traveling between Tombstone and Bisbee.

Virgil thought the whole thing ill-advised. He was upset with his headstrong sibling for going ahead with the plan without first talking it over with himself and their older brother James. Meanwhile, the family, although still annoyed with Wyatt, gained a solid victory elsewhere. Ben Sippy, pressured by some to resign his city marshal's badge, was granted a two-week leave of absence on June 6. The council immediately named Virgil Earp acting police chief.[32] After Sippy failed to return—"his interests elsewhere requiring his undivided attention"[33]—Virgil's appointment became permanent on June 28, along with its then-monthly salary of $125. There were no dissenting council votes.[34]

The changing character of that office was dramatic enough for the *Nugget* to remark: "The business before Judge Wallace has been unusually dull for some weeks, but yesterday it began to look up again. Probably the

[32]Minute Book A, Common Council, Village of Tombstone, June 6, 1881, 79.

[33]*Tombstone Epitaph,* June 22, 1881.

[34]Minute Book A, Common Council, Village of Tombstone, June 28, 1881, 80.

new Chief of Police has something to do with the revival of business."[35] Earp kept busy tightening the recent slackness in Sippy's office. He not only hauled in the usual drunks but also chased more sophisticated offenders. The night before his appointment, burglars had broken into the rear of the Golden Eagle Restaurant and drugged a young woman living there with chloroform before stealing $500 stashed under her bed. The new marshal took the case, and his diligence impressed the locals.

None of this surprised those who knew Virgil Earp. Just weeks before, as a deputy U.S. marshal, he came to the aid of Michael Kelleher surveying the Wheeler Mine for a government patent. Others, claiming much of the same ground, harassed the unarmed federal official. But seeing Earp coming up the trail these would-be toughs fled, not wishing to confront such a formidable fighting man.

The Earps saw Virgil's appointment as a good omen, one they welcomed in the face of those recurring rumors claiming Doc Holliday's involvement in the attempted Benson stage holdup, as well as his recent grand jury indictment for shooting up the Oriental back in October. The brothers not only had trouble with Holliday but with one of Wyatt's erstwhile partners as well. Ike Clanton argued with a gambler called Denny McCann (also known as Daniel Burns) in an Allen Street saloon. Both men retired to arm themselves after McCann slapped Clanton's face. They met again in front of the Wells Fargo office and began pulling their weapons. Virgil Earp, aided by Constable Haggerty, "stepped between them and spoiled a good local item. They are both determined men, and but for the interference of the officer there would doubtless have been a funeral, perhaps two."[36]

Wyatt's scheme to capture Leonard, Head and Crane looked good to him—despite James and Virgil's contrary opinion—even though it required specific performance from unreliable allies. But as the Earps waited patiently for Ike Clanton and Joe Hill to deliver on their promised betrayal, Wyatt's scheme began to unravel. Two of the men he hoped to snare met their end in New Mexico at the hands of the Haslett brothers, one-time Cowboys trying to reform.

As a witness from New Mexico explained: "Ike Haslett and his brother Bill have a ranch in the Animas valley, the best one in it, and old man Gray,

[35] *Tombstone Daily Nugget,* June 9, 1881.

[36] *Tombstone Epitaph,* June 9, 1881.

of Tombstone fame, has one on each side of it that he bought from Curley Bill and his gang and he wanted the one belonging to the Haslett boys, so some of the cow-boys were going to turn the H. boys out of the country or kill them."[37] Among those sent to pressure the Hasletts were Bill Leonard and Harry Head. They rendezvoused at a small store and supply center, located near a mine owned by E. P. Parker—the same man shot in the toe by Doc Holliday at the Oriental.

While sharing a bottle, the heavily armed men boasted their intentions. Unfortunately for these braggarts word reached the brothers, who now planned an ambush of their own. A corral stood some distance from the store. At daybreak the Hasletts hid themselves there behind a low fence. They watched as Leonard and Head, their brains still swimming in alcohol, moved slowly up the road. Only Leonard was mounted. At fifty yards the Hasletts rose and opened fire, striking Bill Leonard below the heart. A shot hit Harry Head in the abdomen before half-a-dozen other bullets found their mark. Both men lived a few hours more. The thirty-one-year-old Leonard begged to be put out of his misery. He still carried festering wounds in the groin from Bob Paul's March 15 shotgun blast. Both were buried at sundown.

In less than a fortnight the San Simon outlaws retaliated against the Hasletts. "They were attacked and slaughtered in their own home," the *Epitaph* informed its anxious readers. "Two or three of the cow-boys are reported wounded in the fight. At this rate the gang will soon be exterminated much to the joy of all law-abiding citizens."[38] Wyatt Earp felt no joy. Events were interrupting his plans. If necessary, he would have preferred killing Bill Leonard and Harry Head himself, if it meant more votes in the November 1882 election. Nor did he seem overly concerned that Leonard, a former jeweler from Las Vegas, New Mexico, had been an old friend of Doc Holliday's.

But the people of Tombstone spent little time pondering the fate of dead and wounded men in New Mexico. They soon faced a more serious disaster closer to home.

---

[37]Ibid., June 18, 1881.

[38]Ibid., June 21, 1881.

Chapter 10

Fire and Other Distractions

When John Clum finally returned on June 22, 1881, he could see from the Benson depot that Tombstone was burning. Horrified, he rushed to find transportation to his stricken city. Nervous residents, their town crowded with wooden buildings, had long feared a serious fire. That afternoon at Alexander and Thompson's Arcade Saloon, three doors east of the Oriental, a condemned barrel of whiskey had exploded while being carelessly tapped by an employee lighting a cigar. The blast spread burning alcohol in every direction as customers stampeded out the back door.

Flames, feeding on bone-dry timber and cloth ceilings, spread in a flash. Within three hours the four blocks bordered by Fifth, Fremont, Seventh, and Toughnut were almost totally destroyed. Tasker and Pridham and J. Myers & Bro. refused pleas from those begging blankets to fight the advancing disaster. Ironically, the Myers brothers soon watched their store reduced to ashes. Tasker and Pridham—who later tried disavowing the story of turning away volunteers—sustained far less damage. Much of their property, and that of others lining Fifth Street between Allen and Toughnut, was saved by the heroic efforts of workers from the Tombstone Mill and Mining Company's nearby hoisting works.

At the Oriental, Milt Joyce frantically tried to empty his safe but abandoned $1,200 in greenbacks as the heat drove him out before he could unlock the inner door. Milton Clapp and other employees of Safford, Hudson & Company's bank stuffed their safe with cash and carried records out the back through falling plaster. Fearing for his friend, George Parsons made a valiant attempt at rescue but failed to get inside the building. He was overjoyed to learn of Clapp's escape. Parsons went on to help others before being struck in the face by falling debris at the Abbott House, the worst injury of the day.[1]

[1]Parsons, *Private Journal*, June 22, 1881, 236-237.

The intense heat badly scorched the Fifth Street side of the Wehrfritz Building, even igniting small fires along its veranda and wooden roof. Luckily these were snuffed out or another crowded block might have been lost. A hastily organized bucket brigade kept the two-story landmark out of danger. Others tore down the saloon's overhang and awning posts. Similar tactics kept the flames at bay elsewhere.

Even with the fire company doing all it could, the blaze roared up Allen Street for two blocks. Frank Walker of the Sycamore Springs Water Company opened the main valve to full pressure and ordered his wagons to haul water from company hydrants. One lady on Fremont Street let her belongings burn but saved her two kittens. John Young, who had survived the great Virginia City fire of 1875, stoically stood by as his elegant new restaurant disappeared—a $10,000 loss. Others helped businessmen remove ledgers, money, and valuable merchandise before disaster could overtake them. Marshal Earp and Sheriff Behan mobilized their departments to control curious onlookers and help those in the fire zone. Citizens cheered.

In all, nearly seventy businesses were lost, including the newly refurbished Vizina and Cook Block, Tritle and Murray's brokerage office, newcomers Ritter and Ream's undertaking parlor, the Star Restaurant, Cadwell and Stanford's store, the sheriff's office and county courtroom, Wells Spicer's place, and numerous saloons. Damage at the uninsured Oriental totaled $10,000. Others, such as grocers and general merchandisers McKean and Knight, sustained losses as high as $25,000. As in the case of Milt Joyce, few were covered by insurance. Several heavy losers that held policies, however, were paid off within days.

To assist his seven-man force, Virgil Earp called in twenty-one special policeman—including his brother Warren and Texas Jack Vermillion—at $4 a day to patrol the roped-off area and guard against looting. To keep the peace he ordered all remaining saloons temporarily closed. The ruins smoldered throughout the night. But even at the fire's peak harried businessmen elbowed their way into the crowded telegraph office to scribble orders for fresh stocks of merchandise. Insured or not, they looked upon these events as just another example of life's daily gamble.

Vizina and Cook announced the hiring of contractor John Hanlon to supervise a hundred workmen rebuilding their corner to rival the original. Safford, Hudson & Company moved their banking operations in

An image from late in the year, many weeks after the fire. The explosive nature of Tombstone's growth is still strikingly evident—especially when compared with that day in late 1879 when Harry Buehman first captured the embryo community on his fragile glass plate negatives.

Courtesy, Special Collections, University of Arizona Library.

with Wells Fargo, while negotiating temporary space with Wehrfritz and Tribolet. Brokers Tritle and Murray appropriated a convenient corner of the telegraph office. Grocers McKean and Knight posted notice of plans to reopen as soon as things cooled off.

Amid the chaos contestants argued anew over town lot titles in the burnt district. After conferring with Clum and the city council, Virgil Earp announced he would use his police powers to enforce the status quo, leaving it to the courts to sort out that mess. His calm steadfastness brought with it feelings of relief: "The action of Marshal Earp cannot be too highly praised, for in all probability it saved much bad blood and possible bloodshed."[2] Two days later it was said: "Mr. Sippy, late Chief of Police, did the public of Tombstone one very great service when he asked for [a] leave of absence . . . and insisted that Mr. Virgil Earp be appointed to fill the vacancy. . . . The peace and good order of Tombstone has never before been so perfect as now, and that too under the most trying circumstances that have ever occurred in this community."[3]

Understandably, citizens demanded more fire protection. The city council had passed an ordinance two months before defining a fire prevention zone, but now such proposals were taken more seriously—at least for a couple of weeks.[4] In New York Clum had ordered a pumping engine, two carts, and hose measuring a thousand feet from L. Button & Son of Waterford. The Sycamore Springs Water Company agreed to install ten additional hydrants and offered the city free water to fight fires.

Within the week a second fire company was organized. Members elected A. O. Wallace, a justice of the peace and former proprietor of the popular San Diego Keg House on Fifth Street, as president and named their organization the Rescue Hook & Ladder Company. Among its other officers were Dr. Daniel McSwegan, gambler Denny McCann (a Sacramento Fire Department veteran who in four months became chief of the Tombstone Fire Department), saloonman Robert S. Hatch, and Councilman George Pridham—perhaps feeling guilty over the blanket controversy. Milt Joyce, William Breakenridge, and Virgil Earp became standing members.

[2] *Tombstone Epitaph,* June 24, 1881.

[3] Ibid., June 28, 1881.

[4] Minute Book A, Common Council, Village of Tombstone, Ordinance No. 8, April 18, 1881, 62

Allen Street by mid-June 1881. One the left is the Bank Exchange saloon, and at mid-block the Grand Hotel. To the right is the Wehrfritz Building, still officially called the Golden Eagle Brewery. The man in the foreground is selling watermelons. Beyond is the Cosmopolitan, with its covered balcony, followed by the upper portion of Charlie Brown's building.
Courtesy, Arizona Historical Society.

At the new company's regular meeting on July 8, Virgil, Albert Bilicke, and R. J. Campbell, the young clerk of the board of supervisors, found themselves on the finance committee soliciting subscriptions for equipment. Even Wyatt eventually donated $2.50. David Scannell, chief engineer of the San Francisco Fire Department, had already come to their rescue by offering a $3,000 piece of fire-fighting equipment at one-tenth value. It had been ruled obsolete because of the height of many buildings in the Bay City. Tombstone gladly accepted Scannell's generosity. Local officials eventually furnished a building on Fourth Street above Toughnut as home for the new fire company.

Stunned victims welcomed reports outlining plans for the Huachuca Water Company and of the new Tombstone Water, Mill and Lumber Company's organization.[5] It proposed a reservoir atop Comstock Hill to

[5]Ordinance No. 24, Huachuca Water Company Franchise, September 9, 1881, original in the author's collection; Minute Book A, Common Council, Village of Tombstone, September 9, 1881, 109; Articles of Incorporation of the Tombstone Water, Mill and Lumber Company, San Francisco, June 15, 1881; filed with the Secretary of State, No. 13263, June 16, 1881, Book 36, 229, California State Archives; and *Langley's San Francisco Directory*, 1882.

be supplied from wells at Watervale, expanded to a daily yield of 100,000 gallons, and from Granite Springs at the base of the Dragoons. Claiming they held an exclusive, officials at the Sycamore Springs Water Company balked, threatening not to install the promised hydrants or allow free water to the city. The council ignored their complaints and granted a joint franchise. Ill feeling persisted. Weeks later it was said: "The hydrants purchased by the Sycamore Water Company for the city are still lying idle in front of their office. Perhaps they were intended for that purpose."[6]

But all that was for the future. Several hundred workmen and teamsters now swarmed over the area hauling away mounds of rubble and starting new construction. Burned adobes were ground into mortar for fresh use. Virgil insisted that property owners reinforce freestanding walls to limit the chance of injury. When they ignored his warning he ordered those structures destroyed. Within thirty days laborers prepared foundations for all the new buildings along Fifth Street north of Allen. A. P. K. Safford jumped at the chance to buy much of the southeast corner of Fifth and Fremont, replacing the burned-out Abbott House with a one-story adobe partitioned for stores. Wells Spicer owned the small bottom portion of that lot. He ordered adobe used to rebuild his destroyed office and justice court. Leslie F. Blackburn moved in within weeks as Spicer's new tenant.

This time around Vizina and Cook left off the second story in their rush to reopen. Merchants Charles Glover & Company and H. Keyser & Company's Pioneer Boot and Shoe reoccupied the site first. The Oriental, now run by Lou Rickabaugh (Milt Joyce having retired to the world of politics), did not reopen for many weeks. Expensive interior details caused the delay, including a thirty-foot bar required to maintain the saloon's elegant reputation. The new Oriental expanded back the full 120-foot lot size, absorbing for storage the old space occupied by Safford, Hudson & Company's bank. The bank, in turn, relocated just to the north in a larger building.

Unlike other victims, McKean and Knight learned their lesson and rebuilt on the northwest corner of Sixth and Allen in brick. Workers poured four inches of earth above the ceiling before installing a heavy tin roof surrounded by a fire wall. Iron doors from the Tombstone Foundry

[6] *Tombstone Epitaph,* August 11, 1881.

were ordered installed to close over windows in case of emergency. Thus the partners claimed the only fireproof building in town. Others, even those with adobe walls, continued using wood for roofs. Some were sealed with asphaltum shipped in from Los Angeles tar pits. That was a good product against rain, but in a major blaze like that of June 22 it could ignite, gutting property from above.

Pioneer merchants Cadwell and Stanford reopened for business on July 11, the Arizona Brewery the next day, with others, such as the Star Restaurant, not far behind. Tritle and Murray planned to occupy their refurbished site within days. The Myers brothers bought Danner and Owens's Bank Exchange for $7,500 to convert that popular corner into a two-story mercantile. Never losing sight of a dollar, the brothers moved the old building up to Seventh for other uses. The saloon business itself, under new proprietors Sultan and Arnold, reopened on Fourth Street.

City Attorney Marcus Hayne and Virgil Earp sent notice to those rebuilding not to ignore provisions of the city's new fire code against frame buildings in the main business district. Many dismissed the warning in their mad rush to reopen and make money. Again Tombstone merchants showed that self-interest overshadowed community concerns. The city then made the mistake of granting limited exceptions, prompting other businessmen to ignore the ordinance. With more wood being used on Allen Street, it was said: "As matters stand now, the council seems to have gotten themselves into something very [much] like a dilemma."[7]

The fire worried county dignitaries, who felt lucky losing only the sheriff's office and courtroom. They had ordered wagons and willing volunteers rushed to other locations to pack up all official papers if the flames spread. After months of unheeded complaints, citizens, thinking primarily of real estate records, now loudly demanded protection for public documents still shelved in dangerously inadequate wood-frame buildings.

Supervisors heeded the call and began looking for a safer spot. Plans to build a permanent courthouse were already mired in that old controversy over legal title to the proposed ground at Third and Toughnut. It

[7] *Tombstone Daily Nugget,* July 7, 1881.

delayed construction for more than a year. Instead, eyeing the convenience of locating opposite the recorder's office and the refreshing delights of The Grotto, the board leased the Mining Exchange on Fremont Street. Officials hoped that that building, nestled between the adobe *Epitaph* and Gird Block, would protect county property from future fires. They reserved the ground floor as a courtroom, with everyone else, including Sheriff Behan, finding space above.

New construction and fear of fires proved contagious. P. W. Smith decided to rebuild John B. Allen's old store. In late September he temporarily moved his bank into the corner lot across Fourth Street—the original site of Edwards's California Variety Store—then occupied by Ike Levi's Bonanza House Saloon. Levi moved up the street and took over the Palace. Smith demolished his old wood-frame and began rebuilding in adobe. The bank would occupy the corner, surrounded on the south and west by his general merchandise emporium. Throughout October workers dug a cellar along Fourth, using much of the excavated material to fill depressions on Allen Street caused by late summer rains.

When completed the new building would be the most elaborate for its size in Tombstone. It boasted the luxury of French plate glass windows and a $7,500 fireproof safe, shipped in from San Francisco suppliers after the Farmers and Merchants Bank of Los Angeles picked another design. People saw Smith's efforts as further proof that Tombstone was still a good place to make money. Smith certainly thought so. Weeks before he had joined with John Clum, E. B. Gage, Milton Clapp, and Charles D. Reppy to operate the *Epitaph* as a joint-stock company. Smith, always keeping his eyes open for profit, saw the newspaper as a moneymaker and an ideal way to promote Tombstone to outside investors. It was all good business.

Ignoring all the confusion caused by fire damage, thirteen plaintiffs filed suit over surface rights to the Mountain Maid mining claim. Those now before the court included such major players as Vizina and Cook, Charlie Brown, P. W. Smith, Wehrfritz and Tribolet, Tasker and Pridham, Comstock and Brown, and Montgomery and Benson of the O.K. Corral. Owners of other disputed properties watched the maneuvering with heightened interest. The case prompted the *Nugget* to concede: "It is . . . gratifying to observe an honest effort to rescue a portion of the town property from the existing chaotic condition into which town titles have so unfortunately fallen."[8]

Troublemakers kept Virgil Earp and his police force busy. In that first month they made forty-eight arrests, representing more than just scooping up drunks on every shift. Their efficiency, contrasting with that of Ben Sippy, convinced the city council it could save money by reducing the number of officers to five—Earp, James Flynn, A. J. Bronk, George Magee, and Alex Young.[9] Virgil dismissed George Bridge, George W. Chapman, and T. J. Cornelison, the latter soon hauled before the grand jury charged with larceny. Officers caught T. J. rifling a woman's trunk.

Then, after nearly sixty days in office, Virgil suggested his force be reduced further to include only himself, Flynn and Bronk. Delighted by any plan promising austerity, the councilmen agreed,[10] although George Magee was soon reappointed. From time to time they also called up special officers to serve without pay. Even so the police made sixty arrests in July, shoveling badly needed coin and currency into the city treasury.

As a deputy U.S. marshal Virgil arrested an army deserter named Peter Schulzhauser, a disgruntled member of the Sixth Cavalry band at Fort Lowell, whose companion had robbed him of $150. The army seemed satisfied with Virgil's resolve. He and Wyatt then set a trap to catch some outlaws who had robbed the Hermosillo stage, but heavy rains washed out the trail in the Sulphur Springs Valley. Earp had no difficulty filling both roles, federal officer and city marshal, a common frontier practice. One of his own policeman, A. J. Bronk, was appointed a deputy sheriff by Johnny Behan that summer.

Amusing police and townspeople, two women brawled in late July. They put on a satisfactory show: "About fifty young men were as much diverted with the scene as they would have been with a circus." Virgil stopped the fun by arresting both combatants for disturbing the peace. Judge Wallace heard the cases of the "Kilkenny Cats," fining Ellen McKenna, who "like a veteran . . . plead guilty," $35.50 including costs. Her opponent, Margaret Kegan, walked free after City Attorney Hayne determined "she had been more sinned against than sinning."[11] It all

[8]Ibid., June 30, 1881.

[9]Minute Book A, Common Council, Village of Tombstone, July 7, 1881, 85.

[10]Ibid., August 1, 1881, 97.

[11]*Tombstone Epitaph*, July 26, 27, 1881.

gave some relief to Earp, who had just chased down Carlos Friejo for stabbing a fellow Mexican.

The police then picked up that old offender and lot jumper Red Mike Langdon, this time for stealing $20 from one of Lou Rickabaugh's faro games at the Oriental. When Virgil transferred him to the county lockup, Mike suggested they stop off for a drink, confessing he planned to pay with cash plucked from his city jail cellmate—a charge he later denied. In any case, Earp refused. On motion of the district attorney, Judge Wallace discharged a less-than-repentant Red Mike. He was then arrested for carrying a concealed weapon and again later for stealing a watch. Unable to pay the fine, Langdon spent fifteen days in jail on the weapons charge.

Virgil also formed prisoners into chain gangs to clean the streets and other areas littered with trash. Officer Bronk arrested Mrs. Jessie Brown, the proprietress of the Grand Hotel, for allowing debris to collect behind the building. Failing to do anything about it, she was rearrested and fined $15, after which she hired a crew to clean her back lot.

Tombstone's streets always suffered from swarms of hungry rats, garbage, fecal matter, and dead animals. The stench was not helped by several slaughterhouses just outside town, or by the many corrals and stables scattered throughout the city. Untreated outhouses, as well as the practice of keeping chickens and swine in back lots, bolstered complaints. "The men engaged in hauling slops and garbage out of the city do not take it far enough away, and the result is that as one is reclining on his divan the breezes that are wafted through the lace curtains are anything but sweet smelling," echoed a recurring theme. "The police should see to this. There is a dead horse at the foot of Second street that don't [*sic*] improve the smell any."[12]

But on those hot, sweltering days of summer no defense could be found against swarms of flies. "It is a wee fragile thing," Harry Woods observed, "but weighs about five tons when gauged by its persistency. . . . it flies on the top of the dish, and with 7,000 of its near relatives laughs at our wary endeavors." In another column he reported, "A drunk stretched out on Fremont street yesterday afternoon was covered with three million flies. There may have been a few more, but no less."[13] Every-

[12]Ibid., August 5, 1881.

[13]*Tombstone Daily Nugget*, August 5, 1881.

one recalled this curse. George Parsons commented the year before: "The flies are simply abominable. They are worse here it is said than anywhere else. Our meals are most uncomfortable ones. Flies by the billion. The old miners never knew them so bad before anywhere. It's a curiosity to gaze at the beams at dinner. They're black."[14]

Someone bemoaned conditions at Dr. Goodfellow's hospital: "To a person passing within a block of that structure, particularly upon its northern side, the effluvia emanating therefrom is . . . suggestive of a dead horse or a bone boiling establishment."[15] Supervisors Joyce, Tasker, and Dyer, Virgil Earp, Officer Magee and a reporter from the *Epitaph* all visited the building and declared that description overstated. George Goodfellow may have hoped to turn the hospital over to the Sisters of Mercy, but in retaliation the council eliminated his position as city health officer. In his place they named the marshal, much to the delight of editor Harry Woods: "Chief Earp believes cleanliness is next to godliness, and is making the streets of Tombstone a model."[16]

Woods exaggerated. Earp's best efforts aside, the streets remained corridors of finely ground dust. As a main thoroughfare for ore wagons from the Grand Central, Contention, and Tombstone Mill and Mining Company, Fremont Street stood out as particularly offensive. City officials, worried about cost, repeatedly ignored pleas to water the streets. Some businessmen paid to have the road fronting their stores treated, but with each gust of wind Tombstone found itself once again clothed in dirt.

Rain helped, but on occasion it proved more troublesome than the dust. A particularly violent thunderstorm in early August 1881 caused the recorder's office to cave in. It seems water cascading off the pitched roof of the Papago Cash Store ate away the adobes. Crashing into The Grotto, fragments also fell "upon a shed of the O.K. Corral . . . under which were tied a number of horses."[17] Parsons wrote: "several horses had to be shot from injuries received."[18] Repairs were done quickly, but Clement Eschman had had enough. He moved his subterranean deli-

[14]Parsons, *Private Journal,* July 4; 1880, 135.

[15]*Tombstone Epitaph,* July 10, 1881.

[16]Minute Book A, Common Council, Village of Tombstone, July 9, 1881, 87-88; and *Tombstone Daily Nugget,* August 3, 1881.

[17]*Tombstone Daily Nugget,* August 12, 1881.

[18]Parsons, *Private Journal,* August 11, 1881, 248.

catessen and watering hole into the basement of the Grand Hotel, much to the regret of hungry and thirsty county employees.

In contrast to natural catastrophes, the police spent days chasing down an "accordion fiend" performing late-night concerts on Fremont Street, upsetting property owners who feared loss of value. No sooner was that hated musician arrested than a trombone man took his place with unwelcomed serenades. Even with such seemingly frivolous duties, Virgil stayed busy trying to build his reputation as a resolute officer.

A blemish against the Earps' carefully crafted image came with Sheriff Behan arresting Doc Holliday on July 5. Warrants sworn out before Wells Spicer charged the Georgia dentist with murdering Bud Philpot and trying to rob the United States Mail. Both warrants were issued on the strength of an affidavit signed by the gambler's common-law wife, then using the name Kate Elder. Spicer heard the first charge as a justice of the peace and the second as a United States commissioner. He set bail at $5,000. Wyatt Earp and two others posted Holliday's bond.

Doc was now in serious trouble, not helped by his admitted friendship with the late Bill Leonard. Kate distrusted Holliday's ties to the Earps. Years later she confessed seeing "all manner of disguises" in a trunk Wyatt opened on their trip to Prescott in 1879. Virgil's wife, too, was disturbed by these beards, wigs, and false mustaches after seeing them at Wyatt's house in Tombstone.[19] Kate never claimed Doc used them, despite her own suspicions and a tough closed-door grilling by Spicer—who wanted to know how Holliday acted that night, if any of the Earps had come to her for his rifle, if Doc had changed his clothes that afternoon, what he and Warren Earp had talked about, and how long had she known the Earps. The angry jurist tossed out the questions as if checking off a string of damaged pearls. Propped in a corner chair, Sheriff Behan listened in silence. Kate said nothing but left shaken, convinced Wells Spicer "felt sure that the Earps and Holliday were in that hold-up."[20]

Worrying how Doc's predicament might undermine their own efforts, the brothers pressured Kate to disavow her affidavit. On Holliday's insistence she was arrested and locked in the Earps' private room at

[19]Memoirs of Mary Katherine [Holliday] Cummings; and Waters, *The Earp Brothers of Tombstone*, 109.

[20]Memoirs of Mary Katherine [Holliday] Cummings.

the Cosmopolitan, before paying $12.50 at the recorder's court on two counts of drunk and disorderly. Rearrested, she was then found guilty in Judge Felter's court for making threats. Fighting back, Kate hired Spicer who appealed her case to T. J. Drum, commissioner of the first judicial district. He granted a writ of habeas corpus and discharged the frustrated defendant, claiming the case resulted from "a warrant sworn out by an enraged and intoxicated woman."[21]

Despite her victory, Kate now feared for her safety if she stayed in Tombstone and wisely refused to testify against Doc. At an early morning hearing on July 9, a clearly dissatisfied Judge Spicer dismissed all charges against Holliday after the district attorney concluded "there was not the slightest evidence to show the guilt of the defendant; that the statements of the witnesses did not even amount to a suspicion" of wrongdoing.[22] But others, including Bob Paul, would always believe that Doc Holliday was one of the men on the road that night near Drew's Station.[23]

Then, as if to counterbalance Holliday's weak reputation for honesty, Virgil Earp came upon a drunken Michael Ryan leaning half-asleep against an adobe wall, surrounded by three opportunists poised to rob him. Earp scattered the would-be thieves and took the man in tow, removing a wad of bills from his overalls for safekeeping. At the *Epitaph*'s editorial office the marshal asked a reporter to count the cash as a witness. To their surprise they discovered Ryan had been staggering around town with $930. Virgil locked the money in his safe, explaining to the intoxicated Irish miner that he could reclaim it when sober. "No doubt he is thankful . . . that Tombstone has a vigilant and honest Chief of Police," mused Harry Woods.[24]

The *Nugget*'s editor had missed the mark. Rather than thankful, Ryan scrounged three $100 bills from another corner of his pocket and "abused the Marshal as only a senseless, drunken fellow would," taunting, "You thought you were d—d smart, but you didn't get it all."[25] Virgil ignored the remark and took him home. Unguarded, Ryan reemerged and stumbled into the Eagle Brewery Saloon, winning $80 at faro. After

[21] *Tombstone Daily Nugget*, July 9, 1881.
[22] Ibid., July 10, 1881.
[23] *Arizona Republican*, June 26, 1892.
[24] *Tombstone Daily Nugget*, July 14, 1881.
[25] *Tombstone Epitaph*, July 15, 1881.

he ordered a round of drinks in celebration, a quick-fingered patron robbed him of $200. Earp searched unsuccessfully for the thief before jailing Ryan as drunk and disorderly.

Further investigation tentatively identified William "Red Billy" Freeman, a refugee of San Francisco's Barbary Coast, and former policeman James Bennett as those responsible for Ryan's loss. Freeman had already hired a team and skipped out. The marshal sent word to the deputy at Charleston, but Red Billy was already on his way to Benson. Virgil then telegraphed Morgan—who often gambled there, finding easy pickings among railroad passengers—to make the arrest. Morgan eventually grabbed his man, but Freeman broke loose during the confusion surrounding the arrival of the eastbound train. Fearing he had no real authority to use force, Morgan insisted, he hesitated to open fire on the fleeing fugitive. At Tombstone Virgil arrested Bennett, but Judge Wallace turned him loose for lack of evidence.

None of this helped the unrepentant Mike Ryan. Wallace had fined him $7.50 during the morning session. With most of his cash safely returned, Ryan stomped off mumbling about life's inequities. More importantly, however, in light of the growing embarrassment surrounding Doc Holliday, Virgil Earp had regained the confidence of less skeptical citizens.

Morgan Earp spent so much time gambling and visiting a lady friend at Benson that Deputy Sheriff McComas, fearing trouble from the notorious John J. Harlan (well known in Kansas and New Mexico as the "Off Wheeler"), telegraphed Behan asking if he could hire Morgan as a temporary deputy. The sheriff told him to use his own judgment. Thus Morgan Earp briefly became one of Johnny Behan's deputies just as the Off Wheeler started arguing with an Indian. The incident ended with both men exchanging gunfire, along with a local constable named McCarty who mistakenly came to Harlan's aid. Morgan helped arrest all the offenders. At Tombstone Wells Spicer punished their public indiscretions with heavy fines.

Johnny Behan could have done worse than temporarily name Morgan Earp a special deputy. In fact he did, appointing the likes of Frank Leslie and Frank Stilwell as Cochise County officers. Many more citizens began having serious doubts about that office. Then his man Billy

Breakenridge embarrassed them all by losing his revolver. Reminiscent of Ben Sippy, he went to the *Nugget* and ordered a printed notice offering $5 for its return. Walking to Spangenberg's Fourth Street gun shop, the unarmed deputy bought another six-shooter, but by then Charlie Brown had found the original.

What happened next did not amuse Harry Woods: "He returned to the store and gave back the pistol just purchased, thereby beating the proprietor out of his profit on it. Then on around to the *Nugget* office, and stopped the publication of the notice, beating the office out of the advertisement, and in addition beating the party finding the pistol out of the $5 reward he intended giving."[26] Chuckling at the deputy's expense turned out to be short-lived. Questioning the competence of local peace officers replaced the fun of needling the red-faced Breakenridge.

During that summer of 1881 the Cowboys became especially belligerent as their smuggling and cattle rustling forays into Mexico increased, both in terms of frequency and levels of violence. Profits stayed high as anxious buyers in Arizona and New Mexico competed with one another to do business with the outlaw bands. For the likes of Curly Bill Brocius and his friends it seemed a golden time. But gloom replaced joy in outlaw camps in mid-August with news of two dozen or more Mexicans ambushing sixty-five-year-old Newman Haynes Clanton and his companions Richard "Dixie Lee" Gray, William Lang, Charles Snow, William Beyers, Harry Ernshaw, and Jim Crane as they camped in Guadalupe Canyon near the New Mexico border. Five of these men died, attacked in the early morning as they awoke. Beyers escaped with a belly wound, and Ernshaw dashed to safety after a single bullet grazed his nose.

One of the victims turned out to be Mike Gray's youngest son, a bookkeeper, nineteen years old and lame. The Grays were friends of the Clantons. Another son, John P., had left for his father's New Mexico ranch just three weeks earlier, accompanied by Ike Clanton and Phillippi S. Montague, a young man from Nova Scotia then in the real estate and brokerage business with Dick Gray. William Lang and his father had come to Arizona from Kansas only five months before. They were men

---

[26] *Tombstone Daily Nugget*, July 30, 1881.

of means, then organizing a stock ranch in the Sulphur Springs Valley. At the time of the attack young Lang planned to bring cattle into the Tombstone market. By chance the third known participant in the abortive Benson stage affair, Jim Crane, was also caught unawares that Saturday morning. Crane, a victim of poor timing, had just joined the group the night before. "He was a fugitive from justice and an outlaw," Clum remarked without sympathy, "and the six bullets that struck him were certainly well expended."[27]

In all probability the attack at Guadalupe Canyon marked revenge for Cowboy raids in Sonora. Just two weeks before, twenty of the Arizona outlaws ambushed sixteen Mexicans leading a caravan near Fronteras. The Mexican consul later verified four killed and two wounded. The Cowboys stole all their animals, merchandise, and some $4,000 in cash. At Fronteras survivors begged authorities for assistance. The local military commander explained he had no authority to cross the border. In southern Arizona, however, almost everyone believed Mexican soldiers dressed as civilians had carried out the raid killing Clanton and the others.

No records supporting these suspicions have yet surfaced in Mexican archives. But an interesting cache of documents unearthed at Hermosillo details just how exasperated authorities there had become over Cowboy raids and the seeming inability of the U.S. government to stop them.[28] They tried embarrassing officials at Washington to take action: "Is it possible that within the boundaries of the best organized Government on the planet a few outlaws, the whole number probably not exceeding one hundred, can band together, defy the civil authorities, and while taking advantage of the security our soil affords, reach out and paralyze the industries of a neighboring State?"[29]

At the height of this crisis J. W. Evans—a one-armed deputy U.S. marshal—warned his boss, after conferring with the Mexican consul at Tucson, that they "express great dissatisfaction at the seeming neglect of our Gov't and threaten to take vengeance on all Americans in Sonora."[30]

---

[27]*Tombstone Epitaph,* August 16, 1881.

[28]"Campana contra los Tejanos o Cow-boys," Biblioteca y Museo de Sonora Archivo Historico, Hermosillo, Mexico.

[29]Legation of Mexico, Washington, to James G. Blaine, Secretary of State, April 13, 1881, Records of the Office of the Secretary of the Interior, Record Group 48, National Archives and Records Service.

[30]J. W. Evans to C. P. Dake, August 11, 1881, General Records of the Department of Justice, Record Group 60, National Archives and Records Service.

Perhaps they decided instead to take vengeance in Guadalupe Canyon. Buckskin Frank Leslie, returning to Tombstone after several weeks in southwestern New Mexico, rode past the massacre site. Just to the west at a place called Blackwater he found the bodies of three recently killed Mexicans, "which would indicate," Harry Woods surmised, "that the work of retribution has already commenced."[31]

In Tombstone George Parsons sided with his neighbors to the south: "This killing business by the Mexicans, in my mind, was perfectly justifiable as it was in retaliation for killing of several of them and their robbery by cow-boys recently[,] this same Crane being one of the number. Am glad they killed him, as for the others—if not guilty of cattle stealing—they had no business to be found in such bad company."[32] Interestingly, Parsons failed to mention Newman H. Clanton by name. But even John Clum, in his account of the massacre, did not list him as a leader of the Cowboys. Instead, he simply referred to his old acquaintance as "the senior Clanton."[33]

The Mexicans had good reason to complain. Long before Schieffelin's discovery smugglers ravaged the border. In late 1878 Governor Frémont admitted as much in a letter to the secretary of the interior. But less than a month later he changed his position, claiming, "Except for the occasional crimes . . . no mention has come to me of any disturbances on the frontier."[34] That must have been news to those living along the line. Then in mid-February 1879 Frémont reported, after speaking with cattleman and legislator Walter Vail, a twenty-six-year-old native of Nova Scotia: "He tells me that he found nothing to show the existence of any such bands."[35] Of course, Vail, one of the owners of the sprawling Empire Ranch and an early employer of the McLaury brothers, was himself later accused of involvement with cattle rustlers. Frémont proved easy to deceive, but even he finally admitted, after a year of convincing: "I have been able to learn the number of men in the band known as Robert Martin's is about one hundred and twenty."[36] Frémont's information was already hopelessly out of date.

[31] *Tombstone Daily Nugget*, August 20, 1881.

[32] Parsons, *Private Journal*, August 17, 1881, 249.

[33] *Tombstone Epitaph*, August 16, 1881.

[34] Gov. J. C. Frémont to Carl Schurz, Secretary of the Interior, January 6, 1879, Records of the Office of the Secretary of the Interior, Record Group 48, National Archives and Records Service.

[35] Ibid., February 15, 1879.

[36] Ibid., January 26, 1881.

Other federal agencies were better informed than the governor's office. The United States Treasury Department and Customs Service both sent undercover agents into the Tombstone area to study conditions for themselves. They drew a clearer picture of circumstances than had Arizona's beleaguered chief executive. Lamenting the absence of customs stations, one agent reported: "This is a very unguarded space of 230 miles with several active mining camps on the border. And then there is the San Simon Valley, through which I believe there have been driven in the past 2 years, not less than 8000 or 9000 head of beef cattle that paid no duties. . . ."[37]

Much of this stock was routed through Arizona and New Mexico for sale to the San Carlos Apache Agency. Thomas Benton Catron, a New Mexico land baron, Republican power broker, and boss of the so-called Santa Fe Ring, was believed at the time deeply involved. Agents identified sixteen or seventeen offenders but regarded Catron "as chief among them."[38] On the surface the forty-year-old Missourian and Confederate army veteran seemed an odd candidate for criminal infamy. A dozen years before, Catron had been named attorney general for the New Mexico Territory, resigning only after President Grant appointed him United States attorney. Unproven accusations linking him to the Cowboys made little or no impact on his subsequent career. With statehood in 1912, Thomas Catron became one of New Mexico's first United States senators. They even named a county after him.

Along with suspicions about Catron, government agents also singled out Walter Vail's "crooked transactions," although they considered him a gentleman.[39] But it was the loss of revenue that drove the agents, one of whom, H. L. Williams, reported: "I am of the opinion that not over 25% of the duties have heretofore been collected upon cattle, not one per cent upon liquors, and little or nothing upon articles taken in by Mexicans on waggons [*sic*] and sold in mining camps. . . ."[40] The chief items smuggled were cattle, cigars, tobacco, and mescal.

[37]R. M. Moore, Special Agent, Customs Service to John Sherman, Secretary of the Treasury, April 30, 1880, United States Customs Service, Special Agents Reports, Record Group 36, National Archives and Records Service.

[38]Ibid., December 1, 1880. For more on this man's rather remarkable life, see Westphall, *Thomas Benton Catron.*

[39]R. M. Moore to A. K. Tingle, Supervisor Special Agents, Washington, April 5, 1881, United States Customs Service, Special Agents Reports, Record Group 36, National Archives and Records Service.

[40]H. L. Williams, Special Agent Treasury Department to William Windom, Secretary of the Treasury, June 27, 1881, ibid.

Federal agents were realistic about the enemy facing them in southern Arizona. Williams pointedly identified that area east of the Dragoon and Mule mountains:

> here lies the celebrated Sulphur Spring[s] Valley, one of the most extensive and richest in grass and water along the Mexican line, and through it are driven more smuggled cattle than through all other points added together. It is inhabited and controlled by "Cow Boys," a class more difficult and dangerous to contend with than Indians—caring nothing for their own lives, and less for that of others, committing murder without hesitation and in many instances for the mere sport of the thing. They go into Mexico and steal large herds of cattle, drive them into this Valley upon their ranches, and dispose of the cattle at their leisure to butchers at the different towns and mining camps and it is said that many of these cattle pass into the hands of Indian Contractors through the means of agents.

But what could be done? For as Special Agent Williams concluded: "It would be sure death for anyone, or even a small party of men to attempt to seize any of these cattle, there is said to be in the Valley as many as 200 of these desperadoes, who would kill upon the first favorable opportunity anyone who interfered with their plans."[41]

The people of Tombstone understood the Cowboy menace. True, they generally avoided the town while on their murderous forays, but outlying areas suffered. Having avoided any responsibility for the accidental shooting of Marshal White, Curly Bill Brocius turned around and terrorized Contention City. Repeating such an outlandish performance in Tombstone would have been impossible. The place was simply too large and filled with enough determined men unlikely to stand idly by. But the Cowboys had little to fear raiding smaller communities, isolated ranches, and lone travelers on public thoroughfares. Stage passengers still felt reasonably safe, but this too was about to change.

Area residents had long complained about the Cowboys, as well as the ineffective efforts of those officers charged with driving them from power. Both Sheriff Behan, whose jurisdiction covered the outlaws' range, and Deputy U.S. Marshal Virgil Earp felt the criticism each time the Cowboys made news. The episode at Guadalupe Canyon succeeded in pulling these deep-rooted concerns to the surface and holding them there.

---

[41]H. L. Williams to H. F. French, Acting Secretary of the Interior, November 9, 1881, ibid..

The death of Newman Haynes Clanton further challenged the Earp brothers. It had always fallen to the older man to control Ike's unpredictable behavior. The father's sense of political balance had saved his son much grief in both California and Arizona. The Earps had gotten along rather well with Newman; all their occasional meetings conducted in an openly friendly manner. Virgil particularly admired the old man's grit. But Ike had begun to resent Wyatt, whom he suspected of betraying their mutual scheme to capture Leonard, Head, and Crane. Wyatt had not done so directly, but Marshall Williams had figured out the request to Wells Fargo about the rewards made payable dead or alive. Ike feared retaliation if the story got around.

Nor did the Earps fade into the background for Ike Clanton's benefit. Virgil enforced Tombstone's ordinances with stern indifference to anyone's reputation, so much so that it could be reported with sarcasm: "Cow-boys don't seem to visit our city very much. Don't they like the climate? We feel slighted."[42] The outlaws may have worried about Virgil's position as a federal officer, but he offered them little discomfort on that score. The marshal busied himself instead with Tombstone. Besides normal duties Virgil raided Chinese opium dens from time to time, usually a $15 fine, and then arrested John Clum, on a complaint filed by Milt Joyce, for riding too fast. The embarrassed mayor quietly paid the $5 assessed by Judge Wallace. Earp soon confronted more serious offenders.

Returning from Tucson on September 9, Sheriff Behan told Virgil that Bob Paul had arrested Pony Diehl, wanted by Wells Fargo for helping rob their coach from Globe back in February. Another suspect, Sherman McMasters, was known to be in Tombstone. Earlier Paul had asked Earp to make no arrest, as he hoped to grab both suspects together. Now, learning that Diehl was in custody, Virgil telegraphed Paul around six o'clock that evening asking if he still wanted McMasters. Awaiting word in the doorway of the telegraph office, Earp was approached by someone claiming that John Ringo (who had earlier robbed a poker game at Galeyville) had just ridden into town. Asked where Ringo was, the man answered, "With McMasters." Virgil, still with no reply from the sheriff of Pima County, hesitated to act.

---

[42] *Tombstone Epitaph,* August 16, 1881.

Over two hours later a telegram from Bob Paul arrived for Marshall Williams. Williams was out of town with Wyatt and Morgan Earp chasing stage robbers when the message was delivered. But a clerk named Clayton went looking for Virgil. The marshal had been standing on Allen Street as McMasters walked by, apparently headed toward the O.K. Corral. Following Clayton back to Wells Fargo, Earp read the message: "Tell V. W. Earp, to-night, that I want McMasters."[43] Unarmed, the marshal asked for a revolver. The clerk had none but quickly borrowed one from a friend. Virgil called on his brother James for support. After the older man armed himself the two walked down Allen Street searching all the saloons.

As they reached the corral someone rode out on McMasters's horse. Virgil ordered a halt, but the rider ignored him. Earp got his attention by firing a shot. Since the man turned out to be a stranger the marshal let him go. Realizing his mistake, and suddenly suspecting McMasters had sent the man to retrieve his horse, Virgil raced on foot to the corner of Third and Safford. It appeared the horseman was searching for someone. Two blocks east Earp stopped the rider a second time, ordering him to dismount or be shot from the saddle. As Virgil reached for the stirrup, McMasters bolted from some brush about a hundred feet away and jumped a fence. The marshal emptied his borrowed revolver at the shadowy figure but missed. McMasters ran through an arroyo and escaped into the twilight.

Hearing the shots, Behan rushed over. Together he and Earp searched for more than half an hour without success. McMasters then reportedly stole two horses from the superintendent of the Contention and made good his escape. Behan saddled up but lost the trail when it got mixed up with that of a cattle herd.

Of the whole affair the *Nugget* complained, "The worst mistake was in Paul not telegraphing immediately to Earp or the Sheriff, and the error was in the officers not arresting him when they first saw him in the street."[44] Clum remarked that McMasters left behind "a splendid mare

[43]Ibid., September 10, 1881. At first glance it seems odd that the sheriff addressed this telegram to Williams rather than the city marshal's office. In all likelihood he declined to send the message in the clear, but instead picked a Wells Fargo code. Paul frequently used one of the company's code books when communicating directly with its San Francisco headquarters. Thus Marshall Williams, or another trusted employee, needed to see the telegram first in order to decode the text.

[44]*Tombstone Daily Nugget*, September 10, 1881.

and one of the finest saddles ever seen in this part of the country. He will miss his outfit very much, but does not want it bad enough to come back after it."[45] This would not be the last the Earps heard of McMasters, or of Pony Diehl or John Ringo for that matter.

The day before McMasters's run-in with Virgil, masked gunmen had robbed the stage between Tombstone and Bisbee at a point beyond Hereford about a dozen miles from its destination. The outlaws got away with $2,500 from Wells Fargo's treasure box and passengers' valuables, including $8 and a gold watch from one and nearly $600 from another. They fled after rifling the mail sacks. Two messengers rushed to Tombstone and alerted Agent Williams.

Undersheriff Woods rode to the scene, followed closely by Williams, Deputy Breakenridge, Wyatt and Morgan Earp, and Fred Dodge. Together they began tracking the outlaws over a cold trail. Three days later the two Earps, Williams, and Breakenridge returned with a pair of suspects—Peter Spencer, then supporting himself as a gambler and wood cutter, and his friend Frank Stilwell, Johnny Behan's deputy at Bisbee. (Two months before Stilwell mortgaged some land along the Babocomari to Spencer, together with some personal property, for $3,000.) Because of this latest incident the *Epitaph* suggested, "it would seem to be in order for Sheriff Behan to appoint another deputy."[46]

The sheriff would have done better not appointing Frank in the first place. That twenty-five-year-old officer, younger brother of the celebrated army scout Simpson E. "Comanche Jack" Stilwell, hardly boasted a reputation recommending him as a lawman. Some years earlier, working for George Young near Miller's ranch outside Prescott, Stilwell shot Jesus Bega through the lung and nearly killed him during an argument over the newly hired Mexican cook serving tea instead of coffee on his first day at work.[47] Stilwell and Jack Cassidy then escaped a grand jury indictment for the murder of J. Van Houton in late 1879. It was also rumored Stilwell had raped a young woman at his stable in Charleston. Not normal items on the résumé of a budding peace officer.

Now he found himself arrested by the Earps and lodged in the

---

[45] *Tombstone Epitaph,* September 17, 1881.

[46] Ibid., September 13, 1881.

[47] *Weekly Arizona Miner,* October 19, 26, 1877.

county jail awaiting a federal indictment for robbing the mail. Justice of the Peace Spicer, also acting as a United States commissioner, set bond for each man at $2,000 for the robbery itself and $5,000 on federal charges. Ike Clanton, Billy Allen, and Charles H. Light, for whom Stilwell had once worked as a teamster in Mohave County, posted bail. Clanton and Stilwell had remained friends despite vague rumors of Ike's deal with Wyatt Earp over the Benson stage business.

In a strange twist, after being released on bond, Peter Spencer asked Marshall Williams for a loan to buy provisions for his family. The Wells Fargo agent replied he had no money to lend. The next day, however, he accompanied Spencer and Virgil Earp to P. W. Smith's store. There Earp and Williams gave security for $25 worth of supplies.

The arrest of his two friends may have disturbed Ike Clanton, but Johnny Behan was both irritated and embarrassed over the Earps jailing his deputy. The sheriff took it personally. The month before Behan might have allowed Morgan to serve as a special deputy, but those days were over. Feelings of animosity proved mutual. Later, when Acting Governor John J. Gosper questioned Behan about the Cowboys, he claimed Virgil Earp "seemed unwilling to heartily cooperate with him (the Sheriff) in capturing and bringing to justice these outlaws." But then, Gosper added, "In conversation with the Deputy U.S. Marshal, Mr. Earp, I found precisely the same spirit of complaint existing against Mr. Behan . . . and his deputies."

Gosper went on to level a more serious charge: "The opinion in Tombstone and elsewhere in that part of the Territory is quite prevalent that the civil officers are quite largely in league with the leaders of this disturbing and dangerous element."[48] Gosper was not alone; Clum at one point charged rather bluntly, "There is altogether too much good feeling between the Sheriff's office and the outlaws infesting this county. . . ."[49]

Four months later a newcomer from Massachusetts, Cambridge theology student Endicott Peabody, deduced in a more evenhanded fashion, "The misfortune is that the cowboys are countenanced by the

[48]John J. Gosper, Acting Governor, to James G. Blaine, Secretary of State, Washington, September 30, 1881, Records of the Office of the Secretary of the Interior, Record Group 48, National Archives and Records Service.

[49]*Tombstone Epitaph,* August 19, 1881.

sheriff for political reasons and the marshal's party on the other hand is not quite above suspicion."[50]

Disgusted by events, Gosper later explained,

> the underlying cause of all the disturbances of the peace . . . is the fact that all men of every shade of character in that new and rapidly developed section . . . have grossly neglected local self government, until the more lazy and lawless elements of society have undertaken to prey upon the more industrious and honorable classes. . . . The civil officers of the county of Cochise, and City of Tombstone partaking of the general reckless spirit of rapid accumulation of money and property, is another cause of public disturbances, in as much as they have seemed to "wink at crime" and to have neglected a prompt discharge of duty, for the hope and sake of gain.

As for the Cowboys themselves, Gosper continued, "when apprehended or detected by the officers of the law, [they] have in many cases no doubt, *purchased* their liberty, or have paid well to be left unmolested."[51]

Despite accusations of official impropriety, Johnny Behan's troubles with the Earps, especially Wyatt, took many political and personal turns. An intriguing element to this drama involved both men's infatuation with the same woman. A self-styled "actress," Josephine Sarah Marcus actually exhibited talents of a more physical nature. Although frumpy and disagreeable in old age, in those early years, Sadie, as she was then called (rather than "Josie," as is generally believed today), possessed the charm to turn men's heads. At Tombstone she moved in with Johnny as his teenage paramour, presenting herself as Mrs. Behan. Earp first met the vivacious young temptress during the late summer of 1881, while buying newspapers at Sol Israel's Union News Depot on Fourth Street. During the 1930s Sadie tried—with some success—to mislead a gullible public, while at the same time enlivening her own standing in the scheme of things, by falsely claiming that her relationship with Wyatt actually started in 1880.

Except for Virgil, the Earp boys were not particularly monogamous. Morgan visited his lady friend at Benson, and Wyatt dallied with at least

---

[50]Letter from Endicott Peabody, January 30, 1882, as quoted in Ashburn, *Peabody of Groton,* 51. Peabody, who never forgot his adventurous days in Tombstone, went on to become headmaster of Groton. Later he performed the marriage ceremony between Eleanor and Franklin D. Roosevelt.

[51]Acting Governor John J. Gosper to Hon. C. P. Dake, U.S. Marshal, November 28, 1881, General Records of the Department of Justice, Record Group 60, National Archives and Records Service.

two others in Tombstone, one of whom, a married woman, he continued to see years later in Los Angeles, much to Sadie's annoyance. Ignoring his own wife, who committed suicide in 1888, but understanding the political implications of any openly adulterous relationship, Wyatt succeeded fairly well in keeping these indiscretions from public view. Sadie, however, foolishly broke the news of her latest infatuation to Johnny Behan. Her romantic interest in his political rival left the sheriff deeply disillusioned. His distrust of Wyatt and Sadie soon included the whole Earp family.

Wyatt would later claim he married Miss Marcus in 1885. If true it is a curious admission considering the death date of Mattie Blaylock, whom he deserted but never divorced. In fairness, however, no marriage record has yet been found between Wyatt and Mattie, so perhaps there was no legal bond requiring divorce proceedings. Of course, there is no proof he ever married Sadie either. Still, not even her charms could control Earp's roving instincts. He enjoyed the company of many attractive women over the years. Sadie made him pay dearly for these transgressions, while conveniently ignoring her own.

By mid-September 1881 Judge Spicer dropped the charges against Peter Spencer. Then, in early October, Frank Stilwell also walked free. Spicer claimed the evidence insufficient to convict. Yet, even with this favorable turn of fortune, Stilwell and Spencer faced continuing troubles with the persistent Earp brothers.

Adding to the confusion over the proper use of authority, Fred Dodge, a self-styled capitalist and occasional Wells Fargo operative described locally as "a well-known sport," had gotten into an argument over a monte game at the Alhambra. Dodge pulled a revolver and fired one poorly aimed shot. Officer Bronk made no arrests. The incident refocused old concerns: "We have not heard of the ordinance prohibiting the carrying or firing of firearms within the city limits being repealed. It is possible, however, that it only applies to laboring men, and is a case of 'special legislation' exempting a certain class."[52] Bronk took the heat. He also faced a complaint of blackmailing a prostitute, extorting money from her as a private watchman. In a move that surprised no

[52] *Tombstone Daily Nugget*, September 14, 1881.

one, the city council dismissed all charges during a hastily assembled special session, and Bronk continued on the police force.[53]

Following a breakout from the San Carlos Apache Agency, Tombstone witnessed a mild Indian scare in early October. Raiders, either Apaches or Cowboys, hit outlying areas stealing horses, mules, and some cattle. The McLaurys lost fourteen horses from their ranch at Soldiers Hole, beyond the Turquoise Mining District at the southern end of the Dragoons.

Thirty volunteers, including John Clum, Charles D. Reppy, Ward Priest, and two of the Earp brothers took to the saddle chasing down rumors of pitched battles with hostiles in the Sulphur Springs Valley. The men elected Johnny Behan their captain and awarded Virgil Earp the rank of lieutenant. While camped at McLaury's ranch George Parsons was startled to see Curly Bill and two confederates: "I will say that our present Marshal and said 'C Bill' shook each other warmly by the hand and hobnobbed together some time. . . ." Parsons went on to record, however, that Brocius "was polite and considerate."[54]

The whole affair became a rather humorous and futile forty-eight-hour exercise. Some townspeople may have taken the reports seriously, but Tombstone was never in any danger. Others saw the absurdity of it all, Harry Woods capturing the mood:

> Frank Ingolsby has sent from San Francisco to a friend in this place a crayon drawing entitled, "Tombstone to the Rescue". . . . Coming to the assistance of the troops are some fifteen or twenty horsemen, and so well has the work been done that many of those who went out with Sheriff Behan and Marshal Earp may be readily recognized. . . . We understand it will be photographed, when its merits and absurdities may be appreciated by the public.[55]

Wyatt rode with his brother knowing it was good politics. He was always looking for votes. His obsession with becoming sheriff had led to his joining the fire company, getting his name in the papers as a resolute officer, occasionally donating a few dollars to high-profile causes, and helping local political figures. He joined Joseph Tasker, Sylvester B.

---

[53]Minute Book A, Common Council, Village of Tombstone, September 14, 1881, 111.

[54]Parsons, *Private Journal,* October 6, 1881, 263-264.

[55]*Tombstone Daily Nugget,* October 20, 1881.

Comstock, and Benjamin Fickas as sureties for the city attorney's $5,000 bond.[56] It all made perfect sense to Wyatt Earp as he sat pondering the money to be made if only he could convince enough voters to elect him sheriff. Earlier the board of supervisors approved a 10 percent fee for collected taxes. Clum objected, but Wyatt thought it was just fine.

On the afternoon of October 8, about four miles north of Contention City, five men robbed the coach making its daily run between Benson and Tombstone. The driver, perhaps recalling the fate of Bud Philpot, deserted the coach and scampered off into the brush. Even the horses escaped after running the stage into an earthen embankment and breaking loose. The robbers picked the eleven abandoned passengers clean, getting in all about $800. Sheriff Behan and Marshall Williams rode out to track the outlaws. The board of supervisors, growing annoyed over stage robberies, authorized a $100 reward for the arrest and conviction of each of the highwaymen.

Then, just five days later, Wyatt Earp arrested Frank Stilwell in Tombstone. Overhearing their prisoner asking a man to inform Peter Spencer, Virgil and Wyatt rushed to Charleston and took care of that chore themselves, nabbing Spencer as he rode into town. The *Nugget* declared both men were in custody on "a charge of robbing the Benson stage last Saturday afternoon." The *Epitaph* suggested the same, saying Stilwell and Spencer had been arrested "for complicity in the late Contention stage robbery."[57] It is interesting how quickly both editors assumed the worst. Actually the Earps had simply rearrested the two for robbing the Bisbee stage six weeks before. Citing new evidence, Virgil Earp escorted them to Tucson on October 13—sidestepping an overly lenient Wells Spicer—for an appearance before a U.S. commissioner: "Sheriff Paul received them at the depot."[58]

Virgil returned to Tombstone the next day. He soon helped Officer Bronk arrest two men for robbery and snagged Tom Corrigan on an assault and battery charge. Tom was then tending bar at the Occidental, having long since sold his interest in the Alhambra. On October 18 Vir-

---

[56]Minute Book A, Common Council, Village of Tombstone, September 9, 1881, *Tombstone Epitaph,* September 11, 1881.

[57]*Tombstone Daily Nugget* and *Tombstone Epitaph,* October 14, 1881.

[58]*Arizona Weekly Star,* October 20, 1881.

gil Earp, who had been subpoenaing more witnesses to the Bisbee stage affair, traveled to Tucson with Johnny Behan. Both planned to attend the preliminary hearing for Stilwell and Spencer scheduled to open on the morning of October 20.

While at Tucson, Virgil later explained, "Wyatt Earp had been sworn in to act in my place . . . and on my return his saloon was opened [actually just the gaming concession at the refurbished Oriental] and I appointed him a 'special,' to keep the peace, with power to make arrests. . . ."[59] He had already sworn in Morgan as a special policeman about a month before. Morgan went to Tucson the day of the hearing, then came back two days later with Doc and Kate Holliday. Those two had reconciled after her boarding house at Globe burned down, destroying all her possessions. Doc borrowed and then promptly lost her last $75 bucking faro at the Congress Hall Saloon.

No sooner had the brothers returned than three prisoners escaped from the county lockup after one of the jailers, William Soule, went uptown late on the afternoon of October 24. The jail itself, located near the northeast corner of First and Toughnut, was "about twenty feet square by ten feet high. The sides are composed of pieces of scantling, two by four inches, spiked down one over the other on the flat, thus making the wall four inches thick. There are high, narrow grated windows on each of three sides, and on the fourth is the only door. This opens into a little office, occupied usually by the jailers."[60]

In this instance, Milt Hicks, arrested only two days before by Deputy Breakenridge for possession of stolen cattle; Charlie "Yank" Thompson, serving 150 days for grand larceny; and a man named Sharp, charged with killing a Mexican at Charleston, simply overpowered jailer Charles Mason after he opened their cell to let the hired boy remove the slop buckets. In the struggle, Mason, aided by Jerry Barton—accused of wounding Jesus Gamboa at his saloon at Charleston nearly two weeks before—tried preventing the escape. Throwing the padlock, the frustrated jailer cut Charlie Thompson's cheek. But Sharp and a bleeding Thompson managed to lock the unfortunate officer in one of his own

[59]Deposition of Virgil W. Earp, Territory of Arizona vs. Morgan Earp, et al, Defendants, Document No. 94, In Justice Court, Township No. 1, Cochise County, A. T. (Hayhurst typescript), 147.

[60]*Tombstone Epitaph,* October 25, 1881.

cells before fleeing. Hicks had already made a run for it. Then, as Mason explained, "I halloed for some time, but as we occasionally have noisy customers, the neighbors paid no attention to me. Finally I called a Chinaman who lives near by, and when he came I sent for Soule, who returned in a few minutes and let me out."[61]

Sheriff Behan and his deputies, including Breakenridge and Frank Leslie, grabbed their guns and rushed to the scene. They were immediately joined by Virgil, Wyatt, and Morgan Earp. Spreading out, the men searched for the fugitives until well after dark. As it turned out the three escapees enjoyed too much of a head start, and the officers' efforts proved fruitless.

With all the troubles the Earps faced trying to get convictions for the Bisbee stage holdup, together with long hours in the saddle chasing elusive Apache warriors, trips to Tucson, and now searching darkened arroyos for departed prisoners from Sheriff Behan's jail, their tempers were short concerning anything to do with Cochise County's outlaw element. All this frustration helped set fire to a powder keg that was about to explode.

[61] *Tombstone Daily Nugget,* October 25, 1881.

Looking southeast from Comstock Hill during that fateful autumn of 1881. Rocked by a series of desperate encounters, Tombstone soon stumbled into a murky world of myth and legend. Even history failed to escape the capricious appetite of popular culture.

Courtesy, Arizona Historical Society.

Chapter 11

AN ARIZONA SHOWDOWN

Late in the morning of October 25, 1881, Ike Clanton and Tom McLaury rode into Tombstone on a commonplace utility wagon. Ike had been away from his family's ranch on the San Pedro for three days and was headed home. On the road from the Sulphur Springs Valley he happened upon the two McLaurys and his younger brother Billy. They talked and rode together several miles before agreeing to meet again in a day or two, as Frank and Tom had some business in town. After breakfast at John Chandler's milk ranch, about nine miles east of Tombstone off the Chiricahua road, Tom McLaury decided to go on ahead with Ike.

Putting up their wagon and team and checking their guns at the West End Corral—on the southwest corner of Second and Fremont—the two men whiled away the afternoon sampling whiskey. That evening Clanton became embroiled in an ugly dispute with Doc Holliday. A few weeks before at Vogan's saloon Ike had accused Wyatt of having told the unpredictable southerner about their secret deal to capture Leonard, Head, and Crane. Earp denied the charge, assuring Clanton that when Holliday returned from Tucson he would prove it. Ike dismissed Wyatt's promise, insisting Doc had already told him all about it.

Now, standing along the back wall of the crowded Alhambra, Ike again whispered his explosive accusation. Openly bewildered by the course of the conversation, the gambling dentist demanded answers. Hearing Clanton's version, Holliday replied by calling him a liar and "a son of a bitch of a cow boy." Doc's less-than-gentlemanly demeanor was not helped by learning of Ike's plan to betray his old friend Bill Leonard. As Clanton later testified, Holliday "told me to get my gun out and get to work. I told him I had no gun. He said I was a damned liar and had threatened the Earps. I told him I had not, to bring whoever said so to me and I would convince him that I had not. He told me again to pull out my gun and if there was any grit in me to go to fighting. All the time he was talking he had his hand in his bosom and I supposed on his pistol."[1]

Wyatt, not wishing to interrupt his dinner, called out to Morgan to use his authority as a special policeman and separate the two noisy combatants. Cutting short his conversation with the bartender, Morgan climbed over the lunch counter, took Doc's arm and led him outside. Clanton followed closely behind and continued the debate. Morgan joined the quarrel, calling Ike a son of a bitch, saying he could have all the fight he wanted. Holliday and Clanton were still at it when Wyatt stepped outside. Overhearing the commotion from next door, Virgil left the Occidental Saloon and threatened to arrest them all if they did not stop arguing.

That seemed to work and everyone separated. Holliday walked toward the Oriental, Morgan headed home, Virgil returned to the Occidental, and Ike crossed over to the Grand Hotel. At Wehrfritz and Tribolet's Wyatt checked on one of the many faro games he had scattered around town. After a few minutes he stepped outside and saw Ike. Clanton wanted to talk. The two men moved halfway down the Fifth Street side of the building. Ike kept insisting that when Holliday insulted him he was unarmed, but with all the loose talk and wild rumors he would be ready for him in the morning. Wyatt brushed him off saying he was not interested in fighting "because there was no money in it."[2] Clanton mumbled he would soon be ready for them all.

Wyatt crossed over to the Oriental, followed closely by Ike who took a drink and continued making threats over Holliday's behavior. Before long Earp's faro dealer brought in the night's receipts. As Lou Rickabaugh's gambling partner, Wyatt locked the money in the safe and withdrew, leaving Ike to nurse his clouded complaints with alcohol. Finding Holliday between the Alhambra and Occidental, the two walked together; Doc to visit Kate's room at C. S. Fly's boarding house, Earp to his residence on the northeast corner of First and Fremont.

---

[1]Deposition of Joseph I. Clanton, Territory of Arizona vs. Morgan Earp, et al, Defendants, Document No. 94, In Justice Court, Township No. 1, Cochise County, A. T. (Hayhurst typescript), 99. Hereafter cited as: "Hayhurst typescript, Document No. 94."

[2]Deposition of Wyatt S. Earp, ibid., 139.

Clanton should have done likewise. Instead he joined Virgil, Johnny Behan, and Tom McLaury in an all-night poker game in the back room of the Occidental. Ike lost just enough money that his mood stood little chance of improving. He took greater offense after the game when he discovered Virgil had been playing with a loaded revolver in his lap. Catching up with Earp near the Cosmopolitan, Ike berated him about the six-shooter, accusing the marshal of siding with those who wanted to murder him the night before.

Clanton growled that Holliday would have to fight, and the sooner the better. Clearly annoyed, Virgil warned him, "Ike, I am an officer and I don't want to hear you talking that way, at all. I am going down home now. . . . I don't want you to raise any disturbance while I am in bed."[3] As Virgil walked away in the early morning light, Ike demanded his message be relayed to Holliday and the others. Earp refused. Clanton again threatened there would be a fight before they knew it.

Around eight o'clock Ike Clanton retrieved his Colt revolver and Winchester carbine from the West End Corral, claiming, "I had those weapons about my person for self-defense."[4] But he now paraded through downtown Tombstone telling several people what he planned for Holliday and the Earp brothers. Ike freely admitted, "I . . . got my Winchester expecting to meet Doc Holliday on the street, but never saw him. . . ."[5]

Edward F. "Ned" Boyle, a laborer and part-time bartender at the Oriental, saw Ike in front of the telegraph office that Wednesday morning of October 26. He noticed Clanton armed and overheard him say that as soon as the Earps and Holliday made their appearance he planned to settle things. Boyle suggested he go to bed. Instead Clanton stomped over to Kelly's Wine House, then located west of the Grand Hotel at 418 Allen. Boyle "went down to Wyatt Earp's house and told him that Ike Clanton had threatened that when him and his brothers and Doc Holliday showed themselves on the street that the ball would open."[6]

---

[3]Deposition of Virgil W. Earp, ibid., 148.

[4]Deposition of Joseph I. Clanton, ibid., 109.

[5]Statement of Joseph I. Clanton, Coroner's Inquest, Wm. Clanton, F. & T. McLowery [*sic*], Document No. 48, Cochise County Recorder's Office; filed December 1, 1881. Hereafter cited as: "Coroner's Inquest, Document No. 48."

[6]Testimony of E. F. "Ned" Boyle, *Tombstone Daily Nugget*, November 24, 1881.

At Kelly's saloon Clanton ordered drinks for himself and Joseph Stumpf, a thirty-four-year-old Bavarian baker turned part-time silver miner. Kelly saw Ike's carbine and overheard his tale of woe. He cautioned Clanton "against having any trouble, as I believed the other side would also fight if it came to the point. . . ."[7] Just past noon Ike voiced similar threats to saloonman R. F. Hafford, saying "that Holliday and the Earps had insulted him the night before; that he was unarmed at the time. Said he was looking for Holliday or the Earps. . . ." Hafford advised the armed and exhausted Cowboy, "you had better go home. There will be none of it."[8]

But Ike had already been looking for Holliday at C. S. Fly's Fremont Street boarding house. He missed him only because Doc was still asleep. A worried Mollie Fly found Kate standing in the rear gallery looking at photographs. She hurriedly explained that Ike Clanton, rifle in hand, had been asking about her husband. Kate rushed to her small room and awoke him with the news. Holliday seemed unconcerned, saying, "If God lets me live long enough to get my clothes on, he shall see me."[9]

Not only had Wyatt heard about Ike's threats from Ned Boyle, but Virgil, too, began receiving ominous reports from downtown. As he later recalled: "I don't know how long I had been in bed. It must have been between 9 and 10 o'clock when one of the policeman [Officer A. J. Bronk] came and told me to get up, as there was liable to be hell." Bronk added, "Ike Clanton has threatened to kill Holliday as soon as he gets up. He's counting you fellows in too."[10] Virgil ignored the situation for the next couple of hours before finally walking downtown with Morgan soon after Ike's conversation with Colonel Hafford.

Not yet aware of any trouble, John Clum left his *Epitaph* office for a late lunch at the Grand Hotel. Seeing Ike with a Winchester near the post office he asked innocently, "Hello Ike! Any new war?" Clanton glanced up and mumbled, "Oh, nothing in particular."[11] Thinking no more about it, Clum walked down Fourth Street. Running into Charlie Shibell, the editor stopped to ask if he had any fresh items of interest for the paper and promptly forgot about Clanton.

---

[7]Deposition of Julius A. Kelly, Hayhurst typescript, Document No. 94, 81.

[8]Deposition of R. F. Hafford, ibid., 60. [9]Memoirs of Mary Katherine [Holliday] Cummings.

[10]Deposition of Virgil W. Earp, Hayhurst typescript, Document No. 94, 148, 151.

[11]John P. Clum, "It All Happened in Tombstone," 46.

At the Oriental Harry B. Jones had just spoken with Wyatt, reporting, "Ike Clanton is hunting you boys with a Winchester rifle and a six-shooter." Not overly concerned at the time, Earp replied, "I will go down and find him and see what he wants."[12] Running into Virgil and Morgan, who had already heard warnings from their brother James, Wyatt passed on the information from Jones. As Wyatt walked down Allen his two brothers went up Fifth and turned down Fremont. They found Clanton standing on Fourth Street talking with a forty-five-year-old mining man named William Stilwell (no relation to Frank).

Ike claimed he was going from the Capitol Saloon to the Pima County Bank—"on the left hand sidewalk going from Fremont to Allen"[13]—when the two Earp brothers slipped up behind him undetected. As Virgil described the resulting struggle: "I found Ike Clanton . . . with a Winchester rifle in his hand and a six-shooting stuck down in his breeches. I walked up and grabbed the rifle in my left hand. He let loose and started to draw his six-shooter. I hit him over the head with mine and knocked him to his knees and took his six-shooter from him. . . . I asked him if he was hunting for me. He said he was, and if he had seen me a second sooner he would have killed me."[14]

Clum and Charlie Shibell saw it all. Shibell asked, "What does this mean?" Worried, John Clum replied, "Looks like real trouble."[15]

Virgil and Morgan hauled Clanton down the street. Crossing Allen, they guided their stumbling prisoner past P. W. Smith's temporary bank site, Mrs. Frary's millinery, and Hawkins, Boarman & Company's wholesale liquors to Judge A. O. Wallace's new police court at 114 Fourth Street. Not finding the judge, who was uptown performing a wedding, Virgil left Ike in Morgan's custody and went looking for Wallace as Wyatt walked in. In the small, crowded courtroom a heated exchange exploded between Ike Clanton and the two Earp brothers.

Rezin J. Campbell, a deputy sheriff and clerk of the board of supervisors, had followed Wyatt into the room. "He took a seat on a bench inside of the railing," Campbell explained. "Ike Clanton was sitting on the outside of the railing." Morgan Earp stood behind Ike. Wyatt stared at Clanton several minutes before saying, "You have threatened my life

[12]Deposition of Wyatt S. Earp, Hayhurst typescript, Document No. 94, 140.

[13]Deposition of Joseph I. Clanton, ibid., 109.

[14]Deposition of Virgil W. Earp, ibid., 148.

[15]Clum, "It All Happened in Tombstone," 47.

two or three times and I have got the best evidence to prove it and I want this thing stopped." Moments later Wyatt leaned against the railing, challenging Ike: "You cattle thieving son of a bitch, and you know that I know you are a cattle thieving son of a bitch. You've threatened my life enough and you've got to fight."[16]

Feeling brave, Clanton responded, "Fight is my racket, and all I want is four feet of ground." Turning to Morgan he added contemptuously, "If you fellows had been a second later, I would have furnished a coroner's inquest for the town."[17] Losing his temper, Morgan offered to return Ike's revolver. Campbell intervened, pushing Clanton into a chair and separating the two men. Virgil walked in with Wallace and the situation cooled.

Wyatt stomped out. On the sidewalk he bumped into Tom McLaury, who had heard about Ike's troubles and was coming to see what had happened. One witness said McLaury stood with both hands in his pockets. Wyatt asked if he was armed. Tom replied that he carried no weapons. McLaury then foolishly tossed out an obscene insult. Enraged, Earp slapped his face. Tom raised both hands to ward off other blows, but Wyatt pistol-whipped him into the street. As Earp turned to leave, butcher and real estate developer Appolinar Bauer heard him mumble, "I would kill the son of a bitch."[18] Thomas Keefe, a carpenter working on the P. W. Smith remodeling, saw it all. As Wyatt moved away Keefe kept his eyes on Tom, who "got up and staggered and walked toward the sidewalk and picked up a silver band or roll, to put on his hat again, that was knocked off."[19] An old man had come over and helped McLaury to his feet.

Fined $25 by Judge Wallace for carrying weapons, Clanton walked out only to find a fresh victim of Earp violence. With his friend William Claiborne, a buggy driver and self-styled badman then working part-time at the Neptune Mining Company's smelter near Hereford, Ike guided Tom McLaury toward Dr. C. F. Gillingham's Fourth Street office. That English-born physician, formerly of Virginia City, hurriedly cleaned and bandaged both men's head wounds. Meanwhile, Virgil Earp took Clanton's guns to the Grand Hotel's bar, handing them to J. H. All-

---

[16]Deposition of R. J. Campbell, Hayhurst typescript, Document No. 94, 82.

[17]Ibid., 82, 83.

[18]Deposition of A. Bauer, ibid., 68.

[19]Deposition of Thomas Keefe, ibid., 63.

man, one of the proprietors. Clanton's revolver went behind the bar and his carbine was propped up behind the front door. Allman later turned both weapons over to policeman George Magee.

Between 1:30 and 2:00 that afternoon, unaware of the day's troubles, Frank McLaury, Billy Clanton, and J. R. Frink, an elderly Sulphur Springs Valley cattle dealer known as Major Frink, rode into Tombstone from Antelope Springs, a dozen miles east toward the Dragoons. Frank and Billy had gone there after having breakfast with Ike and Tom at Chandler's ranch the day before. The three riders reined in at the Grand Hotel. According to witness William Allen, Doc Holliday approached Billy and asked, "How are you?" as he shook young Clanton's hand.[20] The two had never met.

Holliday walked away as the others entered the hotel's bar followed by the forty-year-old Allen, who just weeks before helped post bond for Frank Stilwell and Peter Spencer. Taking McLaury to one side, he explained what had happened. Frank asked, "What did he hit Tom for?" Allen replied he did not know. Lowering his glass McLaury added, "We won't drink." Back on the street he promised, "I will get the boys out of town."[21]

Frank and Billy led their horses down toward the Dexter Livery Stable, then owned by John O. Dunbar and Sheriff Behan, nearly opposite the O.K. Corral. Almost immediately they ran into Claiborne. Billy asked about Ike and for the moment behaved with a clear head. As Claiborne recalled, Billy simply wanted to find Ike and get back to their ranch. Young Clanton said he did not come into Tombstone to fight anyone, "and no one don't want to fight me." When Ike finally showed up he promised his brother "that he would go directly."[22] Seeing the stableman from the West End Corral, Ike asked him to hitch up his team as he would be leaving town.

Still walking their horses, Frank McLaury and Billy Clanton followed Ike up Fourth Street to George Spangenberg's gun shop, located behind Brown's Hotel. Smoking a cigar at the door of Hafford's saloon, Wyatt Earp decided to investigate. Frank McLaury's horse suddenly stepped

[20]Deposition of William Allen, ibid., 23; and Statement of Joseph I. Clanton, Coroner's Inquest, Document No. 48.

[21]Deposition of William Allen, Hayhurst typescript, Document No. 94, 23.

[22]Statement of Joseph I. Clanton, Coroner's Inquest, Document No. 48.

onto the sidewalk and thrust its head into the building. "I took the horse by the bit, as I was deputy city marshal," Wyatt explained,

> and commenced to back him off the sidewalk. Tom and Frank McLoury [*sic*] and Billy Clanton came to the door. Billy Clanton laid his hand on his six shooter. Frank McLoury took hold of the horse's bridle, and I said, "You will have to get this horse off the sidewalk." He backed him off into the street. Ike Clanton came up about this time and they all walked into the gun-shop [actually Ike had been there all along]. I saw them in the gun-shop changing cartridges into their belts.[23]

A number of people, some hearing of the earlier trouble, had gathered to witness this exchange. One of them, saloonman Bob Hatch, raced off to find Virgil, who was at Wells Fargo picking up a ten-gauge double-barreled shotgun. Virgil claimed the weapon had "been in my service for six months. No one handed it to me at the time. I got it myself."[24] Hatch bolted through the door, blurting out nervously, "For God's sake, hurry down there to the gun shop, for they are all down there and Wyatt is all alone. . . . They are liable to kill him before you get there."[25] Hurrying along, Virgil rounded the corner just in time to see Frank backing his horse off the sidewalk.

Making his way to the post office around two o'clock that afternoon, a twenty-eight-year-old laundryman named Patrick Henry Fallehy noticed what he described as "a large crowd of people" gathered in front of Spangenberg's. Curious about the situation, he glanced in and saw Ike Clanton "with his hand under his jaw kneeling on the counter."[26] Trying to explain his presence at a gun shop, Ike later testified, "I often frequent the gunshop every day that I am in town, almost. I went in there and asked for a pistol. The gentleman that runs the shop remarked that my head was bleeding, that I had been in trouble, and he would not let me have it."[27]

Leaving Spangenberg's the Cowboys walked past Virgil Earp. As Fallehy recalled:

> I see the marshall [*sic*] in the door way of a vacant store with a double barrel shot gun. He had the shot gun in his left hand and passing up the street

[23]Deposition of Wyatt S. Earp, Hayhurst typescript, Document No. 94, 141.

[24]Deposition of Virgil W. Earp, ibid., 158. [25]Ibid., 152.

[26]Statement of P. H. Fallehy, Coroner's Inquest, Document No. 48.

[27]Deposition of Joseph I. Clanton, Hayhurst typescript, Document No. 94, 113.

> I saw Ike Clanton passing by where the marshall was standing. Ike Clanton took no notice of the marshall. When he passed some stranger asked Ike Clanton what is the trouble he says I don't think that there will be any trouble and he kept on walking. . . .[28]

Ike went straight to John Doling's saloon next to the Dexter Livery Stable for another drink.

In all probability Tom McLaury had not been at Spangenberg's with the others. Ike Clanton remembered, "I am very sure that Tom McLoury [*sic*] was not there—Frank McLoury came in the shop and asked where Tom was."[29]

Instead Albert Bilicke watched as Tom entered Everhardy's Eagle Market opposite the Cosmopolitan Hotel. As McLaury came out, Bilicke remembered, "his pants pocket protruded, as if there was a revolver. . . ."[30] Army surgeon J. B. W. Gardiner stood beside Bilicke: "I observed . . . that I was sorry to see Tom McLowry [*sic*] had gotten his pistol; saw no pistol, but supposed at the time on seeing the right hand pocket of his pants extended outwards, that he had got his pistol."[31] Yet Ernst Storm, the forty-five-year-old German butcher who managed the place, stated that although Tom "was bleeding on the side of the head when he came in. . . . He had no arms on his person and did not get any in there that I saw."[32] When asked if anyone even mentioned firearms while McLaury was in the shop, Storm shook his head no.

Yet Tom had been carrying a pistol with him at one point. Saloonman Andrew J. Mehan claimed McLaury turned over a loaded revolver early that afternoon. Mehan locked it in his safe. Ike Clanton testified this was the gun his friend brought with him to Tombstone: "I know it by the guard being sprung and by its general appearance."[33]

Having risen late after the all-night poker game, Sheriff Behan first heard of the trouble while having a shave at William Baron's barber shop and bath house near the corner of Fourth, across Allen Street from Hafford's and the Cosmopolitan. He asked Baron to hurry: "I wanted to get

[28]Statement of P. H. Fallehy, Coroner's Inquest, Document No. 48.

[29]Deposition of Joseph I. Clanton, Hayhurst typescript, Document No. 94, 111.

[30]Deposition of Albert Bilicke, ibid., 89.

[31]Testimony of Dr. J. B. W. Gardiner, *Tombstone Daily Nugget*, November 29, 1881; and Hayhurst typescript, Document No. 94, 90.

[32]Deposition of Ernst Storm, Hayhurst typescript, Document No. 94, 92.

[33]Deposition of Joseph I. Clanton, ibid, 97.

out and disarm the parties. I meant all of them—everybody who had arms except the officers."[34] Seeing Virgil Earp and Doc Holliday standing in the intersection, Behan crossed over with Charlie Shibell. Both Shibell and Bob Paul had arrived in Tombstone two days before.

Behan claimed Virgil Earp told him there were "a lot of sons of bitches in town looking for a fight," and that he explained to the marshal, "you had better disarm the crowd." Behan remembered Virgil saying "he would not that he would give them a chance to make a fight."[35] Others heard the same threatening tone. Behan invited Earp to join him for a drink. The two men disappeared into Hafford's. Later Behan could not recall if Virgil actually took a drink or not.

Inside the saloon, stock and mining broker William B. Murray motioned Earp to the far end of the bar. Murray offered himself and twenty-five other armed men to help arrest the Cowboys. Virgil declined, but Murray reiterated, "You can count on me, if there is any danger."[36] After another brief conversation, Behan left Hafford's, telling Virgil, "They won't hurt me. I will go down alone and see if I can disarm them." Earp replied, "I told him that was all I wanted them to do—to lay off their arms while they were in town."[37]

Virgil crossed Fourth Street and was approached by John L. Fonck, a furniture maker and proprietor of the Mining Exchange Saloon, who made an offer similar to Murray's. Again Earp refused, "I told him I would not bother them as long as they were in the corral—if they showed up on the street, I would disarm them."[38] Friendly with the Earps, having been a neighbor of James and his family, the older man would have made a good ally. During the Civil War Fonck had worked as a government agent and later served four years as a captain with the Los Angeles police department. Perhaps a man like Virgil Earp felt he needed no help. He dismissed Ike Clanton as a loud-mouthed drunk. Billy was a youngster, and physically both McLaurys were small men. In a common street brawl he could very well have handled all four.

Reuben F. Coleman, the forty-seven-year-old Englishman who had served on the coroner's jury with Ned Boyle and E. L. Bradshaw during

[34]Statement of J. H. Behan, Coroner's Inquest, Document No. 48.

[35]Deposition of J. H. Behan, Hayhurst typescript, Document No. 94, 29.

[36]Deposition of Virgil W. Earp, ibid., 152.

[37]Ibid., 149.

[38]Ibid., 152.

the Mike Killeen inquest and later played the organ at Marshal White's funeral, witnessed most of the events of this tragic day. He comes across as an incorrigible busybody just waiting around for something bad to happen. Earlier Coleman watched Virgil arrest Ike Clanton, then followed them to Wallace's justice court and witnessed the trouble there. Now, standing near the entrance of the O.K. Corral, he spotted the Cowboys huddled together in one of the stalls of the Dexter Livery Stable. Soon they crossed over, with Billy Clanton in the saddle and Frank McLaury in the lead, walking his horse. Ike and Tom brought up the rear. Before disappearing into the O.K. Corral, Billy remarked that it was very cold and asked directions to the West End Corral. Coleman responded by giving the young man the Second and Fremont Street location.

Coleman now convinced himself there could be serious trouble. Walking east, he saw Sheriff Behan leaving Hafford's and suggested he disarm those men. Behan asked where they had gone and Coleman obliged. Seeing Virgil Earp in the saloon's doorway, Coleman rushed over and repeated his concerns. He said the marshal told him "he did not intend to disarm them."[39] Suddenly excited by the prospect of violence, Coleman started through the O.K. Corral with William Allen to be on hand for whatever happened next, telling his companion, "come on, let's go see it."[40] Allen later claimed he had no interest, but conceded he went along with Coleman anyway.

H. F. Sills, a furloughed Santa Fe Railroad engineer from Las Vegas, New Mexico, watched the Cowboys in front of the O.K. Corral. They talked "of some trouble they had had with Virgil Earp," Sills testified, "and they made threats at that time, that on meeting him they would kill him on sight." New to Tombstone, having just arrived the day before on a Wells Fargo bullion wagon, Sills had no idea who they were talking about. "I then walked up the street and made inquiries to know who Virgil Earp and the Earps were," Sills recounted. "A man on the street pointed out Virgil Earp to me and told me that he was the city marshal. I went over and called him to one side and told him the threats I had overheard this party make."[41]

[39]Statement of R. F. Coleman, Coroner's Inquest, Document No. 48.

[40]Deposition of William Allen, Hayhurst typescript, Document No. 94, 25.

[41]Deposition of H. F. Sills, ibid., 76.

Later Ike Clanton tried to justify all the strong language with the weak explanation, "I made no worse threats against them than they did against me."[42] Following the Earps, Sills would witness the fight from the entrance of a narrow hallway along the west side of the Mining Exchange.

The always dapper Doc Holliday, wearing a long gray overcoat and carrying a silver-knobbed cane, stood with the three Earp brothers at Hafford's corner. Quietly Virgil pondered his duty as an officer against the seriousness of the challenge. Was any of this for real? Or was it simply men listening to their own voices, impressing themselves by threatening others with violence? With frightening resolve, Virgil suddenly decided the time had come to answer those questions. "Holliday had a large overcoat on," Earp recalled, "and I told him to let me have his cane and he take the shot-gun—that I did not want to create any excitement going down the street with a shot-gun in my hand. When we made the exchange, I said, 'Come along.'"[43] Morgan asked, "They have horses. Had we not better get some horses ourselves, so that if they make a running fight we can catch them?" "No," Wyatt calmly replied. "If they try to make a running fight we can kill their horses and then capture them."[44]

It was nearing two-thirty in the afternoon. With Virgil and Doc Holliday in front, and Wyatt and Morgan Earp bringing up the rear, they started from Hafford's saloon, north along Charlie Brown's property to James Howard's small law office. Walking under the large rifle-shaped sign that straddled the sidewalk at Spangenberg's, they crossed a narrow space between the buildings before passing Sultan and Arnold's saloon, the office of Drs. McSwegan and Dunn, A. M. Robertson's small book shop, the cramped quarters of Drs. Gillingham and F. V. B. Gildersleeve, to Sol Israel's Union News Depot. Before reaching the post office the four men crossed over toward Thomas Moses's Capitol Saloon on the southwest corner of Fourth and Fremont.

Meanwhile, Sheriff Behan had stumbled upon the two McLaurys standing in front of the Union Meat & Poultry Market at 318 Fremont

[42]Statement of Joseph I. Clanton, Coroner's Inquest, Document No. 48.

[43]Deposition of Virgil W. Earp, Hayhurst typescript, Document No. 94, 158-159.

[44]Deposition of Wyatt S. Earp, ibid., 142.

Street, west of the O.K. Corral's rear entrance. As the four Cowboys and William Claiborne, who had followed his friends from the Dexter Livery Stable, came through the corral the two McLaurys stopped at the Union Market to talk with James Kehoe. The brothers and Major Frink had recently sold the firm large numbers of cattle. Frank was in debt to the owners and hoped to straighten out the accounts. They still owed Tom money. But this short meeting with Kehoe stalled their departure. It proved to be a tragic delay.

With their friends talking business, Claiborne and the two Clantons had walked slowly on ahead toward the West End Corral, stopping only beyond earshot. On this breezy autumn afternoon the three men found temporary shelter between Camillus Fly's boarding house and photo gallery and William Harwood's dwelling near the corner of Third. To the rear of this eighteen-foot-wide space stood a small vacant building slated for demolition. Recently two dressmakers had occupied the site, before moving next to the Union Market. Once off the street the Clantons and Billy Claiborne huddled together against the cold.

Back at the meat market Behan exchanged a few words with the two McLaurys before Tom walked off to join his friends. James Kehoe then heard Behan say, "Frank, I want you to give up your arms." McLaury replied, "Johnny, as long as the people of Tombstone act so, I will not give up my arms." In no mood to argue, Behan told Frank he "would take him to the sheriff's office to lay off his arms." McLaury suddenly reconsidered, "You need not take me I will go." "Well, come on," Behan said as they started walking away, with Frank leading his horse, to relay the same message to the Clantons and Tom McLaury.[45]

As Johnny Behan reached the vacant space beyond Fly's he asked how many there were. The Cowboys replied "four." William Claiborne insisted he "was not one of the party."[46] Nor was he armed, having left his weapons at Kellogg's saloon on his arrival in town the day before. Behan had good reason to establish Claiborne's presence. The young man was awaiting the grand jury's decision over his shooting James Hickey to death during a saloon quarrel at Charleston three weeks before.

[45]Deposition of James Kehoe, ibid., 52.

[46]Deposition of J. H. Behan, ibid., 30.

The sheriff told the others they would have to cross over to his office and surrender their weapons. Both Ike Clanton and Tom McLaury insisted they were unarmed. Although Tom pulled open his coat by the lapels, Behan admitted he "might have had a pistol and I not know it."[47] It was not easy to tell. James Kehoe remembered that Tom "had on a dark blue blouse of light (weight) material and dark pants and vest. The blouse came down about the length of his arms. It was outside of his pants."[48] Aware of the day's threats, Behan ran his hand around Ike's waist, convincing himself Clanton was unarmed. Later Behan explained: "No one refused to give up their arms, except Frank McLowery [*sic*]. He said that he came on business and did not want any row. He never refused to go to my office."[49] Frank did say, however, that "he would go out of town, but did not want to give his arms up till after the party that hit his brother was disarmed."[50]

One of James Kehoe's customers, Mrs. Martha J. King, saw a crowd of men on the sidewalk as she approached the market. "When I first went in the shop," she said, "the parties who keep the shop seemed to be excited and did not want to wait on me. I inquired what was the matter and they said there was about to be a fight between the Earp boys and the cow boys. . . ."[51] Later, as the butcher began filling her order, someone yelled, "Here they come," and Mrs. King rushed to the door with everybody else. Peering out, she saw four men coming down the sidewalk. She knew Doc Holliday, but only recognized the Earp brothers as a group. Nor could she identify Sheriff Behan by sight.

The Earps and Holliday had just rounded the corner from Fourth Street, passing the Capitol Saloon, Dr. Henry Hatch's small office, a narrow vacant store front, the *Tombstone Nugget*, and the county recorder's office. At the Papago Cash Store Virgil acknowledged co-proprietor Frank B. Austin with a nod before crossing in front of the gate, made of wooden slats, marking the O.K. Corral's rear entrance. As they stepped under the awning at the Union Market, Mrs. King overheard Morgan Earp say to Holliday, "Let them have it." Doc replied, "All right." She saw "Mr. Holliday," as she called him, trying to conceal "a gun, not a pistol,"

[47]Ibid., 40. [48]Deposition of James Kehoe, ibid., 52.

[49]Statement of J. H. Behan, Coroner's Inquest, Document No. 48.

[50]Deposition of Joseph I. Clanton, Hayhurst typescript, Document No. 94, 94.

[51]Deposition of Mrs. Martha J. King, ibid., 51.

under his overcoat "on the left side, with his arm thrown over it. . . ."[52] She explained, "The way I noticed the gun was that his coat would blow open and he tried to keep it covered."[53] Fearing trouble, Mrs. King retreated toward the back of the store before any shooting started and saw nothing more.

Seeing the Earps and Holliday coming down the sidewalk, Sheriff Behan turned from the Cowboys and hurried to intercept the marshal's party. Approaching Virgil and his three companions as they passed the Union Market, Behan considered the Cowboys under arrest; although even he admitted, "I doubt whether they considered themselves under arrest or not, after I turned to meet the other parties."[54] Ike Clanton later said he did not think so as he was leaving town.

Having come through the O.K. Corral with William Allen, R. F. Coleman, the nosy Englishman, stood near Fly's front door as Johnny Behan left the Cowboys and walked toward the Earps. Only moments before William A. Cuddy, a clerk at the Cosmopolitan as well as a part-time actor and theatrical manager (who later became an evangelical preacher), heard Behan tell the Cowboys, whom Cuddy described as farmers, "I won't have no fighting, you must give me your fire arms or leave town immediately."[55] Coleman insisted one of the Clanton party replied, "You need not be afraid, Johnny, we are not going to have any trouble."[56]

William Cuddy heard essentially the same thing. He had gone to Fly's after talking with dry goods and clothing merchant Edward Dillon at the post office. Dillon told him of "some trouble between the Earps and other men which he termed Cow Boys." As he approached them, Billy Clanton "put his hand on his pistol, as if in fear of somebody. Then he recognized me [and] he removed his hand."[57] Having seen enough, and saying good day to Behan, Cuddy walked off toward Allen Street.

Coleman made it back to the Union Market just as the Earps and Holliday passed by. He thought he heard Behan say, "hold boys. I don't want you to go any farther."[58] Other witnesses seemed to think Behan

[52]Ibid., 50.

[53]Statement of Mrs. Martha J. King, Coroner's Inquest, Document No. 48.

[54]Statement of J. H. Behan, ibid.

[55]Statement of W. A. Cuddy, ibid.

[56]Statement of R. J. Coleman, ibid.

[57]Statement of W. A. Cuddy, ibid.

[58]Statement of R. F. Coleman, ibid.

Fremont Street looking west toward Third, years after its prime. Beyond City Hall—the earlier site of the county recorder's office—can be seen the Papago Cash Store and the old Union Meat & Poultry Market. The last building, with its false-front and wooden awning, is C.S. Fly's boarding house, beyond which the famous gunfight took place in late October 1881.
Courtesy, John D. Gilchriese Collection.

told the Earps that he was going to disarm them—meaning the Clantons and McLaurys. Virgil and Wyatt Earp testified that Behan exclaimed, "For God's sake, don't go down there or they will murder you." Virgil replied, "Johnny, I am going down to disarm them."[59] The Earps thought the sheriff answered, "I have disarmed them," or "disarmed them all."[60]

They must have believed that remark more than some warning about getting murdered—words no other witness testified hearing—since both Virgil and Wyatt relaxed their vigilance. Virgil recalled, "When he said that, I had a walking stick in my left hand and my right hand was on my six-shooter, in my waist pants, and when he said he had disarmed them I shoved it clean around to my left hip and changed my walking stick to my right hand."[61] Wyatt remembered: "When I and Morgan Earp came up to Behan he said, 'I have disarmed them.' When he said

[59]Deposition of Virgil W. Earp, Hayhurst typescript, Document No. 94, 150.

[60]Depositions of Wyatt S. Earp and Virgil W. Earp, ibid., 142, 150.

[61]Deposition of Virgil W. Earp, ibid., 150.

this, I took out my pistol, which I had in my hand, under my coat, and put it in my overcoat pocket."[62]

Whatever Behan actually said the brothers gave no further reply as they brushed past the startled sheriff. By now the four Cowboys had retreated out of sight. Hoping to see their adversaries from a better angle, the Earps and Holliday moved into the street. Passing Fly's they spotted Billy Clanton's horse, then each of the four men in turn. The three grim-faced Earps, determined to clear the air, stepped up into Harwood's lot. Only Doc Holliday held back, standing at the edge of the wooden sidewalk holding the Wells Fargo shotgun.

Approaching the Cowboys, Virgil called out, "Boys, throw up your hands. I want your guns."[63] Instead it appeared to the Earps as if Frank McLaury and Billy Clanton were reaching for their weapons. Wyatt shouted, "You sons-of-bitches, you have been looking for a fight, and now you can have it."[64] Hearing those words, Billy yelled, "Don't shoot me. I don't want to fight."[65] Tom McLaury, pulling back his coat, spoke nervously, "I have nothing," or "I am not armed."[66] With guns coming into play, Virgil Earp, still holding up Doc Holliday's cane in his right hand as a point of emphasis, demanded, "Hold on, I don't want that."[67]

Exactly what happened next is clouded by contradictory eyewitness testimony, theoretical posturing based on little more than fertilized imagination, and old-fashioned mythmaking. There has never been anyone, then or now, who could declare with absolute certainty in what sequence the various shots were fired or by whom. Despite the modern belief that Tombstone and surrounding Cochise County echoed with gossip about the Earps, many citizens could not identify one brother from another. When they walked the streets few took particular notice. All this makes it even more difficult to try and sort out the moves during their gunfight with the Clantons and McLaurys.

"When I saw Billy Clanton and Frank McLoury [*sic*] draw their pistols," Wyatt Earp described, "I drew my pistol. Billy Clanton levelled his pistol on me, but I did not take aim at him. I knew that Frank McLoury had the reputation of being a good shot and a dangerous man, and I

[62]Deposition of Wyatt S. Earp, ibid., 142.
[63]Deposition of Virgil W. Earp, ibid., 150.
[64]Deposition of J. H. Behan, ibid., 31.
[65]Ibid.
[66]Ibid.
[67]Deposition of Virgil W. Earp, ibid., 150.

aimed at Frank McLoury. The two shots were fired by Billy Clanton and myself, he shooting at me and I shooting at Frank McLoury. I don't know which was fired first. We fired almost together. The fight then became general."[68]

At that opening volley William Allen, who claimed Billy Clanton held out his hands after Virgil's command, ran from his position in front of Fly's and ducked for cover along the east side of the boarding house: "I got in between the buildings after the first shots were fired and did not see any more of the shooting."[69] At the end of the fight, however, he did see Frank McLaury fall on the north side of Fremont.

So did saloonman Bob Hatch. By chance he and jailer William Soule had been walking unawares down Fremont Street some distance behind the Earps and Holliday. After the first two or three shots Hatch turned to the deputy and warned, "This is none of our fight. We had better get away from here." Retreating, Hatch ducked into the Union Market, going as far back as the butcher blocks before coming forward soon enough to see the battle's end from "the door on the west side of the building."[70]

Having seen the Earp brothers standing at Hafford's armed to the teeth, gambler Wesley Fuller—a friend of both Holliday and the Cowboys—approached Harwood's lot from the south hoping to warn Billy Clanton to leave town. Fuller carried a grudge against the city marshal's office. More than a month before, police had arrested him for smashing a bottle over a man's head and knocking a woman down. He was taken into custody only after a struggle, thus adding the charge of resisting arrest. His combined indiscretions cost him $37.

Now Fuller got sidetracked passing Mattie Webb's Starlight Saloon (formerly the Red Light at 307 Allen). He stopped to chat briefly with the thirty-four-year-old proprietress at the rear of her building. This delayed his arrival to the space behind Fly's photo gallery until after the fight broke out. "I did move after the shooting commenced," Fuller

---

[68]Deposition of Wyatt S. Earp, ibid., 143.

[69]Deposition of William Allen, ibid., 24.

[70]Deposition of Robert Hatch, ibid., 72.

[71]Deposition of Wesley Fuller, ibid., 58; and *Tombstone Daily Nugget,* November 8, 1881.

recalled, "and was dodging about some, as bullets were flying around there."[71]

With those opening shots echoing between the wooden buildings, Sheriff Behan passed through the hallway of Fly's boarding house to the narrow stoop separating it from the gallery. There he found Billy Claiborne, who had just backed away from his friends. Behan pushed the frightened young man into Fly's building, telling him to stay there until he came back. This may have saved Claiborne's life: he noticed afterward a bullet hole through his trousers near the knee. Even Behan admitted that during the battle, "I did not stand still. I moved around pretty lively." As the fight ended C. S. Fly tried pushing Claiborne out, but Behan stopped him, saying, "Let him stay there. He was not to blame and might get killed."[72]

As the air reverberated with the heavy sound of gunfire and Harwood's lot filled with dust stirred up by the Cowboys' two terrified horses and the smoke from discharging black powder cartridges, Fremont Street pedestrians jumped for cover. Others, such as Charles H. Light, standing motionless at his window at the Aztec Boarding House, saw everything. Visiting California attorney William Morris Stewart, in town representing the Contention in a particularly complicated lawsuit, recalled with some relief years later, "I witnessed this fight—from a safe distance."[73] Luckily no bystanders were hit by stray shots, of which there were many. One bullet neatly pierced a window at the *Epitaph*, while another struck a wagon parked in front of the Union Market, fired by the Earp party at Ike Clanton one witness testified.

Of course most residents saw nothing of this fight or even knew of the trouble brewing between the Earps and the Cowboys. Yet common curiosity caused them to eagerly grab copies of both newspapers the next day to get some idea of what had happened and why. Considering the short time to prepare detailed accounts, both editors did quite well in outlining the history of the trouble. Describing the actual battle, however, the *Epitaph* unfortunately printed only a garbled version supplied by R. F. Coleman. The *Nugget*'s account, though not error free,

[72]Deposition of J. H. Behan, Hayhurst typescript, Document No. 94, 37.

[73]Brown, ed., *Reminiscences of Senator William M. Stewart*, 266.

remains far closer to the truth than its rival's. Concerning the actual event, this is what anxious citizens read in the *Daily Nugget* on October 27, 1881:

> the Sheriff stepped out and said:
>
> "HOLD UP BOYS.
>
> Don't go down there or there will be trouble; I have been down there to disarm them." But they passed on, and when within a few feet of them the Marshal said to the Clantons and McLowrys [*sic*] "Throw up your hands, boys, I intend to disarm you." As he spoke Frank McLowry made a motion to draw his revolver, when Wyatt Earp pulled his and shot him, the ball striking on the right side of his abdomen. About the same time Doc Holliday shot Tom McLowry in the right side, using a short shotgun, such as is carried by Wells, Fargo & Co.'s messengers. In the meantime Billy Clanton had shot Morgan Earp, the ball passing through the point of the left shoulder blade across his back, just grazing the backbone and coming out at the shoulder, the ball remaining inside of his shirt. He fell to the ground, but in an instant gathered himself, and raising in a sitting position fired at Frank McLowry as he crossed Fremont street, and at the same instant Doc Holliday shot at him, both balls taking effect, either of which would have proved fatal, as one struck him in the right temple and the other in the left breast. As he started across the street, however, he pulled his gun down on Holliday saying, "I've got you now." "Blaze away! You're a daisy if you have," replied Doc. This shot of McLowry's passed through Holliday's pistol pocket, just grazing the skin. While this was going on
>
> BILLY CLANTON HAD SHOT
>
> Virgil Earp in the right leg, the ball passing through the calf, inflicting a severe flesh wound. In turn he had been shot by Morg Earp in the right side of the abdomen, and twice by Virgil Earp, once in the right wrist and once in the left breast. Soon after the shooting commenced Ike Clanton ran through the O.K. Corral, across Allen street into Kellogg's saloon, and thence into Toughnut street, where he was arrested and taken to the county jail. The firing altogether didn't occupy more than twenty-five seconds, during which time fully thirty shots were fired. After the fight was over Billy Clanton, who, with wonderful vitality, survived his wounds for fully an hour, was carried by the editor and foreman of the *Nugget* into a house near where he lay, and everything possible done to make his last moments easy. He was "game" to the last, never uttering a word of complaint, and just before breathing his last he said, "Goodbye, boys, go away and let me die."

As the shooting stopped, Billy Clanton tried to extract empty cartridges from his revolver. Photographer Camillus S. Fly, carrying a Henry rifle—an unloaded prop from his gallery as it turned out—pointed toward Billy and called out to no one in particular, "Take that pistol away from that man." Saloonman Bob Hatch replied, "Go get it yourself if you want it."[74] The two then nodded sheepishly, agreeing to go together. Without saying a word Fly reached down and pulled the pistol from young Clanton's grasp as the dying man pleaded for more cartridges. The next day Fly surrendered the gun to County Coroner Henry M. Matthews, who had already retrieved Billy's cartridge belt and nickel-plated watch and chain.

The red-haired laundryman Patrick Fallehy, leaving his vantage point near Summerfield's store on the first-floor corner of the Gird Block, ran to Frank McLaury offering to help him up. "He never spoke," Fallehy remembered, "his lips only moved. I picked up a revolver that was lying 5 ft. from him and laid it at his side." Holliday approached McLaury, saying loud enough for Fallehy to hear, "The son of a bitch has shot me, and I mean to kill him."[75] Doc was too late: McLaury was already dead.

Coroner Matthews took possession of Frank's revolver as well as a cartridge belt. William Soule retrieved the Cowboys' two horses, now calmly milling about in front of the Union Market. Both animals carried Winchester rifles in their scabbards; Billy's was fully loaded and Frank's five or six cartridges short of a full load. Soule walked the horses to the Dexter Livery Stable. When asked later what became of the rifles the deputy confessed giving them to parties guarding the jail: "I don't know who I gave them to."[76]

As the fight ended Sheriff Behan cautiously emerged from behind Fly's boarding house and told the Earps and Holliday they were under arrest. For a moment Wyatt stared back coldly, then responded, "No one could arrest me now . . . you threw us Johnny you told us they were disarmed."[77] Behan replied defensively, "I did nothing of the kind. . . ."[78] Someone spoke up saying there was no need to argue the point now and the sheriff let it pass.

---

[74]Deposition of Robert Hatch, Hayhurst typescript, Document No. 94, 73.

[75]Statement of P. H. Fallehy, Coroner's Inquest, Document No. 48.

[76]Deposition of William H. Soule, Hayhurst typescript, Document No. 94, 22.

[77]Statement of W. A. Cuddy, Coroner's Inquest, Document No. 48.

[78]Statement of J. H. Behan, ibid.

Watching Ike Clanton run into a Mexican dance hall on Allen Street as the shooting stopped, actor William Cuddy, perhaps allowing his theatrical background to get the better of him, began yelling that the sheriff had been killed. Regaining his composure, Cuddy rushed to view the bodies. He overheard Behan tell Wyatt that he must submit to arrest and Earp's refusal. Cuddy then claimed he heard businessman and local politician Sylvester B. Comstock say, "there is no hurry in arresting this man he done just right in killing them and the people will uphold them." To those words of support Wyatt added, "You bet we did right, we had to do it. . . ."[79]

With all the gunshots sounding an alarm, the prearranged signal of the month-old Citizens' Safety Committee opened with short blasts from the Vizina Mine whistle. The Tough Nut took up the call as dozens of miners came to the surface. They hurried into town and armed themselves for any emergency.

As the *Nugget* reported the next day:

> almost simultaneously a large number of citizens appeared on the streets, armed with rifles and a belt of cartridges around their waists. These men formed in line and offered their services to the peace officers to preserve order, in case an attempt at disturbance was made, or any interference offered to the authorities of the law. However, no hostile move was made by anyone, and quiet and order was fully restored, and in a short time the excitement died away.

Still, extra policemen were called out and a force of ten men placed around the county jail.

People gathered from every direction. James Y. Eccleston heard the gunfire from his post at Safford, Hudson & Company's bank on Fifth Street behind the Oriental. Together with other curious citizens, he now stood in the cold staring at the bodies. Later he remembered that "just before three o'clock, closing time, it commenced to snow and sleet."[80]

Billy Clanton, crying out with pain despite the *Nugget*'s assurances to the contrary, sprawled in the dirt near the corner of Harwood's building. Still breathing but making no sound, Tom McLaury slumped around the base of a two-span telegraph pole at the corner of Third. His brother Frank lay dead near the sidewalk on the north side of Fre-

[79]Statement of W. A. Cuddy, ibid.

[80]James Y. Eccleston, "A Day of Excitement in Tombstone," unpublished manuscript.

mont. But Ike Clanton, who had done more than anyone to bring on this fight, was found by Sheriff Behan cowering in Judge Lucas's old Toughnut Street law office.

As the shooting started Ike had run up to Wyatt and grabbed his arm, claiming no weapons and begging not to be shot. Even under the pressure of battle Wyatt knew if he ruthlessly gunned down an unarmed man at point-blank range it would destroy any chance of being elected sheriff. He simply pushed Clanton aside with the warning, "The fight has commenced. Go to fighting or get away."[81] Ike sprinted to safety through Fly's boarding house, leaving his younger brother and two friends to be shot to pieces. So much for his earlier boast of needing only four feet of ground because fight was his racket. After more than thirty hours without sleep, his alcohol-supported courage during that desperate day finally deserted him on Fremont Street.

Onlookers, including William Allen, Wesley Fuller, Thomas Keefe, and the two men from the *Nugget,* carried Billy Clanton and Tom McLaury into a small building on the corner of Third—"The second house below Fly's gallery," Fuller said.[82] They laid both men on the carpet, with Clanton's feet toward the door. Someone got McLaury a pillow. Keefe unbuttoned Tom's clothes, pulled off his boots, and gave him some water. He and William Allen carefully lifted McLaury's shirt to see where he was hit. Struck by a dozen buckshot under the right arm, Tom made no sound, and "as soon as Dr. Matthews came, we searched the body and did not find any arms. . . . we only found money on him."[83] Counting cash, checks, and certificates of deposit, Tom had been carrying $2,923.45.[84]

Henry Matthews, who needed to be told the name of each victim, examined Tom McLaury's wounds, grouped together between the third and fifth ribs: "laid the palm of my hand over them, it would cover the whole of them, about four inches in space; the wound penetrated straight into the body; he had on a blouse, vest and pants. . . ." As for brother Frank, he found that one wound entered "the cranium, beneath the right ear; another penetrating the abdomen one inch to the left of

[81]Deposition of Wyatt S. Earp, Hayhurst typescript, Document No. 94, 143.

[82]Deposition of Wesley Fuller, ibid., 58.

[83]Deposition of Thomas Keefe, ibid., 64.

[84]Report of H. M. Matthews to the Cochise County Board of Supervisors, November 8, 1881; accepted and filed, January 5, 1882.

the navel." Matthews added, "The only thing I can recollect about Frank McLowry's [*sic*] clothing was a buckskin pair of pants. . . . The buckskin pants were worn over overalls. . . ."[85]

Thomas Keefe noted Billy Clanton's wounds in some detail: "Shot through the arm and right wrist; he was shot in the right side of the belly; he was shot below the left nipple, and the lung was oozing out of the wound; he was shot through the pants of the right leg."[86] Clanton pleaded with Wesley Fuller, "Get a doctor and give me something to put me to sleep."[87] Dr. Nelson S. Giberson examined Billy and announced somewhat callously, "Nothing will do him any good, he is dying."[88] Keefe helped hold Clanton down while Dr. William Miller injected two syringes of morphine. He "was turning, and kicking, and twisting in every manner with pain," the carpenter remembered. "They have murdered me," Billy kept insisting, "drive the crowd away from the door and give me air."[89] Ten to fifteen minutes later he was dead. He was the last to die.

Two blocks west the sound of gunfire had startled the Earp women. Allie and Morgan's wife, Louisa, were sewing a carpet when the shots exploded. "The noise was awful," Allie remembered, "Lou laid down her hands in her lap and bent her head. I jumped up and ran out the door. I knew it had come at last. Mattie was outdoors. Her hair was done up in curlers and she was ashamed to have people see them so she ran back inside the house."[90]

Allie rushed up the street and pushed her way through the crowd. Harry B. Jones kept telling her that Virgil was all right, but she needed to see for herself: "I knelt down beside Virge. The doctor was bendin' over his legs, probing for the bullet. Virge was gettin' madder and madder from the pain. At last the doctor gave it up. A good thing he did—the bullet wasn't there. It had gone clear through the calf. Then the men loaded him in a hack and drove us home. Behind us came another one. I never asked which one was inside. I seemed to know it was Morg."[91]

[85]Deposition of Coroner H. M. Matthews, *Tombstone Daily Nugget,* November 1, 1881.

[86]Deposition of Thomas Keefe, ibid., November 12, 1881.

[87]Deposition of Wesley Fuller, Hayhurst typescript, Document No. 94, 57.

[88]Deposition of J. H. Behan, ibid., 33.

[89]Deposition of Thomas Keefe, *Tombstone Daily Nugget,* November 11, 1881.

[90]Waters, *The Earp Brothers of Tombstone,* 158.

[91] Ibid., 158-159.

Holliday joined Kate, who had seen much of the fight from a window at Fly's boarding house. She treated Doc's grazed hip before he ran out to check on Virgil and Morgan. Holliday was deeply troubled, realizing the seriousness of this bloody encounter. According to Kate, "he regretted that . . . affair to the end of his life."[92]

Another hack finally arrived to carry the three dead Cowboys to the city undertakers at 613 Allen Street. Proprietors Andrew J. Ritter and William H. Ream had come over from California only seven months before. This enterprising duo did more than simply bury the dead. Their advertisements boasted: "We are prepared to do anything in the line of jobbing, repairing furniture, making desks, bookcases, tables, and setting-up, leveling and repairing billiard tables."[93] Of course the two McLaurys and young Billy Clanton no longer had any use for these services.

Later that day Behan visited the marshal's house at First and Fremont. As the sheriff sat on the edge of the bed Virgil warned him, while three of the Earp wives shuffled about the room, "You better go slow, Behan, and not push this matter too far."[94] Earp accused the sheriff rather bluntly of trying to incite the vigilance committee to hang the brothers. Johnny emphatically denied it. Virgil then repeated Wyatt's claim that Behan had insisted the Cowboys were unarmed. Patiently, the sheriff tried to explain what he had actually said. Behan later claimed he told Earp he had always been his friend and that "seemed to settle the matter about the vigilance committee."[95]

James Earp was there most of the time, along with a lawyer named Winfield S. Williams seated on a small sofa near the head of Virgil's bed and another man never identified. Williams could not recall the sheriff saying anything about being Earp's friend, but he did remember Johnny admitting he heard Virgil say, "Boys, throw up your hands. I have come to disarm you."[96]

That evening Wyatt showed up to check on his brothers and fill them in on the talk uptown. After he left, Virgil's wife later explained, "Morg called me. 'If they come, Al, you'll know they got Wyatt. Take this six-

---

[92]Memoirs of Mary Katherine [Holliday] Cummings.

[93]*Tombstone Daily Nugget*, October 25, 1881.

[94]Deposition of J. H. Behan, Hayhurst typescript, Document No. 94, 46.

[95]Ibid.

[96]Deposition of W. S. Williams, ibid., 53.

shooter and kill me and Virge before they get us.' So Lou and me locked the doors, stacked up mattresses in front of the windows, and sat there with the six-shooter all night. I would have used it too, if they had come to kill Virge and Morg."[97]

As the events of that tragic day sank in, residents began wondering what would follow. George Parsons recorded, "Much excitement in town and people apprehensive and scary. . . . Desperate men in a desperate encounter. Bad blood has been brewing some time and I was not surprised at the outbreak. It is only a wonder it has not happened before. A raid is feared upon the town by the Cowboys and measures have been taken to protect life and property. . . . It has been a bad scare and the worst is not yet over some think."[98]

[97]Waters, *The Earp Brothers of Tombstone,* 159.

[98]Parsons, *Private Journal,* October 27, 1881, 270-271.

Chapter 12

FROM GUNS TO GAVEL

As Virgil and Morgan Earp lay bedridden, the family had little to fear from direct physical violence. They looked upon the possibility of expensive legal entanglements with more apprehension than they did avenging gunfire. Public opinion widened over these killings, running from the extremes of guilt and innocence to more selfish motives suggesting street fights discouraged investors and were thus bad for business. Even among those who did not know the McLaurys or young Clanton personally, doubts festered over the Earps' handling of the affair.

Before the law could run its convoluted course the three dead men needed to be buried. As preparations rushed forward a genuine sense of melancholy settled over the town. The funeral took place on the afternoon of October 27. Undertakers Ritter and Ream placed the carefully dressed bodies in caskets heavily trimmed with silver. On each a silver plate gave the name, age, birthplace, and date of death. In a nod to posterity shortly before the services, a photographer captured the slain men in final repose.

To the sad notes of a funeral dirge, the Tombstone Brass Band led the procession down Allen Street. Wagons, hastily pressed into service, moved the caskets to the graveyard. Billy Clanton traveled alone, while in the second the two McLaurys lay side by side. Carriages filled with family and friends pulled into line. Over three hundred people followed on foot, with two dozen carriages and buggies, a four-horse stagecoach, and individual horsemen bringing up the rear. Mourners and the simply curious crowded sidewalks for nearly four blocks. Some had ridden twenty miles to witness the largest funeral ever held in Tombstone. Harry Woods at the *Nugget* admitted with some surprise, "it was not entirely expected. . . ."[1] Even Clum's normally rambunctious *Epitaph* reluctantly conceded, "It was a most impressive and saddening sight. . . ."[2]

[1] *Tombstone Daily Nugget*, October 28, 1881.

[2] *Tombstone Epitaph*, October 28, 1881.

The McLaurys were buried together in a single grave with Billy Clanton laid close by. The brothers suffered even in death: the newspapers repeatedly misspelled their names. Nor was the information engraved on their coffin plates accurate: Frank was listed as twenty-nine and Tom twenty-five, both natives of Mississippi. The *Epitaph* attempted a partial correction, claiming Iowa as their birthplace. Although the family later settled there, both were actually born in New York. When they died Frank (whose given name was Robert F. McLaury) was thirty-three and his brother Thomas C., twenty-eight.[3] Billy fared better, his silver plate correctly identifying him as a nineteen-year-old native of Texas.

Tombstone's fear of Cowboy raids subsided after hearing reports of Curly Bill Brocius being incarcerated at Lordsburg, New Mexico. By early November John Ringo checked into the Grand Hotel but caused no trouble. He would be arrested for robbery by Deputy Sheriff David Neagle at the end of the month on two indictments handed down by the grand jury. The court finally set bond at $3,000. Ike Clanton stood as one of Ringo's sureties.

To better understand the facts, County Coroner Henry M. Matthews began questioning witnesses on the evening after the fight. Since it was not an official proceeding, no transcript was made. The official inquest, held in the courtroom of the Mining Exchange, opened at ten o'clock on the morning of October 28. With no shorthand reporter present, testimony was written out in longhand. This document has survived.

The twelve-man coroner's jury included two witnesses to parts of the affair under investigation—R. F. Hafford and Sylvester B. Comstock. This shows the casual nature of official proceedings in early Arizona; Hafford later testified at the justice court concerning his conversation that day with Ike Clanton. After hearing eight witnesses the coroner's jury handed down their findings on October 29, conceding the two McLaurys and Billy Clanton had died from the effects of gunshot wounds inflicted by Doc Holliday and the Earp brothers. Of course they refused to offer opinions regarding guilt or innocence. Only the courts could decide those issues.

The *Nugget* ridiculed R. F. Coleman's eyewitness testimony: "We

[3]1880 United States Census, Babocomari Valley, Pima County, Arizona, E.D. 4, Sheet 33, Lines 16-17; and Pima County Great Register, No. 1815, Robert F. McLaury, Precinct No. 12, October 28, 1878.

have seen cross-eyed men, and heard of the wonderful feats performed with their optics, but in this case we think circular eyes must have been used. . . ."[4]

The Earps were not laughing. Following the coroner's verdict, Mayor Clum called a special Saturday afternoon session of the council "to consider grave charges against Chief of Police Earp, and it was ordered that pending investigation of said charges Chief Earp be temporarily suspended and James Flynn act as Chief during such suspension."[5] Clum may have helped push Virgil aside, but Harry Woods assumed, "Marshal Earp . . . in a few weeks will be able to attend to his duties."[6] Instead he lost his badge and monthly $150 paycheck.

Virgil collected his final payment for services on November 12, a city warrant for $135.49. Ten days before, Morgan was paid his last $20 as a special policeman.[7] James Flynn opened his tenure by discharging Dan Lynch for being drunk and striking a prisoner over the head with his revolver. Two weeks later Flynn arrested Sheriff Behan and then Dr. Goodfellow, both apprehended for driving too fast.

On October 29, based on a complaint signed by Ike Clanton, Wells Spicer issued warrants to Sheriff Behan ordering the arrest of the three Earps and Doc Holliday. Due to their injuries Virgil and Morgan were not served, but the sheriff took Wyatt and Doc into temporary custody.

At first Spicer denied bail. Only after defense attorneys submitted affidavits of fact did he reconsider, setting bond for each man at $10,000, a startling sum by Tombstone standards. Eight people, including Wyatt and James Earp, R. J. Winders, and Fred Dodge, came forward on Doc's behalf, raising $15,500. Ten others, again including James Earp, Dodge, and Winders, served as Wyatt's bondsmen. His partner Lou Rickabaugh put in $5,000 and attorney Thomas Fitch was good for another $10,000. Earp collected $27,000—far more than needed.[8] As a sign of overall support, however, this generosity proved a false benchmark. Already confidence in the brothers had begun to slide.

---

[4]*Tombstone Daily Nugget*, October 30, 1881.

[5]Minute Book A, Common Council, Village of Tombstone, October 29, 1881, 131.

[6]*Tombstone Daily Nugget*, October 28, 1881.

[7]Warrant No. 275, V. W. Earp, Services as Chief of Police, November 12, 1881; and Warrant No. 265, Morgan Earp, Services as Special Police, Register of Warrants, City of Tombstone.

[8]*Tombstone Daily Nugget*, October 30, 1881.

Not only had Virgil been maneuvered out of office as city marshal, but on the day after Wyatt and Doc posted bond tax assessments on many Earp holdings, including Wyatt's personal property, were suddenly raised. Surface rights on the Mattie Blaylock and Long Branch mining claims increased $2,500, while seventeen acres of the Northern Extension of the Mountain Maid went up $2,000. And so it went; although values on town lots one through four, Block M, were reduced from $800 to $500.[9]

Immobilized by their injuries, Virgil and Morgan could do little more than watch in angry silence as the legal maneuvering began. Virgil put up with the pain and waited for his leg to heal. Morgan suffered complications. He had been wounded while backing from Harwood's lot and tripping over the partially exposed four-inch pipe of the Sycamore Springs Water Company. Doctors now discovered that Clanton's shot had pulled bits of cloth from Morgan's coat into the wound, forcing them to reopen the shoulder and extract that material along with some previously undetected bullet fragments. It proved a painful procedure and delayed Morgan's recovery.

The case against the Earp brothers and Doc Holliday—listed on the docket as "Territory of Arizona vs. Morgan Earp, et al, Defendants"—opened before Justice of the Peace Wells Spicer at three o'clock on the afternoon of October 31. Dr. Henry M. Matthews appeared as the session's only witness. Normally such proceedings would have taken place in Spicer's small Fifth Street office. But due to the number of lawyers involved—ten for the prosecution and four for the defense—authorities moved the case into the county courtroom at the Mining Exchange. Accepting a defense motion agreed to by the prosecution, Spicer barred public attendance during the early stages, restricting access to the various attorneys and one witness at a time.

In order to transcribe the testimony accurately, county officials sent to Tucson for Eugene W. Risley, the shorthand reporter at the federal court and clerk of the Pima County Board of Supervisors (misidentified in the Tombstone press as "E. J."). He took some notes of Dr. Matthews's testimony, but then, for reasons never explained, prosecutors objected to his services. Since Risley enjoyed strong political ties

[9]Minute Book A, Common Council, Village of Tombstone, October 31, 1881, 132-133.

Fremont Street in the mid-1880s. The *Epitah* is the first two-story building on the left, followed by the Mining Exchange and the Gird Block. Across the intersection stands Schieffelin Hall. Farther up the street is the Vickers Building, identified by its shaded balcony. To the right is City Hall. *Courtesy, Arizona Historical Society.*

throughout Arizona his dismissal remains a mystery. He moved on to other things, among them serving as a Tucson city councilman and representing Pima County in the legislature, before ending his career as a superior court judge in California.[10] Judge Spicer selected the inexperienced Fred W. Craig to replace him. That fifty-five-year-old speculator proved woefully ignorant of procedure and did not know shorthand. Craig's transcript is disappointing. He at times succumbed to the annoying habit of writing down answers without recording the questions.

For decades this document lay undisturbed in the Cochise County Courthouse. During the 1930s Pat Hayhurst, an Arizona journalist, publicist for the State Industrial Commission, and acting state director of the WPA, removed it along with other related papers to type a copy as part of Roosevelt's Federal Writers Project. Unfortunately, he sum-

[10] 1880 United States Census, Tucson, North of Congress Street, Pima County, Arizona, E.D. 5, Sheet 29, Lines 27-28; Wagoner, *Arizona Territory*, 517; Vandor, *History of Fresno County*, 1669; and 1900 United States Census, Third Township, Fresno City, Fresno County, California, E.D. 7, Sheet 3-A, Lines 39-43.

marized much of the material so that his version is not a complete record, nor are witnesses listed in proper order. Later the original document was either lost, stolen, discarded, or destroyed in a house fire, depending on which rumor one wishes to believe. Whatever happened, it was never returned to the courthouse, nor has it resurfaced elsewhere. Hayhurst's typescript is of value, but only when compared alongside the almost verbatim reports of testimony carried in the surviving issues of both newspapers.

The *Nugget* ignored Spicer's gag order and published Dr. Matthews's testimony anyway. John Clum hurriedly dismissed the account as garbled. (Hayhurst only summarized Matthews so no comparison can be made.) Assessing the *Nugget*'s overall accuracy throughout the hearing, however, Clum's complaint seems unfounded. Harry Woods accused the *Epitaph,* which did not carry the testimony, of whining: "If through incompetency of its reporters, indifference of its managers, or bias of its editors it chooses to omit matters of such importance, it concerns us not. But when it resorts to such contemptible means to cover its shortcomings, we enter a protest."[11]

Bowing to public pressure on November 2, defense lawyers withdrew their objections to open hearings. Testimony was now available without restrictions. At times the *Epitaph* summarized its coverage. It appears the *Nugget,* hiring its own shorthand expert, did not. Harry Woods made a point of reminding his readers that he published the testimony in full, whereas, "The 'short-hand' reporter who is furnishing the Epitaph with the testimony of the Earps' examination must have had several of his fingers cut off, as the testimony it publishes at present is so very 'short' that it is barely intelligible."[12]

Such attacks may have amused readers, but the animosity thus reflected was not always evident on the street. John Clum and the *Nugget*'s city editor, Richard Rule, chummed around during a small Thanksgiving Day dinner party at the Maison Doree that November of 1881. Earlier, when the *Epitaph*'s press broke down, Harry Woods offered the use of his so they could publish. After Clum's San Francisco agent denied him telegraphic dispatches for failure to pay in a timely fashion, Woods

[11] *Tombstone Daily Nugget,* November 3, 1881.
[12] Ibid., November 15, 1881.

shared his. The *Epitaph* "refused to return the compliment and then boasts of its enterprise."[13] Still, all their mutual complaining during the Earp-Holliday hearing only helped boost circulation, and in the end that was all that really mattered.

On the evening of November 3 William R. McLaury, the older half-brother of Frank and Tom, arrived from Fort Worth and checked into the Grand Hotel. He had learned of their deaths by Western Union telegram the day after the shootings. McLaury was no stranger to the rigors of life. At nineteen he enlisted for Civil War service as a private in an Iowa infantry regiment, "stationed at the sickly place of Helena, Arkansas. . . ." After the war he returned home, married, and then spent several years in the Dakotas before moving to Texas.[14] As a lawyer himself, McLaury now approached Cochise County's Republican district attorney, Lyttleton Price, and was admitted as an associate council for the prosecution the next morning.

McLaury demanded that bail be revoked, especially after watching Wyatt and Doc strolling "into the court room heavily armed." The older man complained with growing disgust, "The Dist. Atty. was completely 'Cowed,' and after promising me on the 4th to move the court to commit these men without bail he would not do it. . . ." After informing Price that he intended to make the motion himself, the district attorney and others on his staff vigorously tried to dissuade him. "They did not want to see me killed," McLaury confessed. Then, thinking of the two arrogant defendants: "I only hope they would [try] as I would be on my feet and have the first go—and thought I could kill them both before they could get a start. . . ."[15] McLaury deliberately "stood where I could send a knife through their hearts if they made a move and made the motion."[16] "They were as quiet as lambs," he wrote his Texas law partner, "only looked a little scared, it was granted after some discussion."[17]

Earp and Holliday appeared before Probate Judge J. H. Lucas with writs of habeas corpus claiming Spicer had overstepped his authority.

[13]Ibid., November 10, 1881.

[14]*History of Tama County*, 111; Caldwell, ed., *History of Tama County Iowa*, 334; and 1880 United States Census, Fort Worth, Tarrant County, Texas, E.D. 90, Sheet 24, Lines 1-5.

[15]W. R. McLaury to S. P. Greene, Fort Worth, Texas, November 8, 1881, Arizona Historical Society.

[16]W. R. McLaury to D. D. Applegate, Toledo, Iowa, November 9, 1881, New-York Historical Society.

[17]W. R. McLaury to S. P. Greene, Fort Worth, Texas, November 9, 1881, Arizona Historical Society.

Reviewing the documents, Lucas dismissed their claims and remanded them to Sheriff Behan's custody at three o'clock on the afternoon of November 7. This time both were tossed into the county jail, raising the population of that small structure to eighteen. Stuffed into those foul-smelling cells, it must have galled them to learn of Frank Stilwell's release at Tucson on a $2,000 bond paid by, among others, Cochise County Recorder A. T. Jones and his deputy B. A. Fickas.

Cheered by events and the reaction on the street, William McLaury wrote his brother-in-law, also a lawyer: "That evening and night was a perfect hurrah. A large crowd followed me from the court room to the hotel, and at night the hotel was completely thronged with people and they nearly shook my hands off."[18] A couple weeks later he wrote his sister, in an overly optimistic mood as it turned out: "I had little support from the people when I came here. I now have the whole camp, with the exception of a few gamblers."[19] Despite any personal satisfaction, McLaury worried about how all this would effect his seventy-year-old father, still farming in Iowa with his young wife, five stepchildren and three-year-old son.[20]

In the county jail Wyatt Earp and Doc Holliday shared cramped quarters, lousy food, and slop buckets with the likes of Jerry Barton and former policeman T. J. Cornelison. Fearing reprisals, guards heavily armed with Winchester rifles and double-barreled shotguns surrounded "that insecure edifice to prevent anyone breaking in," joked the *Epitaph.* "If the Sheriff had been as active in preventing prisoners' [*sic*] breaking out, there would have been three more gentlemen for the courts to pass upon at the next term."[21]

Inmates complained constantly of the cold. There never seemed to be enough firewood for the jail. With the unusually severe autumn weather, heat was in demand all over town. Metal fabricators Fredericks and Hill had already sold over a hundred stoves from their crowded Allen Street shop. Demand drove the price of wood from $15 to $21 a cord. Mexican families controlled that business locally, but as prices rose Tombstone reacted as one might expect—people simply stole what they needed. Amidst the turmoil prisoners suffered. Wyatt fell ill, interrupt-

[18]W. R. McLaury to D. D. Applegate, Toledo, Iowa, November 9, 1881, New-York Historical Society.

[19]W. R. McLaury to Mrs. M. F. Applegate, Toledo, Iowa, November 17, 1881, ibid.

[20]1880 United States Census, Buffalo Township, Buchanan County, Iowa, E.D. 81, Sheet 11, Lines 32-39.

[21]*Tombstone Epitaph,* November 8, 1881.

ing the hearing for one day. Spicer finally reconsidered and on November 23 admitted Earp and Holliday to bail. Again he set the amount at $10,000 each. This time mining men E. B. Gage and James Vizina stepped forward as bondsmen.

Dr. Matthews headed the list of twenty-nine witnesses who came forward throughout November to give a mass of contradictory testimony, filled with enough ignored revelations of fact and meandering digressions to push the limits of logic and human endurance. From the lawyers' uniformly flaccid examinations—flawed samples of that demanding art—it is clear none of them lived up to their reputations as courtroom advocates. Studying the fragmentary record, one can only wonder if they actually earned their fees, which in the case of Thomas Fitch, Wyatt Earp's chief counsel, proved greater than many of his colleagues could hope to earn during a full year of litigation.

Fitch's dull performance is particularly surprising. His long career as an editor, politician, entrepreneur, lawyer, and lecturer had earned him the title "Silver-Tongued Orator of the Pacific Slope." His high-priced mental powers and well-oiled vocal cords were certainly not on display here. As much as any single feature of the proceedings, his less-than-aggressive posture led to a deeply held suspicion that the fix was on.

Much of the testimony proved self-serving. After all, Wyatt and Virgil were fighting for their lives; neither Morgan Earp nor Doc Holliday testified, although Doc took copious notes. Hoping to assure guilty verdicts, Ike Clanton overstated his case, much of the time finding refuge behind theory not fact. He steadfastly maintained that his knowledge of Wyatt and Doc's involvement in the attempted Benson stage holdup led to all the trouble. It was simply a plot to silence him, he said. Ike also claimed Bill Leonard "told me that if Doc Holliday had not been there and drunk that Philpot would not have been killed."[22]

Although Clanton had admitted to Dr. Matthews during the coroner's inquest, "We had a transaction I mean myself and the Earps but it had nothing to do with the killing of these 3 men," he now denied working with Wyatt on any scheme to capture Leonard, Head, and Crane.[23] Instead, Ike insisted,

[22]Deposition of Joseph I. Clanton, Hayhurst typescript, Document No. 94, 116.

[23]Statement of Joseph I. Clanton, Coroner's Inquest, Document No. 48.

> I then asked him why he was anxious to capture these fellows. He said that his business was such that he could not afford to capture them. . . . He said he and his brother, Morg, had piped off to Doc Holliday and Wm. Leonard the money that was going off on the stage, and he said he could not afford to capture them, and he would have to kill them or leave the country, for they [were] stopping around the country so damned long that he was afraid some of them would be caught and would squeal on him. I then told him I would see him again before I left town. I never talked to Wyatt Earp any more about it.[24]

Later, when asked if he still feared for his life after being taken into Wyatt's confidence, Ike responded, "Well, after the attempt to murder me the other day I do."[25] Such testimony was not convincing, for of all the Cowboys confronting the Earps and Holliday, he was the one Wyatt ordered to fight or get away. Considering the known facts and everyone's position in Harwood's lot, Ike Clanton could have been killed first and easiest had it been the plan.

Wyatt Earp testified, by reading into the record a statement prepared with the help of attorney Thomas Fitch, that trouble with the McLaurys started back in the summer of 1880 with the stolen mules from Camp Rucker. About a month later, Earp insisted, "I met Frank and Tom McLoury [*sic*] in Charleston. They tried to pick a fuss out of me down there and told me if I ever followed them up again as close as I did before they would kill me." The attempted robbery of the Benson stage only heightened tension, Wyatt said. "It was generally understood among officers and those who have information about criminals that Ike Clanton was a sort of chief among the Cow Boys; that the Clantons and McLourys were cattle thieves and generally in [on] the secret of the stage robbery; and that the Clantons' and McLourys' ranches were meeting places and places of shelter for the gang."[26]

Wyatt repeated the story of finding his stolen horse in the possession of Billy Clanton at Charleston, and of his deal with the three Cowboys to capture Leonard, Head, and Crane. He then told of threats made against Morgan by Ike, Frank McLaury, John Ringo, and others in front of the Alhambra over the Earps' involvement in the arrest of Frank Stil-

[24]Deposition of Joseph I. Clanton, Hayhurst typescript, Document No 94, 105-106.

[25]Ibid., 118.

[26]Deposition of Wyatt S. Earp, ibid., 136.

well and Peter Spencer. Virgil Earp related a similar conversation he had had with Frank McLaury. At that time Frank accused him of raising a vigilance committee to hang the Cowboys. Virgil dismissed the implication, but McLaury insisted Johnny Behan had told him so.

Such threats were well known around Tombstone, where a general consensus accepted the arrests of Stilwell and Spencer as the point where serious trouble began. Even the *Nugget* conceded, before any testimony was taken: "The co-operation of the Earps with the Sheriff and his deputies in the arrest causing a number of the cowboys to, it is said, threaten the lives of all [those] interested in the capture."[27] Clum, too, was aware of these developments: "Since the arrest of Stilwell and Spence . . . there have been oft repeated threats conveyed to the Earp brothers . . . that the friends of the accused, or in other words the cowboys, would get even with them for the part they had taken in the pursuit. . . ."[28]

Wyatt and Virgil wisely concealed another possible motive for animosity. The year before, James Earp's teenage stepdaughter had been seeing someone on the sly. Following a night out,

> Wyatt and Virge was there with Jim and Bessie," Allie recalled with some embarrassment. "When Hattie slipped back in, all hell broke loose. Me and Lou standin' in the dark, our noses against the window, could hear Hattie screamin', Bessie yellin', and the men cussin' as they leathered her to a finish. . . . I thought my blood would turn to water. Lou and me looked at each other in the dark at the same time. I knew she was thinkin' the same thing I was: "First Mattie, now Hattie. Is it goin' to be my turn next?"

Both women had a right to wonder. "Finally the strappin' stopped," Allie continued. "I could hear Virge's voice stern but calm. 'That's enough now!' Maybe she told who the cowboy was, and maybe because she was an Earp she never did tell. I hope she didn't! Anyway I never learned. But I always thought she was sweet on one of the McLowery [*sic*] brothers."[29]

When Allie ran from her house to Harwood's lot to find Virgil after the shooting stopped, she saw Tom McLaury lying at the corner of Third. But for no more than a instant did she wonder, "Was he the one

[27] *Tombstone Daily Nugget*, October 27, 1881.

[28] *Tombstone Epitaph*, October 27, 1881.

[29] Waters, *The Earp Brothers of Tombstone*, 123.

Hattie had kissed and hugged in the moonlight? I never stopped runin' past him. All I had a mind for was Virge."[30]

Meanwhile, back in the courtroom, Wells Spicer not only had to try and make some sense out of all the wildly conflicting testimony, but he also took it upon himself to privately interrogate witness Addie Bourland, a milliner who had watched the opening moves of the fight through the front window of her house opposite Harwood's lot.

During a recess, Spicer, who believed she knew more than she had testified to, went down to her place and talked with her alone, unknown to any of the attorneys. That afternoon he recalled Mrs. Bourland, overruling strenuous objections from the prosecution, and questioned her on the record as to whether or not the Cowboys put their hands up. "I didn't see anyone hold up their hands," she replied, "they all seemed to be firing in general, on both sides."[31] The worried prosecutors failed to shake her testimony. Spicer's decision to question a witness that had already been excused by both sides was unusual even for Tombstone.

By the end of November the final witness, butcher Ernst Storm, stepped down. On the last day of the month, and without hearing closing arguments from either side, Judge Spicer—a man never particularly friendly and on occasion openly hostile toward the defendants—announced his decision.

First he outlined the trouble. Because of the confrontation the night before between Doc Holliday and Ike Clanton, as well as Wyatt's early afternoon encounters with Clanton at Wallace's courtroom and then outside with Tom McLaury, Spicer concluded that Virgil's decision to include either one in any attempt to arrest the Cowboys was "an injudicious and censurable act," but "considering the many threats that have been made against the Earps, I can attach no criminality to his unwise act. In fact, as the result plainly proves, he needed the assistance and support of staunch and true friends, upon whose courage, coolness and fidelity he could depend, in case of emergency." Spicer never bothered to ask why Virgil failed to call out the police force. Officer A. J. Bronk and the others were certainly aware of the trouble.

The judge took great pains explaining his position on the question of felonious intent:

---

[30]Ibid., 158.

[31]Deposition of Addie Bourland, Hayhurst typescript, Document No. 94, 88.

> In view of the past history of the country and the generally believed existence at this time of desperate, reckless and lawless men in our midst, banded together for mutual support and living by felonious and predatory pursuits, regarding neither life nor property in their career, and at the same time for men to parade the streets armed with repeating rifles and six shooters and demand that the chief of police and his assistants should be disarmed is a proposition both monstrous and startling.

"It is claimed by the prosecution that their purpose was to leave town," the judge made a point of considering. "Whatever their purpose may have been, it is clear to my mind that Virgil Earp, the chief of police, honestly believed (and from information of threats that day given him, his belief was reasonable), that their true purpose was, if not to attempt the death of himself and brothers, at least to resist with force of arms any attempt on his part to perform his duty as a peace officer by arresting and disarming them."

In all likelihood Tom McLaury and Ike Clanton were both unarmed as they stood listening to Virgil's command. Spicer dismissed this point, saying, "I will not consider this question, because it is not of controlling importance." Later he described the circumstances differently:

> Another fact that rises up preeminent in the consideration of this sad affair is the leading fact that the deceased from the very first inception of the reencounter were standing their ground and fighting back, giving and taking death with unflinching bravery. It does not appear to have been the wanton slaughter of unresisting and unarmed innocents, who were yielding graceful submission to the officers of the law, or surrendering to, or fleeing from their assailants; but armed and defiant men, accepting their wager of battle and succumbing only in death.

Clearly troubled by conflicting accounts, Spicer reluctantly conceded: "Witnesses of credibility testified that each of the deceased or at least two of them yielded to a demand to surrender. Other witnesses of equal credibility testified that Wm. Clanton and Frank McLowery [*sic*] met the demand for surrender by drawing their pistols, and that the discharge of firearms from both sides was almost instantaneous." The prosecution had tried accusing the defendants of criminal haste. "I cannot believe this theory," Spicer patiently explained, "and cannot resist the firm conviction that the Earps acted wisely, discretely and prudentially, to secure their own self-preservation. They saw at once the dire necessity of giving the first shots, to save themselves from certain death."

Spicer supported his decision further by saying: "Was it for Virgil Earp as chief of police to abandon his clear duty as an officer because its performance was likely to be fraught with danger? Or was it not his duty that as such officer he owed to the peaceable and law-abiding citizen of the city, who looked to him to preserve peace and order, and their protection and security, to at once call to his aid sufficient assistance and proceed to arrest and disarm these men?"

Anticipating some disappointment after his releasing the defendants, Spicer cautiously reminded everyone:

> I have the less reluctance in announcing this conclusion because the Grand Jury of this county is now in session, and it is quite within the power of that body, if dissatisfied with my decision, to call witnesses before them or use the depositions taken before me, and which I shall return to the District Court, as by law required, and to thereupon disregard my findings and find an indictment against the defendants, if they think the evidence sufficient to warrant a conviction.[32]

Because of his rulings toward the end of the hearing, the judge's decision surprised no one. Conceding conflicting testimony, many felt the need "that all the circumstances leading up and connected with the affair be thoroughly investigated." Some went so far as to mistakenly question Spicer's personal motives: "the suspicion of reasons of more substantial nature are openly expressed upon the streets, and in the eyes of many the justice does not stand like Caesar's wife, 'not only virtuous but above suspicion.'"[33]

George Parsons concluded: "Earps released today or yesterday. Grand Jury may indict but I doubt it."[34] William McLaury may have hoped, "There will be an indictment against Holliday and, I think, two of the Earps and one Williams for the murder in the attempted robbery," but Parsons understood Tombstone better than did the embittered lawyer from Fort Worth.[35]

The newly impaneled seventeen-member grand jury included William Harwood, David Calisher, Thomas Sorin, Sylvester B. Comstock, and

[32]Judge Wells Spicer's Decision, ibid., 160-167. Also see *Tombstone Epitaph* and *Tombstone Daily Nugget,* December 1, 1881.

[33]*Tombstone Daily Nugget,* December 1, 1881.

[34]Parsons, *Private Journal,* December 1, 1881, 278.

[35]W. R. McLaury to Mrs. M. F. Applegate, Toledo, Iowa, November 9, 1881, New-York Historical Society.

Marshall Williams, with lumberman Lewis W. Blinn acting as foreman. They indicted William Claiborne on December 3 for the Charleston killing, but after Spicer's decision freeing the Earps and Holliday it was reported, "The affair will probably be investigated by the Grand Jury . . . but from the confessed and known bias of a number of its members, it is not probable that an indictment will be found."[36] Predictably the grand jury took no action, but the troubles for the Earps were far from over. Disappointed, William McLaury kept busy trying to settle his brothers' estates while fantasizing about revenge.

Then, as if Tombstone needed another distraction, a fire broke out on the evening of December 9 at the Grand Lodging House on Toughnut. Owned by the proprietors of the Grand Hotel, the twelve-room wood-frame was used as a dormitory. R. F. Coleman lived there with his twenty-three-year-old son Walter. At first it was thought a hall lamp had exploded and spread flames to the cloth ceiling, but an employee of Taylor and Brown's nearby Pioneer Livery Stable saw a Chinese domestic empty a pan of ashes at the rear of the building and assumed this started the blaze. Fire engulfed the lodging house so fast residents could not retrieve their belongings. It then spread quickly westward, igniting the roof and gutting W. T. Richards and Frank Hunter's assay office. Taylor and Brown opened their stalls and drove thirty frightened animals out their Allen Street entrance west of the Grand Hotel. The two men saved carriages and leather goods, before losing their building, hay, and grain.

When the fire started S. H. Lyall, who lived across the street but was then talking with a friend near the Vizina Mining Company's lumberyard, rushed to the hoisting works and sounded the alarm. The fire department responded on the run. Finding coupling problems with hydrants at both ends of Toughnut, they were forced to deliver water to their pumper, hauled to the end of the Russ House, by a hose finally connected to a hydrant at Fifth and Allen. Crowds hurriedly assisted nearby residents and businessmen removing goods and merchandise into the street. The Rescue Hook & Ladder Company pulled down many burning structures and outbuildings in hopes of hindering the fire's advance.

---

[36] *Tombstone Daily Nugget,* December 1, 1881.

When the threat lessened on Toughnut Street firemen hauled their engine to Fourth and Allen, sending out a stream of water that finally stopped the flames at the rear of the Mint Saloon. Everyone cheered the victory. Liquor wholesalers Hawkins and Boarman, whose own building might have been lost with less vigilance, offered free refreshments to exhausted firefighters. Taylor and Brown found their horses grazing on the mesa east of town and drove them to Hiram Tuttle's Tombstone Livery and Feed Stable near the corner of Third and Fremont. No animals had perished, and no one suffered injury in the blaze. Monetary damage was small, with some losses covered by insurance. Everyone felt lucky not to have witnessed a repeat of the June 22 firestorm. City Marshal James Flynn ordered all saloons closed at eleven o'clock, which helped calm the situation. With all its recent troubles, Tombstone had a low threshold for excitement.

Chapter 13

Aftermath of a Tragedy

Decades later the street fight of October 26, 1881, became famous as the *Gunfight at the O.K. Corral*. It sounded so dramatic to satisfied admirers that no one seemed to care that legend got the location wrong. Instead they blindly surrendered to thirty seconds of violence and elevated the Earp brothers and Doc Holliday to the height of popular culture—Western heroes all. The reality was different. The bullets fired that cold afternoon on Fremont Street proved more tragic than heroic.

Popular myth allows that following this fight Tombstone rallied behind the Earps, standing firm against the county Cowboy element. Actually the town fractionalized over this bloody episode. For reasons covering personal, political, and economic considerations, many ordinary citizens found themselves on the far side of the line from the Earps and their allies. The gunfight that later made them so famous now alienated the brothers from the very community they so much wanted to impress. The only civic nod received came with the Rescue Hook & Ladder Company appointing Virgil to the reception committee for its New Year's ball. Otherwise, recriminations filled the air as anonymous threats of further bloodshed surrounded participants and partisans alike. No one in a position of prominence seemed immune from danger.

Amidst all these wild rumors of retaliation, John Clum boarded Sandy Bob's six-horse stage on December 14 for its nightly run to Benson. The mayor planned visiting his older brother George, clerk of the U.S. District Court at Tucson, before going east for the holidays. A bullion wagon, driven by a fellow known as Whistling Dick, closely followed the stage. Below Malcolm's Wells, the last house on the road to Contention City, and nearly four miles outside Tombstone, masked highwaymen stepped out of the brush, yelled "halt," and opened fire on the coach. One misplaced shot hit the off wheeler in the neck (the animal nearest the vehicle on the right). A stray bullet accidently struck Whistling Dick in the leg. Ignoring the pain and repeated commands to stop, he whipped his team and followed stage driver Jimmy Harrington to safety. They made it only a mile or so before the wounded horse stumbled and collapsed from loss of blood. Harrington and his passengers frantically cut the dead animal free before rushing on to Contention.

For a time no one knew what had happened to John Clum. Passengers assumed he had climbed up top with Harrington. They did not realize his absence until reaching Contention City. Sheriff Behan and Charles Reppy began searching for the missing mayor. Learning more about the attack from two teamsters at Malcolm's, they combed the area without success. Enjoying the spectacle, Harry Woods surmised, "The prevailing opinion is that he is still running."[1] The *Epitaph* saw events through a different lens. Claiming the attack "threw the city into the wildest excitement," the paper saw it all as part of a wider conspiracy: "Since the late unfortunate affair, rumors have been rife of the intended assassination of not only the Earp brothers and Holliday, but of Marshall Williams, Mayor Clum, Judge Spicer and Thomas Fitch."[2]

Woods countered by describing how "Tombstone's Chief Magistrate" had "valiantly maintained his position in the bottom of the coach, occupying as little space as possible, until a halt was compelled by the falling of the wounded animal. . . ." Then, after helping Harrington, "like an inspiration, the whole diabolical plot of the would-be assassins was revealed to the mind of our city's ruler. It was not booty they were after . . . it was something far more precious than that; it was his own august person. . . . But our Worthy Mayor was not to be thus ensnared. In short, he was to 'fly.'" Regarding the possibility of ransom, Woods concluded, "this community would have regarded two dollars and fifty cents as rather steep."[3]

Whether he ran away or not, Clum failed to climb back aboard with the others as the coach pulled out. Instead, he walked the seven miles along the ore road to the Grand Central Mill. After resting he borrowed

[1] *Tombstone Daily Nugget,* December 15, 1881.

[2] *Tombstone Epitaph,* December 16, 1881.

[3] *Tombstone Daily Nugget,* December 16, 1881.

a horse and rode into Benson soon after sunrise. Clum never wavered in his assertion that the attack was aimed at him. Five decades later he still reveled in the claim of escaping assassination that night outside Tombstone.[4]

Not everyone took Clum's "Death List" pronouncements seriously, although Thomas Moses of the Capitol Saloon and Frank Leslie (now seen by the *Epitaph* as "law-abiding and peaceable citizens"[5]) received anonymous cards telling them to get out after they publicly criticized the Earps' handling of the Clantons and McLaurys. Lou Rickabaugh was then threatened for voicing approval of Wyatt and his brothers. Until the situation cooled, Moses sent his family to California. Frank Leslie showed steadier resolve. He closed his short-lived saloon business at John Doling's old place and became night manager for Ben Wehrfritz at the Golden Eagle.

Annoyed with rumors predicting assassination, attorney Thomas Fitch wrote the *Epitaph* from Tucson: "Will you permit me the use of your columns to say that . . . I have never received a warning or menace from 'cow-boys' or anybody else, and I have not and never have had occasion to suspect myself to be in the slightest danger from any source whatever."[6]

Actually, hard language came from both sides. Earlier, the Citizens' Safety Committee strongly suggested that certain parties unfriendly to the Earps, including Sheriff Behan, leave town.[7] It is now impossible to judge the seriousness of any of these threats, except toward the Earp brothers and Holliday.

Certainly Tom Fitch was not concerned, nor was Wells Spicer particularly frightened. He admitted receiving a letter calling him a son of a bitch and warning, "you are liable to get a hole through your coat at any moment." The justice may have believed the attack on Sandy Bob's stage was directed at Clum, but as to the source of the threats, he wrote: "And now I will try to do justice to the Clanton brothers by saying that they and men outside the city, living on ranches and engaged in raising cattle

[4]Clum, "It All Happened in Tombstone," 56-62.

[5]*Tombstone Epitaph*, December 16, 1881.

[6]Thomas Fitch to the editor, Tucson, January 11, 1882, ibid., January 13, 1882.

[7]*Tombstone Daily Nugget*, December 15, 1881; and Parsons, *Private Journal*, December 17, 1881, 280-281.

or other lawful pursuit, as heartily condemn these proceedings as any man in our midst, and that they as honestly denounce all such affairs as any man can. That the real evil exists within the limits of our city."

Spicer went on to explain:

> There is a rabble in our city who would like to be thugs, if they had courage; would be proud to be called cow-boys, if people would give them that distinction; but as they can be neither, they do the best they can to show how vile they are, and slander, abuse and threaten everybody they dare to. . . . when they threaten me they do so because they are low-bred, arrant cowards, and know that "fight is not my racket"—if it was they would not dare to do it.

Spicer had no intention of running: "I will be here just where they can find me should they want me, and that myself and others who have been threatened will be here long after all the foul and cowardly liars and slanderers have ceased to infest our city."

The threats against the justice of the peace all implied corruption in his ruling favoring the Earps and Holliday, but as Spicer pointed out in some detail:

> It is but just to myself that I should here assert that neither directly nor indirectly was I even approached in the interest of the defendants, nor have I ever received a favor of any kind from them or for them. Not so the prosecution—in the interest of that side even my friends have been interviewed with the hope of influencing me with money, and hence all this talk by them and those who echo their slanders about corruption. . . . every one who says that I was in any manner improperly influenced is a base and willful liar.[8]

At the Oriental Milt Joyce joked with Virgil Earp about the attempted stage robbery below Malcolm's Wells, saying he "had been expecting something of the sort ever since they (the Earps and Holliday) had been liberated from jail." Angered by the implication, Virgil slapped his face. Seeing himself surrounded by heavily armed Earp partisans, Joyce remarked as he backed toward the door, "a man would be a fool to make a fight single-handed against that crowd." Before stepping onto the sidewalk he tossed out one final taunt: "Your favorite method is to shoot a man in the back, but if you murder me you will be compelled to shoot me in front." Harry Woods concluded Joyce's "coolness

[8]Wells Spicer to the editor, *Tombstone Epitaph,* December 18, 1881.

and good judgment undoubtedly saved Tombstone from the disgrace of another bloody tragedy, all who are cognizant of the peculiar characteristics of the Earp party will readily admit."[9]

These rapidly unfolding events prompted a sneer from James Reilly, who had never forgiven Wyatt for arresting him in his own courtroom back in August of 1880. First he suggested that the plot to assassinate Clum was a hoax, conceding, "I have long been satisfied that Mayor Clum was weak and perhaps fond of notoriety, but I believed him honest and still hope he is so, and that he was not a party to that scheme." After saying "the Earps and Holliday are not good men," Reilly asked, "Is it true that Mayor Clum has given permits to carry arms to the Earps and twenty-five of their determined friends, and that our city police must not interfere with them, while all others are subject to fine and imprisonment if they carry arms?" Reilly believed the trouble between Virgil and Milt Joyce confirmed his suspicions, "and if so, then our city government has followed the precedent set a year ago by the Randall-Gray detestable clique, and placed the lives of all men intelligent and manly enough to have opinions of their own, at the mercy of a band of reckless bravos. . . ."[10]

Ned Boyle, one of the Oriental's bartenders, blunted the attack by claiming the ex-justice of the peace "was charged when he lived in Yuma with being connected in a stage robbery and the taking of Wells-Fargo's box. . . ." He also reminded everyone of Reilly's history as a lot jumper. As for the Earps, Boyle wrote, "I shall speak only of one of them, Wyatt Earp; he is one of the partners of the firm I am working for, and a more liberal and kind-hearted man I never met."[11] Others argued that point privately.

Charges such as these flowed first one way, then the other, becoming the underlying issue for the city elections scheduled for January 3. Mayor Clum and the council had already divided the city into four wards.[12] The First Ward covered everything from the western edge to Third Street; the Second from Third to Fifth; the Third from Fifth to Seventh; and the Fourth Ward extended from Seventh Street to the east-

[9] *Tombstone Daily Nugget,* December 16, 1881.

[10] James Reilly to the editor, December 20, 1881, ibid., December 21, 1881.

[11] "A Card" from E. F. Boyle, *Tombstone Epitaph* (weekly), December 26, 1881.

[12] Ordinance No. 29, Minute Book A, Common Council, Village of Tombstone, December 6, 1881, 149.

ern limits of the city. Done for no other reason than to create bases for political patronage, the measure fooled no one and prompted widespread complaint: "Dividing as small a place as this into so many wards has one compensation, it will give employment to twice as many judges, clerks and inspectors of elections as there is any use for."[13] It was a point well taken since at the time only 913 voters were registered. By election day, however, the official total had risen to 1,441.

As tensions rose so did interest in finding a political solution to Tombstone's many problems. On December 13 a group including John Clum, Lou Rickabaugh, William Harwood, George Spangenberg, Sylvester Comstock, and Marshall Williams joined 170 others signing a petition that informally urged lumberman Lewis W. Blinn to run for mayor. Four days later, somewhat reluctantly, he announced his willingness to do so.

The opposition started late but they got organized first. On the evening of December 24, in what became known as Cuddy's Convention, theatrical manager William A. Cuddy called to order a small political meeting at the Mining Exchange. The group voted to call themselves the People's Independent Ticket. They ostensibly supported a platform dedicated to cleaning up the mess surrounding real estate titles, but a complete change in the direction of city government was their goal. Cuddy nominated Allen Street blacksmith John Carr for mayor. Saloonman Bob Hatch tossed in Blinn's name for consideration. Carr won 24 to 6. George Chapman beat out A. O. Wallace and George Swain for recorder. Heyman Solomon got the nod for treasurer, with J. P. Rafferty carrying the vote to run as city assessor.

Then, in what was described as a "tug of war," votes were counted for the all-important office of city marshal. Eight names went into nomination, including incumbent James Flynn, Behan's deputy David Neagle, Deputy U.S. Marshal Leslie F. Blackburn, and John L. Fonck. On the first go-around Flynn got 9 votes, Neagle 8, and Fonck 2. Blackburn came up with zero. On the third ballot Neagle won the nomination, defeating Flynn 17 to 13. Virgil Earp knew better than to even announce an intention to run, fearful of ending up in the cellar with Leslie Blackburn.

---

[13] *Tombstone Daily Nugget*, December 3, 1881.

The *Nugget* threw its support solidly behind the People's Independent Ticket. The *Epitaph* immediately launched negative attacks: "The fact that the Daily Cow-boy and the Ten-per-cent Ring are advocating the claims of John Carr for mayor and Dave Nagle [*sic*] for city marshal is a sufficient reason why all citizens who would have the city government conducted differently from that of the county should vote against them."[14] And three days later: "The ten-per-cent ticket is losing ground every day. The people don't seem to desire that the corrupt gang who have run Cochise county so dishonestly should get control of our municipal government as well."

The *Epitaph* tried belittling the rotund Irishman John Carr by describing him as the "Village Blacksmith." Actually, he was a man with wide experience who had come to the United States by way of Canada, then crossed the plains from Peoria, Illinois, in 1850. Working as a blacksmith in California, he became one of the pioneer settlers of Trinity County, where he naturalized as a citizen in 1855. Later Carr moved to Humboldt County and served two terms on the Eureka City Council, once as president of the board. He traveled back to Illinois for a visit by way of Panama and Cuba. These experiences and others taught Carr much about the hardships of frontier life. Later he wrote a book about what he had seen in the early days, but unfortunately Tombstone is mentioned only as a reference.[15] When he was picked to run for mayor, Carr had been a resident long enough to clearly understand the troubles plaguing that unhappy camp. And those troubles were about to get worse.

Near 11:30 on the evening of December 28, 1881, Virgil Earp stepped from the Oriental, pausing to glance up and down Allen Street before crossing Fifth toward the Wehrfritz Building. He passed Dr. Goodfellow coming the opposite way. As Virgil reached the sidewalk, three assassins—concealed in an adobe under construction on the southeast corner as a drug store and offices for the Huachuca Water Company—fired ten-gauge shotguns at the unsuspecting deputy U.S. marshal.

[14] *Tombstone Epitaph*, December 25, 1881.

[15] Carr, *Pioneer Days in California*; Pima County Great Register, No. 4219, John Carr, Precinct No. 17, October 11, 1880; Cochise County Great Register, No. 248, John Carr, Tombstone, December 2, 1881; and Irvine, *History of Humboldt County*, 734-737.

Earp took the force of at least one full charge of buckshot, the balls shattering his left elbow and leaving deep wounds above the thigh and along his back. One glanced off the spinal column as others struck his kidneys and liver.[16] Luckily several shots missed, blowing holes instead through the first window along the Fifth Street side of the saloon and passing harmlessly over the heads of startled patrons before burying themselves in the wall beyond. Evidence of nineteen buckshot peppered the outside wall and awning posts.

Miraculously, Virgil was not killed. Despite his wounds he stayed on his feet, turned and started back toward the Oriental. Hearing the heavy shotgun blasts, Wyatt bolted from his gambling perch to find his brother seriously injured. Someone yelled for Dr. Matthews, but Goodfellow was already on the scene. Together with other of Earp's friends, he helped Virgil to the same room at the Cosmopolitan where Kate Holliday had been incarcerated, a site maintained by the brothers as a sanctuary for private card games. There they regularly emptied the pockets of the less skilled and unwary.

At the time of the ambush George Parsons happened to be in Goodfellow's office, one the doctor shared with dentist E. P. Ryder above the saloon on the second floor of the Wehrfritz Building. "Doc G had just left and I tho't couldn't have crossed the street," Parsons wrote, "when four shots were fired in quick succession from very heavily charged guns, making a terrible noise and I tho't were fired under my window under which I quickly dropped, keeping the dobe [*sic*] wall between me and the outside till the fusilade [*sic*] was over. I immediately tho't Doc had been shot and fired in return, remembering a late episode and knowing how pronounced he was on the Earp-cow-boy question."[17]

Once out on the crowded street, Parsons rushed to the hospital at Goodfellow's insistence: "got various things. Hotel well guarded, so much so that I had hard trouble to get to Earps [*sic*] room. He was easy. Told him I was sorry for him. 'It's hell, isn't it!' he said. His wife was troubled, 'Never mind, I've got one arm left to hug you with'. . . ."[18]

Following the shooting, witnesses watched passively as three men ran

---

[16]Virgil W. Earp, Declaration for Invalid Pension, October 14, 1890; and M. F. Price, M.D., Physician's Affidavit, May 5, 1891, Federal Pension Record No. 954186, National Archives and Records Service.

[17]Parsons, *Private Journal*, December 28, 1881, 282.

[18]Ibid., 283.

down Fifth Street. Parsons noted with contempt, "Cries of 'There they go', 'Head them off' were heard but the cowardly apathetic guardians of the peace were not inclined to risk themselves and the other brave men all more or less armed did nothing."[19] Passing the ice house on the corner of Toughnut, one of the gunmen yelled out to someone, "Lock your door." Moments later, from a safe distance a curious miner watched all three run into the gulch below the Vizina hoisting works off Fourth Street and disappear into the darkness.

Tombstone came alive with rumors rather than hard evidence identifying Earp's assailants. Parsons himself offered three candidates: "It is surmised that Ike Clanton, 'Curly Bill' and McLowry [*sic*] did the shooting."[20] The diarist here is referring to William R. McLaury. That same morning the *Nugget* reported his departure two days earlier for a month-long visit to Fort Worth. This does not, however, absolve the Texas lawyer nor does it provide a completely convincing alibi.

Earlier, in letters to his law partner and family, Will McLaury alluded to the possibility of revenge. A week into the Spicer hearing he wrote S. P. Greene at Fort Worth, "I regard it as my duty to myself and family to see that these brutes do not go unwhipped of justice."[21] The next day he wrote his brother-in-law, attorney David D. Applegate, "in the event they escape by any trick or otherwise, then if you read the papers there will be more 'Press Dispatches.'"[22] The day after Wyatt Earp read his carefully prepared statement into the record, McLaury wrote his sister Margaret, "I do not intend that by perjury these men shall escape."[23]

McLaury had grown so suspicious of local intrigues that he sent his sister a short note, soon after Spicer's decision, asking her not to write, as he distrusted the post office.[24] Clearly he suspected Clum or someone else of reading his mail. Although unrelated to this complaint, the editor was replaced as postmaster on December 15 by actor Fred E. Brooks.

With his brother crippled, Wyatt Earp desperately needed some legal authority. This time he did not simply crave status from an officer's badge. Wyatt wanted some way to avenge the cowardly attack on Virgil.

[19] Ibid. [20] Ibid., December 29, 1881, 283.

[21] W. R. McLaury to S. P. Greene, Fort Worth, Texas, November 8, 1881, Arizona Historical Society.

[22] W. R. McLaury to D. D. Applegate, Toledo, Iowa, November 9, 1881, New-York Historical Society.

[23] W. R. McLaury to Mrs. M. F. Applegate, Toledo, Iowa, November 17, 1881, ibid.

[24] Ibid., December 9, 1881.

At two o'clock the following afternoon he sent a telegram to U.S. Marshal Crawley Dake at Prescott: "Virgil was shot by concealed assassins last night. His wounds are fatal. Telegraph me appointment with power to appoint deputies. Local authorities are doing nothing. The lives of other citizens are threatened."[25] From Phoenix, where Western Union operators intercepted Wyatt's message for the visiting marshal's perusal, Dake telegraphed Earp an appointment as deputy U.S. marshal. He then visited Tombstone and confirmed the position in a face-to-face meeting with his new deputy.

Although serious, Virgil's wounds were not fatal, as Wyatt feared, but they did require drastic measures: "The surgeons extracted the bone from Virgil Earp's arm from the elbow about half the way up to the shoulder. While his wounds are painful, and he will lose the use of his arm, they are not necessarily mortal. With a good constitution and plenty of courage he will surely recover. The shot that penetrated the back struck the vertebra and glanced following around to the front without striking any vital spot."[26] George Parsons confided much of this to his journal: "Longitudinal fracture, so elbow joint had to be taken out today and we've got that and some of the shattered bone in room. Patient doing well."[27]

The Earps had fallen out of favor, and most of them were wise enough to know it. A mounting backlash over the Fremont Street gunfight had already begun destroying their plans. They had mistakenly crossed an invisible line, beyond which lay a frightening new reality, one decidedly outside their power to influence events. Warnings to leave came not only from their friends, wives, and brother James, but also from their father. Nicholas Earp hoped his sons would return to California and leave the violence of Arizona behind them. Wyatt resisted them all and insisted on staying, foolishly convincing himself he could ride out the storm and still be elected sheriff. Wyatt Earp seemed blind to the fact that major elements of the local Republican Party had already deserted him.

On the street, as Virgil suffered in his cramped quarters at the Cos-

[25] *Phoenix Herald,* December 30, 1881.

[26] *Tombstone Epitaph,* December 30, 1881.

[27] Parsons, *Private Journal,* December 29, 1881, 283.

mopolitan, the political opposition to John Carr's candidacy for mayor coalesced around Lewis Blinn under the banner of the Citizens' Ticket. Alongside the lumberman, James Flynn ran for city marshal and A. O. Wallace for recorder. Other candidates included Robert Eccleston for Second Ward councilman and T. A. Atchison in the Third. Atchison was assured victory, as was Heyman Solomon for city treasurer, since he was nominated by both tickets.

The campaign was short and hot; a mud-slinging match from the start, instigated by hard feelings at both newspaper offices. The Republican *Epitaph*, supporting the Democrat Lewis Blinn, accused the *Nugget* of being the "Cow-Boy Organ" and backer of the "Ten-Per-Cent-Ring." The Democratic *Nugget*, supporting the Republican John Carr, retaliated by describing the *Epitaph* as the "Daily Strangler" and backers of the "Bulldozer's Ticket" composed of the "Clum-Earp clique."

The *Epitaph* claimed, "The Daily Cow-boy, which is kept alive by means of a ten-per-cent addition to the burdensome taxes of our people, endorses Carr for mayor and Nagle [*sic*] for chief of police. And why not? If the ten-per-cent ring could handle the city finances, as they do those of the county, there would be just that much more to keep the wolf from the door of the Daily Cow-boy."[28] The *Nugget* had its own issue it felt played well with voters: "The election will to-day decide whether Tombstone is to be dominated for another year by the Earps and their strikers. Every vote against the People's Independent Ticket is a vote in favor of the Earps. Miners, business men, and all others having the welfare of our city at heart should remember this."[29]

With the addition of some truly independent candidates, voters swarmed the polls on January 3, 1882. They elected the fifty-three-year-old blacksmith by a vote of 830 to 298. Carr and all the newly elected councilmen were Republicans. David Neagle, another Republican, easily defeated two challengers for chief of police, getting 590 votes against 434 for Flynn and 103 for Leslie Blackburn. Except for Atchison and Solomon, who ran on both slates, the *Epitaph*-supported Citizens' Ticket took only three offices: A. O. Wallace beating out independent candidate A. J. Felter and three others for city recorder by a 45 vote margin;

[28] *Tombstone Epitaph*, December 28, 1881.

[29] *Tombstone Daily Nugget*, January 3, 1882.

O. O. Trantum defeating two challengers by 394 votes for city attorney; and Dennis McCarty topping the list of five with 267 votes more than his closest rival for assessor.

The *Nugget* celebrated victory under the headline "EXEUNT EARPS! —The Alleged 'Ten Per Cent' Ring Captures the Entire Confectionery Outfit."[30] Despite so many Republican victors, the *Epitaph*'s view was understandably subdued: "The result is not just what we should have been pleased to see, and not what we labored for. . . ." Then, in a statement hardly anyone believed, the editor promised, "we acquiesce in the result without a murmur, and have only the kindliest feelings and best wishes for the officers elect, and will support them in every effort for the promotion of the public good."[31]

A case of wine was delivered to both newspaper offices "compliments of the 'Ten-Per-Cent gang.'" Clum's crew accepted with grace, confessing, "The worst wish that the *Epitaph* has for the officers elect is that their official records may be as bright and free from blemish as the glorious wine that made glad the weary toilers on this morning's paper."[32]

Much of the animosity between the two journals lessened after a new regime took over the *Nugget* on February 3. Its city editor Richard Rule, always a favorite among *Epitaph* men, replaced Harry Woods as chief editor. The *Epitaph*'s staff was further pleased by Mayor Carr's reforms in reducing the cost of city government. He and the council eliminated the 10 percent surcharge on tax collections, despite the paper's claims during the campaign that a vote for Carr was a vote for the "Ten-Per-Cent-Ring." The mayor also cancelled all concealed weapons permits and demanded strict enforcement of Ordinance No. 9. They temporarily sidestepped the issue of Clark, Gray & Company, but by then few expected any action on that front. The *Nugget* may have supported Carr's ticket, but the council turned a deaf ear and awarded the city printing contract to Clum's *Epitaph*.[33]

Keeping tensions high in another arena, two more stage robberies took place in quick succession. W. W. Hubbard & Company's four-horse coach left Tombstone for Bisbee on the morning of January 6. It carried

[30]Ibid., January 4, 1882. [31]*Tombstone Epitaph*, January 4, 1882. [32]Ibid.

[33]Minute Book B, Common Council, Village of Tombstone, February 23, 1882, 24.

four passengers plus a $6,500 payroll, mostly coin for the Copper Queen Mine, and traveled without incident along the San Pedro. Five men were seen above the Clanton ranch riding through the hills. At mid-afternoon the stage was attacked some eight miles from its destination. Wells Fargo messenger Charles A. Bartholomew resisted. The driver tried falling back toward Hereford, amidst a sporadic running gun battle, but in the end surrendered the treasure box after one of the horses was hit.

Then two days later, around one o'clock in the morning, Sandy Bob's stage from Benson was robbed at the arroyo about halfway between Contention City and Tombstone. Nine passengers were relieved of small amounts of cash and three pistols. All were taken by surprise. Even so the two highwaymen missed nearly $1,500 in hidden funds. One of the passengers turned out to be James B. Hume, Wells Fargo's chief of detectives, who surrendered two prized revolvers. Marshall Williams offered $300 for the capture of the outlaws even though the stage carried no Wells Fargo consignments. An embarrassed Jim Hume posted a $500 reward plus one-fourth of any booty recovered from the Bisbee holdup. At San Francisco the company raised the initial amount to $2,500. Reacting to losses, Wells Fargo temporarily closed its Bisbee office. Hume went on to confer with Wyatt Earp, still closely tied to the company.

Despite Hume's interest, the Earp brothers had lost what was left of their political support with John Carr's election as mayor. They then surrendered some of their economic base as well. Wyatt's partnership with Lou Rickabaugh in the Oriental's gaming tables had been profitable. The so-called Earp crowd used those premises as their headquarters. On the evening before the voting they took delight in heckling critic James Reilly. Yet Earp, strapped for cash by early January, sold his interest to Rickabaugh, who in turn sold out to Milt Joyce, "one of the best managers of the saloon business ever in Tombstone. . . ."[34]

Within a fortnight both Rickabaugh and Doc Holliday faced more serious and direct confrontations. As George Parsons described both incidents:

[34] *Tombstone Epitaph,* January 10, 1882.

> Much blood in the air this afternoon. Ringo and Doc Holliday came nearly having it with pistols and Ben Maynard and Rickabaugh later tried to kick each others [*sic*] lungs out. Bad time expected with the Cow boy leader and D. H. I passed both not knowing blood was up. One with hand in breast pocket and the other probably ready. Earps just beyond. Crowded street and looked like another battle. Police vigilant for once and both disarmed.[35]

Wyatt remained obsessed with capturing those responsible for attacking Virgil. Using his new position as a federal officer, he pressured authorities for warrants. Mayor Carr issued a proclamation on January 24: "I am informed by His Honor, William H. Stilwell, Judge of the District Court of the First Judicial District, Cochise county, that Wyatt S. Earp, who left the city yesterday with a posse, was intrusted with warrants for the arrest of divers persons charged with criminal offenses. I request the public within this city to abstain from any interference with the execution of said warrants."[36] Carr's decision greatly disappointed his recent supporters at the *Nugget,* who buried his proclamation among other legal notices and advertisements. In contrast, the *Epitaph* gave it prominent display as a column lead.

With Carr's proclamation appearing so soon after the election, the *Nugget* tried shifting blame from the mayor's shoulders onto Judge Stilwell: "The cause of the intense local feeling that exists here is that a large majority of our citizens believe that the recent killing on our streets was a murder in reality, but done under the cover of a city marshal's authority. To again place such power, in a slightly different form, in the hands of the perpetrators of the former act, is an outrage upon the public that could only be committed by a stupid or vicious magistrate."[37]

But the judge was no fool, nor was he easily intimidated; not by a Tombstone editor, the Earp faction, or the Cowboys, as John Ringo would soon discover. Although only in his early thirties, Stilwell had been associated in an Albany, New York, law firm with Rufus W. Peckham (later a justice on the United States Supreme Court). The president then appointed Stilwell an associate justice of the Arizona Territorial Supreme Court in time for the 1881 term. His new duties led him to

[35]Parsons, *Private Journal,* January 17, 1882, 286.

[36]*Tombstone Epitaph,* January 24, 1882; and *Tombstone Daily Nugget,* January 28, 1882.

[37]*Tombstone Daily Nugget,* January 26, 1882.

preside over the First Judicial District. It did not take long for citizens of southern Arizona to appreciate their new jurist.[38]

Wyatt Earp left Tombstone with a posse numbering nearly ten men, including his brother Morgan and Doc Holliday. Two days later he was briefly reinforced by over two dozen more, miners and other volunteers disgusted by the seeming inability of Behan's deputies to corral wrong-doers, be they the persistent stage robbers or those responsible for a recent stock raid against Helm's ranch. In that case county authorities blamed Apaches. Others suspected the outlaw gangs.

Two groups of Cowboys, one rumored under the leadership of John Ringo, followed the Earp party but did not attack. So many heavily armed men were in the saddle they exhausted harnessmaker John J. Patton's supply of rifle scabbards. During a short but controversial visit to Charleston—looking "like a deserted village and as though having undergone a siege," in Parsons's view[39]—Earp conferred with Charles Bartholomew and five others searching for the Wells Fargo raiders.

Residents of Charleston were not amused by this invasion, one of them sending an unsigned telegram to Sheriff Behan:

> Dear Sir: Doc Holliday, the Earps and about forty or fifty more of the filth of Tombstone, are here armed with Winchester rifles and revolvers, and patrolling our streets, as we believe, for no good purpose. Last night and today they have been stopping good, peaceable citizens on all the roads leading to our town, nearly paralyzing the business of our place. We know of no authority under which they are acting. Some of them, we have reason to believe, are thieves, robbers and murderers. Please come here and take them where they belong.[40]

Despite the seriousness of the complaint, Behan decided to do nothing. Against this reluctance, Earp's posse quietly slipped out of sight. No one seemed to know their whereabouts. Rumors of impending bat-

---

[38]Wagoner, *Arizona Territory,* 498, 504. Wagoner mistakenly has Stilwell presiding over the Third District. Both Pima and Cochise counties during this period fell into the First Judicial District. See Murphy, *Laws, Courts, and Lawyers,* 20. For more on Judge Stilwell, see McClintock, *Arizona,* 3:574-576. This volume claims the judge was born in 1853, but on the 1886 Cochise County Great Register he gave his age as thirty-six, and on the 1900 census said he was born May 1848.

[39]Parsons, *Private Journal,* January 30, 1882, 287.

[40]Telegram to J. H. Behan, Sheriff of Cochise County from Charleston, A. T., January 26, 1882, *Tombstone Daily Nugget,* January 27, 1882.

tles between peace officers and Cowboys dominated the conversation of those arguing the merits of the expedition against the factor of personal revenge.

Causing more ill feeling against the Earps, T. J. Cornelison was finally released from the county jail without trial, after the district attorney decided there was "no evidence to warrant a continuance of the case."[41] After languishing behind bars all those many months, the disgruntled former policeman now charged that Virgil Earp had told him he could force a conviction for rifling the woman's trunk, but would hush up the matter if Cornelison paid him $100. Nothing came of the accusation, but all the born skeptics believed it just the same.

Meanwhile, those hunted by Wyatt Earp and his followers feared capture. Worrying about falling into the hands of that band of angered and determined men, Ike Clanton and his brother Phineas rode into Tombstone under the protection of Charlie Bartholomew's posse—guided to the rendezvous by Peter Spencer. They surrendered to Sheriff Behan around 2:30 in the morning of January 30, prepared, they claimed, to answer the charge of involvement in the assault on Virgil. Earlier a cousin of the Clantons did the same, only to discover no warrant had been issued for his arrest.

Not so John Ringo. He had been released on bail for just one of his two robbery indictments. After authorities discovered their error, Judge Stilwell issued a bench warrant for his arrest. None of the area's lawmen made any effort to do so until January 20, even though Ringo made several Tombstone appearances. Then, after Earp's posse left town, and without any authority from the court, Thomas Moses, John O. Dunbar, and A. T. Jones posted bail. Ringo was released a second time. Incensed, Judge Stilwell revoked the so-called bond and again ordered Ringo's arrest. In the early morning of January 24 he was finally apprehended at Charleston where he had gone in pursuit of the Earps. It proved "quite an experience but no shooting," according to George Parsons.[42] Back in Behan's lockup, Ringo languished nearly two weeks before a lawful bond could be arranged to the judge's satisfaction.

On the day the two Clantons surrendered, Sheriff Behan was arrested

[41] *Tombstone Daily Nugget,* January 28, 1882.

[42] Parsons, *Private Journal,* January 25, 1882, 287.

on a complaint filed by Earp partisan Sylvester B. Comstock, charging perjury on several bills submitted to the board of supervisors. Behan, accusing Judge Stilwell of bias, requested a change of venue. The judge denied the motion, but since no statutory violation could be found covering Behan's supposedly errant behavior, he dismissed the case. Stilwell then chastised the prosecution for wasting the court's time with frivolous charges. Johnny's supporters in the crowded courtroom cheered. The *Epitaph* complained about Behan's release on a technicality, but few took the suit seriously.

In retaliation for the sheriff's discomfort, a warrant was issued for Earp posse member Sherman McMasters, the same man Virgil had tried arresting on Safford Street some months earlier. McMasters, who had sidestepped those difficulties, was a small man physically. Born in Wyatt Earp's home state of Illinois twenty-eight years before, he was the son of a gardener; one successful enough, however, to employ an Italian servant.[43] City Marshal David Neagle, still working concurrently as a deputy sheriff, served the warrant at Earp's camp near Pick 'Em Up, the site of a small hotel, corral, and saloon on the Charleston road. McMasters submitted without incident. The Earps, Doc Holliday, and the rest of the posse returned to Tombstone and posted bond in Judge Felter's court.

Judge Stilwell then granted bail to Ike and Phin Clanton at $1,500 each. For their own protection they remained in Sheriff Behan's custody. Of the impending hearing, the *Nugget* remarked offhandedly, "The Earps have had their examination and were acquitted of criminal intent, and we say, let the matter drop. So, too, let the cowboys accused of crime have their trial, and let all abide by the result."[44]

Stilwell heard the case on the evening of February 2. Sherman McMasters testified seeing Clanton at Charleston and asking him about the shooting. McMasters claimed Ike replied he "would have to go back and do the job over."[45] Half a dozen others, however, including J. B. Ayers, a Wells Fargo operative working undercover as a saloonman, all placed Clanton in Charleston at the time of the attack on Virgil Earp.

[43]1870 United States Census, Fourth Ward, City of Rock Island, Rock Island County, Illinois, Sheet 7, Lines 1-7; and *Past and Present of Rock Island*, 294.

[44]*Tombstone Daily Nugget*, January 28, 1882.

[45]"The Clanton Trial," *Tombstone Epitaph*, February 3, 1882.

Considering the less-than-convincing evidence, Stilwell discharged the prisoners.

Earlier, leading Republicans had crowded the courtroom of A. O. Wallace, chairman of the executive committee of the Tombstone Republican Club (John Clum served as vice chairman). They came to meet with United States Marshal Crawley P. Dake, "to take steps toward harmonizing the Republican Party in view of present complications." Essentially they wanted Dake to appoint new deputies. As city council clerk B. C. Quigley put it, "men who were not allied with either faction of the parties who are now distracting our community."[46] Speaker after speaker echoed Quigley's concerns. In the end they formed a five-man committee, headed by Mayor Carr, to suggest suitable candidates for Dake's consideration. Clearly these Republicans wanted persons other than the Earps to serve as local deputy U.S. marshals. To the party elite the brothers had become an embarrassing liability.

Sensing the absence of a solid political base and fearing the disgrace of dismissal, Virgil and Wyatt handed Marshal Dake their joint letter of resignation, penned by lawyer Thomas Fitch, on February 1:

> And while we have a deep sense of obligation to many of the citizens for their hearty cooperation in aiding us to suppress lawlessness, and their faith in our honesty of purpose, we realize that . . . there has arisen so much harsh criticism in relation to our operations, and such a persistent effort having been made to misrepresent and misinterpret our acts, we are led to the conclusion that . . . it is our duty to place our resignations as deputy United States marshals in your hands, which we now do, thanking you for your continued courtesy and confidence in our integrity, and shall remain subject to your orders in the performance of any duties which may be assigned to us, only until our successors are appointed.[47]

Marshal Dake may have listened quietly to all the concerns voiced during the Republican meeting at Wallace's, and was even considering their choices of Silas Bryant and John H. Jackson for appointments as local deputies (Dake stayed with Jackson—already carrying a federal badge while working on the side for mining tycoon George Hearst—and added the services of Willcox deputy A. F. Burke), but he refused to

[46] *Tombstone Epitaph,* February 1, 1882.

[47] Virgil W. Earp and Wyatt S. Earp to Major C. P. Dake, United States Marshal, Grand Hotel, Tombstone, February 1, 1882, ibid., February 2, 1882.

accept the Earps' resignations. Dake had no serious objections to their performance thus far. Indeed, as he described the unfolding events to his superiors: "The Earps have rid Tombstone and neighborhood of the presence of this outlaw element. They killed several Cow boys in Tombstone recently and . . . were vindicated and publicly complimented for their bravery. . . ."[48]

George Parsons seemed to share Dake's confidence: "The Earps are out too on U S business and lively times are anticipated. . . . at last the National Government is taking a hand in the matter of our trouble and by private information I know that no money nor trouble will be spared to cower the lawless element. Our salvation I think is near at hand. It looks like business now when the U S Marshal Dake takes a hand under special orders."[49]

Actually no money came from government sources. Disgusted with official inaction, the Citizens' Safety Committee collected a separate fund, for which no accounting has been found. Marshal Dake, according to Wells Fargo confidential detective John N. Thacker, independently asked the company's superintendent at San Francisco for $3,000 as he "was powerless to do good as U.S. Marshal for want of funds. . . ." That money, after a $15 transaction discount, was deposited in Tombstone with Hudson & Company. Thacker told Wyatt it was for his use as an officer. Earp insisted Dake withdrew $340 for himself after getting drunk on wine, but that he spent the rest on his posse and "never received a dollar for his services as deputy marshal." Yet records suggest Wyatt only withdrew $536.65 from that account.[50]

Posseman O. C. "Charlie" Smith reported that those riding with Earp served a total of thirty days at $5 a day. Each was supplied with a horse, bridle, saddle, carbine, six-shooter, and rations. Dake, who kept poor records and illegally commingled department moneys with his own, later claimed he advanced upwards of $12,000 to suppress the Cowboy raids. Smith disagreed, saying he thought Wyatt Earp got no more than $500 from Crawley Dake. This prolonged controversy over cash totals and

[48]U.S. Marshal C. P. Dake to Hon. S. F. Phillips, Acting Attorney General, Washington, December 3, 1881, General Records Department of Justice, Record Group 60, National Archives and Records Service.

[49]Parsons, *Private Journal,* January 25, 1882, 287.

[50]Leigh Chalmers, Examiner Dept. of Justice to Attorney General A. H. Garland, August 13 and September 3, 1885, General Records Department of Justice, Record Group 60, National Archives and Records Service.

disbursement rates from the various funds was never resolved to the government's satisfaction.

Dake later tried explaining that by then he had lost track of Wyatt's whereabouts. An examiner from the Department of Justice dismissed that argument, pointing out: "His [Earp's] Father and Brother live in Colton [California] and are in constant communication with him at his home in Aspen Colorado and if Mr. Dake or his bondsmen had used that care which prudent men use . . . they might long since have obtained any vouchers which Wyatt Earp could give."[51]

As events heated up in Tombstone, Lou Rickabaugh pulled out and returned temporarily to San Francisco. Resigning from Wells Fargo, Marshall Williams decided to leave early on the morning of February 3. He was soon reported in Pennsylvania—where a false rumor told of his arrest—before showing up in Brooklyn. All traces of his Tombstone years were quickly swept aside. First, Wells Fargo's new agent, J. M. Seibert, moved the office up Allen Street into Tritle and Murray's old brokerage building. Then, the Baron brothers fitted up Williams's tobacco and stationery store as the town's most luxurious barber shop and bath house. Despite gossip swirling around the former agent's business practices, an examination of his accounts showed no serious improprieties. As for personal debts, Williams had arranged for them to be paid from the liquidation of his Arizona assets.

With some of their friends skipping out, Wyatt, Morgan Earp, and Doc Holliday found themselves rearrested on February 10. A warrant had been issued the day before by Justice of the Peace J. B. Smith at Contention City. Although named in the indictment, Virgil was not arrested on account of his wounds. Ike Clanton had refiled charges against them all for murdering his brother and the two McLaurys.[52]

For their defense they hired William Herring, a former assistant district attorney in New York City and until recently manager of the Neptune Mining Company near Hereford. He had been in private practice at Tombstone for just over a month. That evening he filed a hastily drawn writ of habeas corpus with Court Commissioner T. J. Drum. By

[51]Leigh Chalmers to Attorney General A. H. Garland, August 13, 1885, ibid.

[52]Territory of Arizona vs. J. H. Holliday, Wyatt Earp, Morgan Earp and Virgil Earp, Defendants, Arrest Warrant, J. B. Smith, Justice of the Peace, Cochise County, February 9, 1882, Hayhurst typescript, 5, 10.

morning Drum declined to hear the matter since he had served as Doc Holliday's lawyer at the Spicer hearing. With Judge Stilwell out of town the petition went next to Probate Judge J. H. Lucas for action on February 13. Lucas, from his office at city hall, questioned habeas corpus and said he was sorry for the hardship it caused but felt he had no choice but to remand the accused into Sheriff Behan's custody.[53]

Escorted by friends "armed to the teeth," according to George Parsons,[54] the two Earps and Holliday rode to Contention City and appeared in Smith's court. He, in turn, sent the case back to Tombstone, much to the relief of worried citizens at Contention. One of the prosecutors, Briggs Goodrich, grew so disgusted with the farce—seeing it all as another example of bending the law for personal revenge—that he withdrew, leaving his older brother Benjamin, J. S. Robinson, and District Attorney Lyttleton Price to carry on without him.

Justice Smith came up to hear the case in the county courtroom on the morning of February 15. That afternoon attorney Herring filed a second application for habeas corpus with Judge Lucas.[55] This time he freed the prisoners,[56] citing a provision of territorial law giving discretion to magistrates on charges "defectively or unsubstantially set forth."[57] Arguments before Justice Smith danced around Wells Spicer's earlier verdict and an affidavit from Sylvester B. Comstock showing that the grand jury had indeed considered the case but failed to find against the defendants. Herring moved for a dismissal, but Smith took the matter under advisement before following Lucas's example and dropping the charges.

Free again, Wyatt and Morgan Earp, Doc Holliday, Sherman McMasters, Turkey Creek Jack Johnson, and two or three others rode out of Tombstone on February 17. Posse member Dan Tipton had been

---

[53]Petition of Writ of Habeas Corpus, Wyatt Earp, Morgan Earp, and J. H. Holliday to J. H. Lucas, Probate Judge, Cochise County, February 11, 1882, ibid., 11-13; and *Tombstone Daily Nugget* and *Tombstone Epitaph*, February 14, 1882.

[54]Parsons, *Private Journal*, February 15, 1882, 290.

[55]Petition for Writ of Habeas Corpus, Wyatt S. Earp, John H. Holliday and Morgan Earp, Cochise County, February 15, 1882, Hayhurst typescript, 15, 16.

[56]Ex Parte J. H. Holliday, Wyatt Earp and Morgan Earp, Habeas Corpus, no date; Order Discharging Relators from Custody, J. H. Lucas, Probate Judge, Cochise County, no date, Hayhurst typescript, 6-8, 18; and *Tombstone Daily Nugget* and *Tombstone Epitaph*, February 16, 1882.

[57]Hoyt, *Compiled Laws of the Territory of Arizona*, (2221) Sec. 22, 377.

arrested and fined $30 for carrying concealed weapons. As Parsons explained, "Policeman just prevented Ben Maynard and Tipton from shooting one another."[58] It was all in retaliation for Maynard's earlier attack on Lou Rickabaugh.

At Watervale Earp's posse split up, four heading in the direction of the San Simon Valley to try, it was thought, arresting Pony Diehl, wanted in connection with the recent Bisbee stage holdup. Others rode toward Charleston. Citizens pondered the chances of further bloodshed. As Parsons pointed out at the time of the Tipton-Maynard confrontation: "A bad time is expected again in town at any time. Earps on one side of the street with their friends and Ike Clanton and Ringo with theirs on the other side—watching each other. Blood will surely come. Hope no innocents will be killed."[59] Without finding any Cowboys, Earp's posse returned to Tombstone on February 24. By now they mostly hung around the Cosmopolitan to be nearer Virgil. Even Wyatt's wife finally moved there, registering as "Mattie Earp, Cedar Rapids, Iowa."

With Wyatt's return events seemed to cool off temporarily. Then, as if to prove this offered no trend for the future, Frank Leslie argued with James Floyd at the Oriental. The man on night watch tossed both offenders into the street where the disturbance continued. Finally Buckskin Frank pulled out his revolver and beat Floyd about the head. The gun accidently discharged, injuring no one, but Leslie was arrested. Later at the justice court Floyd withdrew his complaint, and Judge Wallace dismissed the case.

During this somewhat quiet period curious bystanders watched as miners trickled in from Virginia City and other towns looking for work. There was none. Richard Rule tried to warn them about conditions in "this already over-populated camp" by writing for the sake of his West Coast exchanges: "A large number of men are now idle in this camp, and every day dozens of such can be found about the leading mines vainly seeking work. This is an expensive place to reach, costly to live in and none too pleasant for men out of work and 'busted.'"[60] Those who did come were forced to seek employment at mines opening in Sonora.

[58]Parsons, *Private Journal,* February 15, 1882, 290.

[59]Ibid.

[60]*Tombstone Daily Nugget,* March 4, 1882.

The only promising jobs for laboring men locally were as ditch diggers and pipe layers with the Huachuca Water Company. Still, only fifty were hired that March. Despite shortages of wagons and mules, teamsters remained the one class of workers still in demand. Even after the railroad extension between Benson and Contention City finally opening on February 4, movement of supplies to widely scattered mining camps and the hauling of ore from mines to mills still required more teamsters and rigs than those available. Contention witnessed a mild railroad boom. To take advantage of these developments, John Dunbar and Sheriff Behan opened a livery stable across from the depot. In time, because the railroad paralleled the river's west bank, much of Contention City moved with it.

In Tombstone events seemed stalemated. The Earps had failed in their attempts to have anyone convicted for the attack on Virgil, and the Cowboys proved equally inept at having the Earps and Holliday again answer charges of murdering young Billy Clanton and the two McLaurys.

On March 18, 1882, Wyatt Earp approached Briggs Goodrich: "I think they were after us last night. Do you know anything about it?" The lawyer replied he had no specific information on that, but when Wyatt asked if he thought there was any immediate cause for alarm, he admitted they "were liable to get it in the neck anytime." Goodrich had seen men that aroused his suspicion: "The presence of these strangers, added to the fact that Frank Stilwell seemed to expect that there would be a fight, led me to make the remark to Wyatt Earp that I did."

Goodrich also passed along a message from John Ringo, who "wanted me to say to you, that if any fight came up between you all, that he wanted you to understand that he would have nothing to do with it. . . . he was going to take care of himself & everybody else could do the same." Later, as they walked up Allen Street around sundown, Goodrich and Ringo saw two heavily armed men pass by. Goodrich thought they were members of Earp's posse taking more guns down to the Cosmopolitan in preparation for a fight, but Ringo shook his head: "I don't think it's any of the Earp party." Seeing Morgan and Doc Holliday later at Schieffelin Hall, the worried lawyer told them what he and Ringo had seen, warning, "You fellows will catch it tonight if you don't look out."[61]

---

[61]Deposition of Briggs Goodrich, Coroner's Inquest, Morgan Earp, Document No. 68, District Court of the First Judicial District, County of Cochise; filed March 22, 1882. (Hereafter cited as: "Coroner's Inquest, Document No. 68.")

Morgan Earp, Holliday, Dan Tipton, and Goodrich were all at Schieffelin Hall for the Saturday night opening of William Horace Lingard and Company's *Stolen Kisses,* a play its promoters advertised as "the Great Comedy of the Age"—guaranteed to supply theatergoers with "2½ Hours of Incessant Laughter."[62] Thinking the theater as safe a place as any, the men willingly paid their dollar for general admission seating.

After the performance Holliday went home. Earp and Tipton walked down Fourth Street. Turning the corner at Allen, Morgan almost collided with his sister-in-law, out to get Virgil some peppermint-flavored taffy. He asked, "How's Virge?" Allie, who always liked the easygoing Morgan, replied that he was coming along. Actually he had taken a brief walk, outside for the first time, just three days before. Morgan's next words proved sadly prophetic: "Wish he'd get better. I'd like to get away from here. Tonight!"[63] Earp escorted his sister-in-law back to the Cosmopolitan before going on to Campbell and Hatch's saloon and billiard parlor, a few doors farther up Allen Street next to Williams's old Wells Fargo office.

At the front door Morgan and Dan Tipton ran into the co-proprietor, forty-year-old Robert S. Hatch, who had also just returned from Schieffelin Hall. Earp casually asked him for a game of pool. Morgan liked the fun-loving saloonman, an amateur thespian who kept a large jar filled with frogs on the premises, claiming the changing noise level helped him predict the weather. The two friends chose an unoccupied table about ten feet from the back door. Tipton grabbed a chair to watch the game. Wyatt Earp and Sherman McMasters sat talking nearby.

The top section of the saloon's rear door held four panes of glass. The bottom two were painted over, but the others offered anyone outside a clear view into the back of the place. This door opened onto two intersecting passageways. One ran completely through the block, starting as a narrow strip along the west side of the saloon and ending between A. D. Otis & Company and the Melrose Restaurant on Fremont Street. Another crossed most of the block from west to east, beginning at Spangenberg's and meandering along the back of those buildings fronting Allen Street to the south side of Henry Hasselgren's

[62] *Tombstone Epitaph,* March 17, 1882; and Willson, *Mimes and Miners,* 136.

[63] Waters, *The Earp Brothers of Tombstone,* 192.

print shop for the *Weekly Commercial Advertiser*—formerly A. O. Wallace's San Diego Keg House—and the back of Wehrfritz and Tribolet's saloon. It ended there and did not extent to Fifth.

It was nearing eleven o'clock in the evening. The two men finished one game and started another. Bob Hatch, his back to the door, leaned forward to take his turn. Morgan Earp stood a foot or more off to the right watching the play. Before the ball was struck, Hatch explained:

> There were two gun or pistol shots, almost simultaneously, at first did not know from what direction they came from. Thought they came from the front. I got out of the range of the door, with my back against the back wall of the building, and just about that time I saw Earp fall. In about 8 or 10 seconds, I passed through the cardroom and out into the back yard but could see no one as it was very dark. I came back into the saloon & [saw] Earp lying on the floor near the cardroom door, he having been moved by his brother & some others from where he fell. . . .[64]

Wyatt, Sherman McMasters, and Dan Tipton had dragged Morgan out of the line of fire. Tipton recalled, "We pulled up his shirt, found he was shot through the body."[65] Hearing gunfire, Dr. William Miller rushed to the scene, followed closely by Drs. Goodfellow and Henry Matthews. Goodfellow later described Morgan's condition: "He was in a state of collapse resulting from a gunshot or pistol wound, entering the body just to the left of the spinal column in the region of the left kidney emerging on the right side of the body in the region of the gall bladder. It certainly injured the great vessels of the body causing hemorrhage. . . . It also injured the spinal column. It passed through the left kidney and also through the liver."[66] A second bullet struck a fifty-one-year-old mining man named George A. B. Berry, warming himself and talking with three friends by the front stove. Berry suffered an ugly wound in the thigh but survived.

After the doctors finished their examination, Morgan was moved into the cardroom and placed on a lounge. As the men pulled him to his feet, Earp pleaded, "Don't, I can't stand it." Made as comfortable as possible, he turned to Bob Hatch: "I have played my last game of pool."[67] He lay

[64]Deposition of Robert S. Hatch, Coroner's Inquest, Document No. 68.

[65]Deposition of D. G. Tipton, ibid.

[66]Deposition of Dr. G. E. Goodfellow, ibid.

[67]Deposition of Robert S. Hatch, ibid.

surrounded by his brothers Wyatt, James, Warren, and Virgil, who upon hearing the gunfire and shouts from bystanders struggled out of bed to be with his dying brother. Allie and James's wife, Bessie, stood by with some of the wounded man's close friends. After a long while Morgan finally opened his eyes. Looking up, Allie remembered, he remarked in a whisper, "It won't be long. Are my legs stretched out straight, and my boots off?"[68] Everyone nodded in reply. Then, thirty to forty-five minutes after Dr. Goodfellow's examination, Morgan Earp died. He was just weeks short of thirty-one.

Of the scene, it was said: "At the front door of the saloon stood a hound, raised by the brothers, who, with that instinct peculiar to animals, seemed to know that his master had been stuck down, and despite . . . threats or entreaties remained whining and moaning, and when the body was taken to the hotel, no sadder heart followed than that of the faithful dog."[69]

Suspicion quickly implicated five men, a point strengthened by testimony at the coroner's inquest from Peter Spencer's wife. (Although the transcript shows "Spence," she signed the document Maria Duarte Spencer.) She linked her husband to Morgan Earp's murder, along with Frank Stilwell, a German teamster she called Freis—who was actually a man named Frederick Bode—and an Indian she identified only as "Charley." Mrs. Spencer described these heavily armed men hanging around her house. Spencer lived in a small "T"-shaped wood-frame on the southeast corner of First and Fremont. Virgil Earp's house stood across First Street. Wyatt had been residing for some time directly north on the other side of Fremont. From Spencer's place the Cowboys observed the comings and goings of the Earps and their allies.

According to Mrs. Spencer: "Four days ago, while mother and myself were standing at Spence's house, talking with Spence & the Indian who came home with him, Morgan Earp passed by, when Spence nudged the Indian and said, 'That's him, that's him.' The Indian then started down the street, & got ahead of him to get a good look at him (Earp)." Afterward, Maria and her husband quarreled violently, but she still found the courage to come forward and testify: "Spence didn't tell me so, but I know he

---

[68]Waters, *The Earp Brothers of Tombstone,* 193.

[69]*Tombstone Daily Nugget,* as quoted in *Weekly Arizona Citizen,* March 26, 1882.

killed Morgan Earp." She and her mother heard the shots, and after Spencer and his friends excitedly rushed back inside, she "judged they had been doing wrong from the condition, white and trembling, in which they arrived. . . . and after hearing the next morning of Earp's death, I came to the conclusion that Spence and the others have done the deed."[70]

After hearing all the testimony a coroner's jury concluded Morgan Earp came to his death from "the effects of a gunshot, or pistol wound, on the night of March 18, 1882, by Peter Spence, Frank Stillwell [*sic*], one Jno Doe Freeze [*sic*], an Indian called Charley and another Indian, name unknown."[71]

With the crippling of Virgil and now the killing of Morgan, even Wyatt finally realized their days in Tombstone were over. As Andrew Ritter prepared Morgan's body for shipment to his parent's home at Colton, California, Virgil and his wife began packing their belongings for the anticipated twelve-hour train ride. Wives of Wyatt and James Earp, too, hoped to be leaving within days.

With Morgan's body sealed in its casket, the funeral cortege started from the Cosmopolitan Hotel toward rail connections at Contention City. Heavily guarded by Wyatt and Warren Earp, together with two or three close friends, the solemn procession moved slowly down Allen Street at about 12:30 on the afternoon of March 19. As they departed, a bell at the fire company rang out the notes for "Earth to Earth, Dust to Dust." In all it was a rather sad and grim spectacle. Riding only as far as Contention, the men placed Morgan's coffin aboard the train and returned to Tombstone.

The next day Wyatt and Warren, along with Doc Holliday, Sherman McMasters, and Turkey Creek Jack Johnson, escorted Virgil and his wife over the same route. Contractor John Hanlon saw them approaching Contention City and recalled, "Some were on horseback and some were in a buggy." Hanlon noticed that all "had rifles and shotguns except Virgil Earp who had a pistol."[72]

Unlike the day before, all of them now boarded the train for Benson and points west. Nathan W. Waite, later a Bisbee saloonkeeper, also got

---

[70]Deposition of Maria Duarte Spencer, Coroner's Inquest, Document No. 68.

[71]Coroner's Certification, Dr. H. M. Matthews, ibid.

[72]Statement of John Hanlon, "The Stilwell Inquest," *Weekly Arizona Citizen*, April 2, 1882.

on at Contention. McMasters told him they had originally planned to go no farther than Benson, but hearing that Frank Stilwell had traveled on ahead, Wyatt changed his mind and decided to see Virgil and Allie safely off at Tucson. Waite recalled: "On the train McMasters particularly inquired as to the arrival and departure of trains at the Tucson depot. All had guns and McMasters had two belts of cartridges."[73]

At Tucson J. W. Evans, the one-armed deputy U.S. marshal, watched Doc Holliday step from the train carrying two shotguns. He stored them temporarily in the railroad office. McMasters would retrieve them later. For dinner the party walked to Asa Porter's hotel, fronting the tracks just northwest of the depot. Afterward Wyatt spoke with Evans outside the dining room. As they talked Virgil and his wife came out. Shaking hands with the deputy marshal, Virgil said goodbye. Wyatt helped his brother and sister-in-law to the back of the sleeping car before joining the others guarding the coach from the station platform.

David Gibson, at the depot with baggage claims, watched the Earps walk down from Porter's Hotel. The railway newsboy had already told him, "I guess there will be hell here to-night." Gibson said the boy explained, "the Earps and Holliday were aboard and were going to stop here as they had told him that the man who killed Morgan Earp was in Tucson."[74]

Gibson was not the only one who saw the Earps leave Porter's. So did Frank Stilwell and Ike Clanton. Stilwell was in town scheduled to appear before a federal grand jury on charges stemming from the September 1881 robbery of the Bisbee stage. Clanton had been in Tucson for the Jerry Barton case since checking into the Palace Hotel on March 6. He came to the depot that evening to meet a witness from Charleston—former Shibell deputy Milton McDowell. With the appearance of the Earps and their friends, Ike quickly withdrew: "I then started towards my room, and [Stilwell] walked down the track between the cars and hotel."[75]

But the Earp party had already seen him and were coming up fast as the cars began to move. R. E. Mellis, one of the engineers on the outgo-

[73]Statement of Nathan W. Waite, ibid.

[74]Statement of David Gibson, ibid.

[75]Statement of Ike Clanton, ibid.

ing train, remembered, "while on the lookout for tramps, saw a man cross in front of the engine, and shortly afterwards four armed men walked down on the west side of the train to where the man was."[76] It was now around 7:15 that Monday evening, March 20, 1882.

Catching his man alongside the tracks a hundred yards or more from Porter's Hotel, Wyatt Earp pushed his double-barreled shotgun close to Stilwell, who begged for mercy while trying to push the gun away. Although armed, Behan's ex-deputy made no move to pull his six-shooter. Wyatt calmly dismissed Stilwell's frantic pleas and pressed one of the triggers of his ten-gauge. Frank screamed. More shots followed. S. A. Bateman, the other engineer, "heard some cheering in the direction in which the shots were fired."[77] Some railroad workers approached rather tentatively, but Earp's posse warned them off.

Hearing the shooting, many assumed it came from those celebrating the arrival that evening of gas lighting to Tucson's streets. Deputy U.S. Marshal Evans heard it and returned to the depot, only to be told that a group of Mexicans had fired the shots. No one seemed anxious to find out what had actually happened. Stilwell's bullet-riddled body lay unattended until a Southern Pacific trackman found it at daybreak. As a final insult, someone had stolen the dead man's watch during the night, leaving behind only a broken vest chain.

Dr. Dexter Lyford examined Stilwell's body at E. J. Smith's undertaking parlor. He described the wounds as coming from a heavy charge of buckshot, fired at close range, shredding the liver and abdomen. Another shotgun blast shattered Stilwell's left leg. A single shot, from a rifle or revolver, had passed through the body from armpit to armpit, going through the upper portion of both lungs, the gunpowder burning his coat. There was also a rifle wound in Stilwell's upper left arm and another in his right thigh. "Frank Stilwell was shot all over, the worst shot-up man that I ever saw," wrote saloonman and later courthouse janitor George Hand.[78] Stilwell's mangled body was hauled to the Tucson cemetery with only one unidentified mourner following behind.

After this killing the Earp party left Tucson on foot. At Papago Sta-

---

[76]Statement of R. E. Mellis, ibid.

[77]Statement of S. A. Bateman, ibid.

[78]Miscellaneous and undated notation relating to early episodes of violence, Papers and Diaries of George O. Hand, 1875-1887, Manuscript Collections, Arizona Historical Society, Tucson.

tion they flagged down an eastbound freight and rode to Benson on their way back to Tombstone. Sheriff Paul immediately wired Behan to arrest Wyatt and Warren Earp, Doc Holliday, Sherman McMasters, and someone called Johnson (Jack's first name then unknown to Pima County authorities). Tucson officials, in a special dispatch making a veiled reference to the Spicer hearing, declared, "if the parties are apprehended there will be no sham examination, but a trial on merits, and the guilty parties, whoever they may be, will suffer the penalty of the law."[79] At Tombstone, diarist George Parsons took a different view: "A quick vengeance and a bad character sent to hell where he will be the chief attraction until a few more accompany him."[80]

Pima County prosecutors stood prepared to move ahead, but before they could seriously consider a trial, on its merits or not, the wanted men needed to be captured. Late in the afternoon of March 21 the Earp party returned to Tombstone and gathered more belongings from those rooms occupied at the Cosmopolitan during Virgil's long convalescence. Around eight o'clock that evening, as they passed through the hotel office, they met the sheriff. The *Epitaph* reported their brief exchange: "Sheriff Behan, who was standing near the door, spoke to Wyatt, saying, as reported by the bystanders, 'Wyatt, I want to see you.' Wyatt replied, 'You can't see me; you have seen me once too often,' or words to that effect."[81]

The heavily armed men brushed past Behan and made their way down Allen Street, retrieving their horses from P. W. Smith's Corral at the southeast corner of Third. They spent the night camped near Watervale. The sheriff and chief of police did nothing to stop them. Later Behan organized a posse, borrowing guns and ammunition from P. W. Smith's, but showed little enthusiasm for taking up the chase. Parsons noted that the Earp party "refused arrest and retired from town, first though waiting for Behan and Neagle to do what they threatened. Bad muss this. Sheriff is awake now that one of his friends is killed. Couldn't do anything before. Things are very rotten in that office. Fine reputation we're getting abroad."[82]

---

[79] *Tombstone Daily Nugget*, March 22, 1882; and *Tombstone Epitaph* (weekly), March 27, 1882.

[80] Parsons, *Private Journal*, March 20, 1882, 297.

[81] *Tombstone Epitaph*, March 22, 1882. In another column the paper gave a slightly different version of this conversation, but the meaning remains the same.

[82] Parsons, *Private Journal*, March 21, 1882, 297.

The *Nugget*, too, conceded Behan's tardiness in responding to Bob Paul's telegram: "The Sheriff, finding that the time consumed in arming and equipping his posse had enabled the other party to secure at least half an hour's start, concluded not to commence the pursuit until this morning at 5 o'clock."[83]

A little before noon on March 22, Wyatt and Warren Earp, together with Holliday, McMasters, Johnson, Texas Jack Vermillion, and two others, rode over a small hill into Peter Spencer's wood camp off the Chiricahua road near Doyle's ranch, below South Pass in the Dragoon Mountains. They stopped first to talk with a Mexican wood cutter named Simon Acosta, but since he spoke no English they moved on to teamster Theodore Judah. Wyatt Earp asked about Spencer's whereabouts and was told he had returned to Tombstone that morning (for a hearing in the justice court, as it turned out, on charges of assaulting one of his in-laws). Wyatt also inquired about Hank Swelling, a minor figure in the Cowboy ranks. Judah replied that he was not there.

Then, the teamster recalled, Earp "asked me my name, and wanted to know if I was not a friend of Pete Spence's and also of Frank Stilwell, to which question I answered that I was. He then turned to the crowd and asked them if they had seen any horses down there with saddles on. They then went off, and passed out of my sight towards the main road to Tombstone."

Curious about Earp's intentions, Judah and Acosta started up a small hill to see for themselves. Looking over after hearing gunfire, they saw the Earp party on another hill across the road. Two or three had dismounted. Soon the two wood camp workers watched as they "came down the hill very leisurely to the road and returned in the direction of the camp. They proceeded but a short distance, and turned around again. Then they went along the road until it makes a sharp turn and kept on in the same direction, easterly, passing into the hills."[84]

Beyond Judah's view Earp's posse had found Florentino Cruz out looking for stray mules. Wyatt was convinced Florentino had been involved in Morgan's murder. For this the unarmed Mexican wood cutter now lost his life. Simon Acosta later testified, through an interpreter:

---

[83] *Tombstone Daily Nugget*, March 22, 1882.

[84] Proceedings of the Inquest Upon the Body of Florentino Cruz, Statement of Theodore Judah, ibid., March 25, 1882.

"I saw this party commence firing; I immediately ran up the hill and saw them shooting at Florentino. I did not see Florentino fall."[85] The frightened workers delayed searching for their missing comrade. The next morning, Judah testified, "I went to the top of the hill where I saw the Earp party after the shooting, and looking around, discovered the body of Florentino, lying under the shade of a tree, a few feet away from the tracks made by the Earp party. He was lying face downwards, with his right arm resting under his head, and his coat was placed over his legs."[86]

Dr. Goodfellow described the wounds, after examining the body in Tombstone at Ritter's undertaking parlor:

> I commenced the examination at his head and followed down. The first shot entered at the right temple, penetrating the brain; the second produced a slight flesh wound in the right shoulder; the third entered on the right side of the body near the liver, and made its exit to the right of the spine. . . . The fourth struck in the left thigh, and made its exit about seven or eight inches above the point of entry. . . . he was lying on the ground after the wounds in the upper part of the body had been received. In my opinion the wound in the thigh was received after he was dead. I form that opinion from the absence of blood around the wound.[87]

The twelve-man coroner's jury, including Mike Gray and Earp's friend Albert Bilicke, found that Florentino came to his death by "gunshot wounds inflicted by Wyatt Earp, Warren Earp, J. H. Holliday, Sherman McMasters, Texas Jack, Johnson, and two men whose names are unknown to the jury."[88]

Samuel Williams, Spencer's half-brother, had heard the shooting and fled to Tombstone without bothering to investigate the cause. After learning of Florentino's death, and fearing his own murder, Peter Spencer surrendered to the sheriff. Behan allowed the frightened prisoner to retain his weapons in case an attempt was made to kill him in his cell. Meanwhile Behan and his posse conducted a rather futile search for the Earp party.

In Tombstone the sheriff detained Dan Tipton and O. C. Smith, claiming they had somehow obstructed his efforts at the Cosmopolitan

[85]Continuation of the Inquest . . . Florentino Cruz, Statement of Simon Acosta, ibid., March 26, 1882.

[86]Proceedings of the Inquest . . . Florentino Cruz, Statement of Theodore Judah, ibid., March 25, 1882.

[87]Statement of Dr. G. E. Goodfellow, ibid.

[88]Continuation of the Inquest . . . Florentino Cruz, ibid., March 26, 1882.

to arrest those charged with killing Frank Stilwell. William Herring acted as their attorney, arranging a $1,000 bail for each man. Earp supporters Frank Walker, superintendent of the Sycamore Springs Water Company, R. J. Winders, Bob Hatch, and James Earp's son-in-law, Thaddeus S. Harris of the Tombstone Foundry, stood as bondsmen. Accepting Herring's arguments about defective warrants, Judge Felter tossed out the charges. Tipton was soon reported near Willcox rejoining the Earps.

With undisguised sarcasm the *Epitaph* kept track of Behan's lack of progress:

> Sheriff Behan left with a posse of some fifteen or sixteen men, among whom were John Ringo, Fin [*sic*] Clanton and several others of the cowboy element, together with some of the permanent residents of Tombstone. They were out as far as the Dragoons where they got track of the party they were after, and tracked them back (so they say) to within four miles of town, where the trail became obliterated by the passing travel. It is supposed by some that they are now within easy reach of Tombstone.[89]

With little difficulty Wyatt Earp avoided Johnny Behan and his band of amateur lawmen. On Friday, March 24, Earp and his men rode toward Iron Springs. As they neared the site, some fifteen miles west of Contention City at the base of the Whetstone Mountains, Wyatt spotted a small group of men already camped there. Riding closer, Earp saw that the man stirring the stew pot was none other than Curly Bill Brocius.

As everyone recognized each other the Cowboys reached for their weapons. At the opening volley, delivered with only thirty yards separating the combatants, the men with Wyatt Earp—including Doc Holliday, McMasters, and Texas Jack Vermillion—retreated to safety. The outlaws' fire killed Vermillion's horse as other bullets cut the leather strap of McMasters's field glasses and peppered his clothing. Only Wyatt stood his ground, even though a bullet shattered the pommel of his saddle and others cut the loose fabric of his coat and trousers. Returning the gunfire, Earp killed Brocius with a double shotgun blast.

One of the *Nugget*'s reporters telegraphed from Charleston at 8:40 that evening: "It is certain that the Earp party have had a fight near the Whetstone mountains. . . . Impossible to ascertain anything in regard to Curly

[89] *Tombstone Epitaph*, March 23, 1882.

Bill." A little more than five hours later a second report arrived from Contention City: "Wyatt, without doubt, is wounded. . . . There is much excitement here, but the report of the killing of Curly Bill is not credited."[90] Ignoring all this, George Parsons wrote, "I got strictly private news [that] 'Curly Bill' has been killed at last—by the Earp party and none of the latter hurt."[91] Despite conflicting rumors Brocius was never seen around Tombstone again. Most assumed his Cowboy followers had buried him somewhere on Frank Patterson's ranch along the Babocomari.

With the situation deteriorating daily, Bessie and Mattie Earp left Tombstone four days after Stilwell's murder. Both papers carried the news. The *Epitaph* printed the item under its local personals column: "Mrs. James Earp and Mrs. Wyatt Earp left to-day for Colton, California, the residence of their husbands' parents. These ladies have the sympathy of all who know them, and for that matter the entire community. Their trials for the last six months have been of the most severe nature."[92] Morgan's wife Louisa—deeply hurt after learning of her husband's sexual indiscretions at Benson—had already left and was not in Tombstone at the time of his death.

After the fight at Iron Springs the movements of Wyatt Earp and his followers, together with the less-than-aggressive pursuit of Johnny Behan and his posse, became a source of speculation and some amusement to the tired citizens of Tombstone. Both groups now did little more than endure the interminable discomfort of traveling about the county on seemingly endless horseback rides.

Hoping to capture those responsible for Frank Stilwell's murder, Pima County Sheriff Bob Paul joined Behan's posse. After observing his dubious companions close up, however, Paul withdrew. At Tucson even the Democratic press supported him: "He refused to pursue the parties, as he felt that they could not be taken with the posse raised, inasmuch as they are most all hostile to the Earps and a meeting simply meant bloodshed, which he believed ought to and could be avoided, and yet get the parties into custody. All things considered, we do not see how Sheriff Paul could have done otherwise."[93]

---

[91]Parsons, *Private Journal,* March 25, 1882, 298.

[92]*Tombstone Epitaph,* March 24, 1882. The next day the *Nugget* remarked somewhat casually: "The wives of James and Wyatt Earp departed yesterday for Colton, Cal., where their husbands' parents reside."

[93]*Arizona Daily Star,* March 28, 1882.

Behan and Harry Woods ignored Bob Paul's rebuff and led their fifteen-man posse to Summit Station in Dragoon Pass. From there they rode north, wondering if their quarry had already boarded a train and fled into New Mexico. But no, the trail led instead to Henry C. Hooker's Sierra Bonita Ranch at the upper end of the Sulphur Springs Valley, across the line in Graham County. Not wishing to be rushed, Behan camped near the Winchester Mining District and did not reach Hooker's place until the morning of March 28. There the sheriff learned that Earp and his followers had left the night before.

Colonel Hooker was not happy to see Behan's posse, it being composed of some of the area's more celebrated Cowboys. He might buy small herds stolen in Mexico, as did every other cattleman in southern Arizona, but he drew the line at socializing with thieves. Hooker greeted Sheriff Behan with cool restraint, deliberately ignoring the others. But he soon shelved the niceties, exclaiming, "D—n the officers and d—n the law."[94] Hooker reluctantly served them food but refused to furnish fresh horses or volunteer information about the Earps.

Behan's supporters accused the feisty cattleman from New Hampshire of having given Wyatt mounts and provisions. Actually, Earp and his men still rode the same horses they left Tombstone with. Defiantly they now camped on a hilltop near Hooker's ranch house. Behan probably knew it but chose to ignore the obvious. After breakfast the sheriff's posse traveled to nearby Fort Grant to try and solicit Apache scouts for the chase. The army belittled the request, explaining those men had all been recently discharged. Behan's offer of a $500 reward was also ignored. Returning to Hooker's, the posse claimed all sign of the wanted men had been lost. After camping at Eureka Springs in Aravaipa Canyon, Behan and his exhausted followers finally returned to Tombstone late in the afternoon of March 30.

Despite his rather pitiful performance, Johnny Behan later boasted to an official from the Department of Justice "that he himself had done more to quell the disturbance than any one else, and that he regarded the whole affair as a fight between two lawless factions. . . ."[95]

With rumors claiming Hooker had paid Wyatt Earp, on behalf of the

[94] *Tombstone Daily Nugget,* March 31, 1882.

[95] Leigh Chalmers, Examiner Dept. of Justice, to Attorney General A. H. Garland, August 13, 1885, General Records of the Department of Justice, Record Group 60, National Archives and Records Service.

Stockman's Protective Association, $1,000 for killing Brocius, Richard Rule downplayed the episode: "it's rough on the party who paid the reward, as the notorious and wily William is beyond question of doubt alive in New Mexico, keeping his weather eye open for a fresh saddle horse."[96] John Clum countered: "Suffice it to say that Curly Bill is as dead as two loads of buckshot can make him, and the man who killed him is entitled to the reward."[97] As it turned out Hooker never paid anyone.

The case against Peter Spencer for killing Morgan Earp had been scheduled for hearing before Justice of the Peace A. O. Wallace on March 24. Meanwhile, Deputy Sheriff Bell had arrested Indian Charlie at Charleston, and Deputy Frank Hereford (a nephew of Governor Tritle) picked up Frederick Bode (John Doe Freis). All avoided trial. After stumbling through a number of false starts, the prosecution refused to proceed after calling Mrs. Spencer and enduring objections from the defense. Claiming he had no choice, Wallace discharged the prisoners.[98] Few were disappointed. Years later, despite Maria Spencer's testimony at the coroner's inquest, even Wyatt remarked somewhat surprisingly, "I am satisfied that Spence had nothing to do with the assassination of Morgan."[99]

The news of Frank Stilwell's death stunned his family. To better understand the confusing circumstances, his famous brother, the old army scout who was then in the cattle business, traveled from Indian Territory to conduct his own investigation. After conferring with Tucson officials, he checked into the Grand Hotel at Tombstone in mid-April as "Jack Stilwell, Fort Sill." He spoke with Ike Clanton and others anxious to give the noted plainsman their version of his younger brother's life in Arizona. Stilwell impressed the locals as a gentleman, one describing him as "a very amiable, good natured man, tall and robust, with remarkably small hands and feet."[100]

Rumors of Jack riding with the Cowboys worried his uncle Jacob. From Baldwin City, Kansas, he wrote to inquire about his nephew's well-

---

[96] *Tombstone Daily Nugget,* March 31, 1882.

[97] *Tombstone Epitaph,* March 31, 1882.

[98] *Tombstone Daily Nugget,* April 4, 1882.

[99] Wyatt S. Earp to Walter Noble Burns, March 15, 1927, Special Collections, University of Arizona Library.

[100] O'B. M. to the editor, *Arizona Daily Star,* April 30, 1882.

being. But Stilwell was more interested in gathering information and settling his brother's affairs than worrying about immediate revenge. Still, decades later, an old cavalry sergeant wrote: "I saw 'Jack' last just after his brother was killed. I met him at Benson, Arizona, and I'm sure he was man hunting then for his brother's murderer. . . ."[101] In the end, however, Stilwell satisfied himself that Frank had in all likelihood killed Morgan. He thus felt no personal obligation to hunt down the Earps or Holliday.

It was just as well. Jack Stilwell carried with him a history of cold courage. He remained a favorite of Nelson A. Miles, who delighted in telling the story of his former scout hiding in some grass to escape roving Cheyennes at the time of the Beecher Island fight in 1868. Surprised by a rattlesnake, Stilwell could neither shoot nor run without exposing his position. Instead, the general revealed, he "spit tobacco juice on his head which caused him to vacate the premises."[102] As one Arizona citizen cautiously pondered Comanche Jack's stolid reputation: "While he is a downright law and order man, I would hate to be in Wyatt Earp's place if they met face to face in the open prairie."[103]

Tombstone was growing weary of this bloody vendetta. Since the Earps seemed to have slipped out of the territory, Cochise County breathed with relief. Nor were many tears shed over the dead Cowboys. Residents had long feared the loosely organized outlaw confederation, but after learning the Earps had killed more of these desperadoes, people greeted the news with a surprising sense of indifference. Yet during the brothers' tenure as peace officers many had objected to their often rough and high-handed methods.

Although the past several months had proved exciting ones for Tombstone, most longed for peace and predictable municipal calm. There were few cries of regret after it became known the Earps had ridden off to Silver City, New Mexico, then to Albuquerque and on into Colorado. Most people, including their friends, prayed they would never return.

Hating to see their bread-and-butter story coming to an end, Arizona journalists kept busy describing efforts by territorial authorities to extradite Wyatt Earp and Doc Holliday from their Rocky Mountain refuge. But ordinary citizens began putting the whole ugly business

[101]Carter, *The Old Sergeant's Story*, 129.

[102]Miles, *Personal Recollections and Observations*, 148.

[103]O'B. M. to the editor, *Arizona Daily Star*, April 30, 1882.

behind them. Based on some rather intriguing machinations among leading Republican politicians, Colorado's governor, Frederick W. Pitkin, finally refused Arizona's request to return the wanted men for trial. At Tucson the opposition press noted, "The Republican party in Arizona has taken shelter under the Earpumbrella," adding, "Governor Tritle to Governor Pitkin: 'My dear brother, let his blow pass!' It passed."[104]

Tombstone had already turned its attention elsewhere. Excited residents began looking forward to the arrival of their most famous visitor.

[104]*Arizona Daily Star,* June 1, 1882.

Chapter 14

The Decline and Fall

The Earps had jailed Frank Stilwell and Peter Spencer for the Bisbee stage robbery only seven months before. Somehow it seemed an eternity; those arrests having led indirectly, in the minds of many, to the October 1881 gunfight and its long, brutal aftermath. Others admitted with some embarrassment of finding the whole thing strangely captivating. Yet as time went by nearly everyone grew weary of repeated disruptions and threats of street violence. Arizona wanted more business and less bloodshed. Distant reports about Wyatt Earp and his followers reflected old news best forgotten. Tired citizens gladly shifted their attention from some future folk legend to the expected arrival of General William Tecumseh Sherman.

Sherman's visit came none too soon. Accounts detailing Tombstone's troubles blanketed the country. With all the negative publicity everyone eagerly searched for signs of civility, anything suggesting visitors—or possible investors—need not fear the place. How better than to host one of the nation's revered heroes? The general and his entourage arrived in two six-horse military ambulances on the evening of April 7, 1882.

Considering the town's wild reputation, a raucous welcome surprised no one, and Sherman

> told, with considerable merriment, his introduction to Tombstone. He said when the carriage he was in was within a few hundred yards of the town a cow-boyish looking individual rode up and asked if General Sherman was there. Being answered in the affirmative, he pulled a pistol and fired two shots in rapid succession. That was the signal for a volley, and for a few minutes the air vibrated with the sharp reports of pistol shots, bursting of anvils and Chinese bombs. He said that Tombstone was a decidedly live town, and that he enjoyed himself hugely there.

Apparently so did everyone else: "The ladies of the party were also much pleased. . . . Miss Poe naively remarked that she was expecting that at least there would be two or three men killed every day, and to her great disappointment, she added, in the suggestive language of the West, 'They hadn't a man for breakfast while they were in Tombstone.'"[1]

On his arrival Sherman joined city dignitaries for an impromptu parade along Allen to Third, then to Fremont and up to Sixth, before returning to Allen Street and the Grand Hotel; decorated for the occasion with flags, bunting, and Chinese lanterns. The brass band blasted out a few martial selections as the general and his party made their way along the street, packed between Fourth and Fifth with crowds "anxious to obtain even a momentary glimpse of the war-worn veteran."[2] Ushered to his rooms, the general and Mayor Carr soon returned, Sherman thanking everyone for the cordial welcome, adding that he was pleased "to find such a number of fine looking, intelligent citizens in this place so badly thought of out side."[3]

Later Sherman and his party crossed over to the Cosmopolitan for dinner at the Maison Doree. There a number of local luminaries including Joseph Tasker, Milton Joyce, Ben Goodrich, John Clum—who could hardly pass up a chance like this—and Colonel R. F. Hafford joined the general's table. The restaurant featured wild game, although patrons could order more traditional offerings of beef, lamb, poultry, and fish. Oysters could also be chosen from its wide menu. An extensive wine list, featuring domestic and European selections, helped enliven the experience, and meal's cost as little as fifty cents.

The next morning Sherman toured the Tombstone Mill and Mining Company and was lowered into the main shaft of the Toughnut. He then repeated the experience at the Grand Central. That afternoon he visited the Girard Company mill site, before returning to his rooms and meeting with local representatives of the Grand Army of the Republic. With dinner behind them the group adjourned to Schieffelin Hall for a crowded reception. There "a great many citizens gathered to shake hands with the 'Hero of Atlanta,' who seemed glad to see everyone. The guests were received upon the platform of the hall."[4]

In celebration of seeing a hero known to them only from their history books, some two-hundred schoolboys petitioned Mayor Carr for per-

[1] *Tombstone Epitaph,* April 13, 1882.

[2] *Tombstone Daily Nugget,* April 8, 1882.

[3] *Tombstone Epitaph,* April 8, 1882.

[4] Ibid., April 10, 1882.

mission to light a bonfire at the Fremont Street intersection. Staring at their eager upturned faces, his resolve weakened. Years later city officials hoped to restrain youthful exuberance: "The action of the City Council in passing an ordinance prohibiting the sale of fire arms, ammunition, or explosives of any character to children under fourteen years of age, is certainly a step in the right direction, and we trust it will become a law and be strictly enforced."[5]

Sherman and his party left Tombstone early Sunday morning and traveled on to Fort Huachuca. Although he spent less than two full days in the notorious and celebrated mining camp, Sherman remembered his welcome. Locals, too, often recalled details years later. After all, William Tecumseh Sherman remained the most famous man that ever visited Tombstone, and his tour did provide a needed antidote to lingering memories of the Earp troubles.

★ ★ ★

During his visit the general did more than tour mines and socialize with locals. While traveling through "the southern & eastern tier of Counties" he studied conditions. The government had long been concerned with problems, primarily international, stemming from Cowboy raids. At Tucson Sherman telegraphed the attorney general at Washington: "The Civil Officers have not sufficient force to make arrests to hold prisoners for trial or punish when convicted." He suggested adopting the governor's plan, calling for the use of $150,000 in federal funds, "to hire [a] suitable posse to aid the sheriff & marshal," or revise the Posse Comitatus Act to allow the use of troops.[6]

President Chester Arthur had tried earlier to have the law changed, but Congress blocked his efforts by tabling the measure in committee. Arthur issued a presidential proclamation on May 3, 1882, essentially reading the riot act to Arizona residents:

> it has been made to appear satisfactorily to me, by information received from the Governor of the Territory of Arizona and from the General of the Army of the United States, and other reliable sources, that in consequence of unlawful combinations of evil disposed persons who are banded together

[5]Ibid., August 24, 1888.

[6]Gen. W. T. Sherman to Hon. B. H. Brewster, Atty. Gen., Washington, telegram, April 12, 1882, General Records of the Department of Justice, Record Group 60, National Archives and Records Service.

> to oppose and obstruct the execution of the laws, it has become impracticable to enforce, by ordinary course of judicial proceedings, the laws of the United States. . . .

As a result the president declared that the law required "whenever it may be necessary, in the judgment of the President, to use the military forces for the purpose of enforcing the faithful execution of the laws of the United States, he shall forthwith, by proclamation, command such insurgents to disperse and retire peaceably to their respective abodes, within a limited period of time."[7]

Reaction in Arizona was mixed; many considered President Arthur's proclamation an insult to the territory. Columns of local newsprint condemned the measure. But other forces, chiefly market pressures, were already at work abating the outlaw threat. As Behan's deputy Billy Breakenridge wrote years later: "A lot of rustlers had been killed off by the Mexicans . . . and in quarrels among themselves when they were drinking. The stockmen had organized for self-protection, and the rustlers got out of the country as fast as possible. With most of the bad men run off there was no more trouble. . . ."[8]

Big-time corporate operations in the Sulphur Springs Valley, primarily the Erie and Chiricahua Cattle companies, did much to stabilize the area by the mid-1880s. Still, law enforcement and government pressures played their role. As James Zabriskie, then United States district attorney for Arizona, outlined in 1885: "Deputy Marshal Earp was very active in his efforts to suppress this 'quasi' insurrection and prevent the violation of United States laws. . . . [He] and his band killed quite a number of these cow-boys. . . . Mr. Earp finally resigned and left the Territory, and subsequently these discordant elements were suppressed, after great exertion, chiefly by the United States Authorities."[9]

★ ★ ★

Times were changing. Even the *Nugget* and *Epitaph*, those partisan stalwarts of the Earp era, would never be the same. The latter changed

[7]Presidential Proclamation, Respecting Disturbances in Arizona, Chester A. Arthur, May 3, 1882, Records of the Office of the Secretary of the Interior, Record Group 48, National Archives and Records Service.

[8]Breakenridge, *Helldorado*, 179.

[9]United States District Attorney Jas. A. Zabriskie to Hon. Benjamin Harris Brewster, Atty. Gen., Washington, January 22, 1885, General Records of the Department of Justice, Record Group 60, National Archives and Records Service.

hands on May 1, 1882. John Clum, Charles Reppy, E. B. Gage, P. W. Smith, and the others associated with the Epitaph Publishing Company, sold out to a group of Democratic investors headed by Milton Joyce. Businessman and longtime political insider Samuel Purdy—son of a former California lieutenant governor and brother of an ex-Egyptian pasha—came up from Yuma to take over editorial control. Later, much to the annoyance of diehard Republicans, Harry Woods became editor and manager,[10] although Reppy did return as publisher. Tombstone now had two Democratic newspapers, but they were no longer across-the-street rivals. Richard Rule had moved the *Nugget* from its niche east of the recorder's office into a space opposite Spangenberg's gun shop. That, too, would not last owning to circumstances beyond Rule's control.

A fire broke out on May 26, 1882, in the water closet of the Tivoli, a saloon and garden restaurant on Allen Street east of the Grand Hotel. Since the Tivoli Garden had a wood frame and canvas roof, it was lost within minutes. The flames spread to nearby shanties; J. Myers & Bro. general merchandise, a Chinese laundry fronting Fifth, and the apartments of the Tombstone Club in the Grand Hotel. The hotel itself was soon engulfed. Within the hour much of that block had disappeared into ruins.

Carried by the wind, flames jumped Allen Street and destroyed such landmarks as the Wehrfritz Building, Campbell and Hatch's billiard parlor, the Alhambra and Occidental saloons, Bilicke's Cosmopolitan, Hafford's corner and Brown's Hotel. Spangenberg's went up amid a colorful display of exploding ammunition and gunpowder. All the buildings north, including the post office, were lost. Crossing Fourth Street, the fire took out half the buildings in that block. The new offices of the *Nugget* were gone, as was the Capitol Saloon and nearly everything else along Fremont to the rear of the O.K. Corral. The recorder's office crashed into the old Grotto's basement after records had been hastily withdrawn. Many buildings on the periphery of this disaster—the *Epitaph*, Mining Exchange, the Gird Block, Schieffelin Hall, Shaffer and Lord's old stand, and the Oriental Saloon—were badly scorched but still standing.

At the Oriental, Milt Joyce, who had lost all in the earlier blaze, again fought to save his property. Aided by Leslie Blackburn and Engine Co.

[10]*Tucson and Tombstone General and Business Directory*, 154, 202.

Dramatic results of the May 1882 firestorm, as seen from the balcony of the Gird Block. In the background, south of the Oriental, Tasker & Pridham's general merchandise stands out as a lonely sentinel surrounded by rubble.
Courtesy, Arizona Historical Society.

No. 1, he tore down the overhang and kept the building wet enough with low-pressure hoses to save it. Following the excitement, Joyce freely distributed all the liquor on hand to exhausted fire department and citizen volunteers. Looking around, everyone saw that Tombstone's historic core was gone. But most owners began rebuilding almost immediately.

The town mourned its loss of three major hotels. Smaller facilities such as the Russ House, then run by Nellie Cashman and her sister, escaped the blaze but were hard-pressed to satisfy demand. Nellie's former partner, Joseph Pascholy, soon joined up with Godfrey Tribolet and built the adobe two-story Occidental on the southeast corner of Fourth and Allen. Opening April 7, 1883: "It is well fitted with all the modern improvements. . . . The hotel is partly surrounded by a veranda, 160 feet in length, which affords a pleasant walk for the guests." With its forty rooms, the new site became "the headquarters for mining experts, capitalists, and mining men generally. . . . A first-class restaurant is attached to the house for the convenience of its guests," along with a billiard parlor and space for private card games. Prices ranged from fifty cents to $2 a day.[11]

★ ★ ★

As if trying to convince themselves that their town could improve despite natural disasters and its violent reputation, residents turned once more to those symbols that had worked so well for them in the past. Again confusing physical monuments with personal accomplishment, be it the Grand Hotel or Schieffelin Hall of earlier days, they decided to end the annoying practice of having scattered county offices. Citizens wanted a courthouse large enough to serve as a center for county government. Early on mining man James Vizina offered land for just that purpose from the surface of his claim at the corner of Third and Toughnut. At first nothing came of the idea. The Townsite Company objected, preferring instead to try and sell the county more expensive real estate near the center of town.

When the Arizona Legislature created Cochise County it had authorized the issuing of bonds, not to exceed $50,000, for a courthouse and jail. But it was not until August 10, 1882, that officials, invited dignitaries, and ordinary citizens gathered at the Vizina site to lay the cornerstone. Speakers delivered their windy orations, after which various items were buried for posterity: some gold and copper coins; a cigar; an ore sample assayed at $2,200; along with a collection of papers, poems, and essays, including one by Milton Joyce entitled, "How We Boss the County." It began raining as ceremonies ended, forcing the band to rush its last few tunes. By early October there were more than 100,000 bricks piled at the site. Telegrams begged for masons as far away as Phoenix. They finished work in January at a cost of $43,000. Today the building is a state museum.

Bids also went out for a new city hall during that fall of 1882. The *Epitaph* reported: "the site selected is that upon which the County Recorder's office was previous to the late fire. The building is to be 30 x 80 two stories, with basement. . . . Connected with the building will be

[11] *History of Arizona Territory*, 241; and *Tucson and Tombstone General and Business Directory*, 174.

The Cochise County Courthouse as it originally appeared, before the rear-wing addition. The 1882 date represents the setting of the cornerstone, a common practice for dating public buildings. Note the wooden braces supporting newly-planted trees.
Courtesy, Arizona State Library.

a jail which will be appropriately fitted up. . . ."[12] At a special meeting the council awarded the contract to William Constable for $11,490.[13] Workers set the cornerstone in late December. This brick building is still Tombstone's city hall and fire station.

☆ ☆ ☆

The town could not escape its sordid past. After surviving the Earps and Doc Holliday, Johnny Behan's troubles continued. He narrowly sidestepped indictments for various improprieties but still failed to hold on to the sheriff's office. At the first Cochise County general election on November 7, 1882, his deputy David Neagle, freighter Jerome L. Ward,

[12] *Tombstone Epitaph*, October 7, 1882.
[13] Minute Book B, Common Council, Village of Tombstone, October 21, 1882, 136.

and Benson grain dealer Larkin W. Carr jostled for that office. Behan knew better than to even enter the race. Ward defeated his two challengers with a margin of 261 votes over Neagle. Carr came in a close third.

A week later, while talking with friends at the Oriental, Frank Leslie had a brief altercation with William Claiborne, who had won an acquittal—after two trials—for killing James Hickey at Charleston back in October of 1881. Physically rebuffed, Claiborne returned with a Winchester rifle loudly determined to put Buckskin Frank in his place. Instead, Leslie stepped from the saloon's center side door and calmly walked down Fifth to the Allen Street corner. Surprised to find his quarry outside, Claiborne got off one quick shot but missed his mark. Frank Leslie fired twice, and Billy Claiborne fell to the sidewalk near a fruit stand at the Oriental's main entrance. The *Epitaph* reported: "LESLIE'S LUCK/ 'Billy the Kid' Takes a Shot at 'Buckskin Frank.'/ The Latter Promptly Replies, and the Former Quietly Turns His Toes Up to the Daisies."[14]

Frank experienced his own troubles, though far less damaging than those inflicted upon the unfortunate Mr. Claiborne. Toward the end of December, "Frank Leslie," the press revealed, "met with a serious, and it is feared, fatal accident. . . . He was going home and had just turned the corner of Fifth and Safford streets, when he slipped, and he was precipited [*sic*] headfirst on to a pile of rock, bruising and gashing his head and face in a frightful manner."[15] Whether alcoholic overindulgence contributed to this mishap was not reported. Leslie suffered but survived. Frank was not the only one out of sorts: "There are a lot of tramps in Tombstone, and it looks as if some of them were just walking around to save funeral expenses."[16]

☆ ☆ ☆

On December 5, 1883, five men robbed Joseph Goldwater's store at Bisbee, killing three men and a woman with indiscriminate gunfire. Outraged citizens, after overcoming delays in rounding up enough horses, finally took up the chase. Suspicion soon pointed to posse member John

[14] *Tombstone Epitaph*, November 18, 1882.

[15] *Weekly Arizona Citizen*, December 24, 1882.

[16] Ibid.

Heith, who seemed overly interested in following the wrong trail. Questioned rather roughly, he admitted planning the robbery. The others were captured and sent to Tombstone for trial. All five killers were convicted and sentenced to the gallows. But John Heith, given a spirited defense by Wyatt Earp's old lawyer, William Herring, received only a prison sentence. On February 22, 1884, a mob rode in from Bisbee, stormed the jail, took Heith from Sheriff Ward's custody, and lynched him from a Toughnut Street telegraph pole. At the inquest Dr. Goodfellow described the cause of death as coming "from emphysema of the lungs—a disease common to high altitudes—which might have been caused by strangulation, self-inflicted or otherwise."[17] On March 18 the five actual holdup men were all legally hanged.

★ ★ ★

Then, in what easily rivaled the Earp-Cowboy controversy as a threat to community well being, labor unrest rocked the Tombstone Mining District that spring of 1884. Years before miners had considered a union. But Richard Gird and the other owners easily crushed those plans with threats of blacklisting and shutdowns. In November 1883 mine owners at Telluride, Colorado, cut wages to $3.50, and the miners struck. Owners broke that effort in little more than a month by importing nonunion labor. The $4 day fell quickly at nearby Silverton. In Arizona superintendents of the Grand Central, Contention, and Tombstone Mill and Mining Company posted notice that beginning May 1 wages would be cut to $3. Other owners sat on the sidelines awaiting reaction. If the three major producers succeeded, those smaller operators planned to follow.

Trapped by decisions beyond their control, workers organized the Tombstone Miners' Union on April 29. Optimism rose as three-fourths of the miners signed up to protect their livelihoods. Organizers, who had watched Sheriff Ward rendered helpless when confronted by determined citizens during the John Heith affair, now proposed to repeat that success by confronting mine owners with men united by common purpose. But

[17]Coroner's Inquest, John Heith, February 1884, Arizona State Museum, Tombstone Courthouse; and Dr. George E. Goodfellow, biographical file, Arizona Historical Society.

those responsible for the mines proved far tougher than officials of Cochise County. In defiance E. B. Gage closed the Grand Central on April 30. Within days the Contention and Tombstone Mill and Mining Company did the same. Superintendents announced that under no circumstances would they reopen until the miners accepted the $3 day.

Gage and the others claimed production no longer supported a $4 wage. But the men who worked the shafts and read the published mill reports assumed the owners were simply sacrificing their wages for higher profits. In a summer report to his stockholders, Tombstone Mill and Mining Company president George Burnham explained:

> the proprietors of mines in Tombstone decided to reduce miners' wages from $4 to $3 per day, the latter being the *highest* rate paid in California or Colorado. The reduction was not accepted by the workmen, and all the mines in the district, excepting two of the smallest, were closed, and are still idle. . . . it is the unanimous opinion of the Board that work on them should not be resumed until the above named reduction in wages shall have been peaceably obtained.[18]

Burnham's words may have satisfied his stockholders, but locals proved harder to convince. Residents had already organized an Allen Street parade supporting the miners. Unlike the Earp-Clanton upheaval, which affected only a handful of citizens directly, the mine closures of 1884 threatened the entire community, from major property owners down to the smallest businessmen and their employees. Serious trouble loomed if miners suffered a 25 percent pay cut.

Major shareholders did not live in Tombstone, nor did they spend their profits there. The district always was of two worlds: the financiers controlling the mines and those living in the city itself. The mines could function without Tombstone, but without the mines and their payrolls the town had no reason to exist. It meant little to distant owners which local businesses succeeded or failed. The mines simply needed ore and a silver price high enough to justify production. Men could always be found. Understandably, Tombstone threw its support behind the miners. During the parade merchants treated strikers to free beer and a thousand cigars. A carnival-like atmosphere captured the community on

[18]Tombstone Mill and Mining Company, President George Burnham, Report to Stockholders, July 15, 1884, 1-2.

Watching local festivities from the corner of Third. Fire had destroyed all the two-story buildings that once graced Allen Street's north side. The O.K. Corral survived. To the right is the adobe wall of P.W. Smith's Corral, the sign of the Dexter Livery Stable—still showing the running horse, and the top of the new Occidental Hotel.
Courtesy, Arizona Historical Society.

Goose Flats as never before. Sympathetic townspeople began collecting money for a strike fund.

Soon more trouble surfaced. Not only did President S. D. Stevens publicly acknowledge the union's inability to support members during a prolonged walkout, but on the very day of the parade Hudson & Company's bank closed its doors while holding $130,000 in deposits. Cashier Milton Clapp had enjoyed his stay, but then, as fellow employee James Eccleston recalled, "Mr. Clapp did the wise thing; he skipped out."[19] Clapp's temporary absence left Eccleston (whose aunt was married to bank president Charles Hudson) all alone to face angry depositors. Or so he thought, until he realized the judge had appointed Frank Leslie to act as his bodyguard. Years later Eccleston explained, "Frank had been selected as the quickest man with a gun in Tombstone, and if

[19]James Y. Eccleston, "Failure of the Bank," unpublished manuscript.

anyone had attempted the familiarity of bulleting me, I am sure he would not have lived to know it."[20]

Public distrust rose dramatically after it became known Sheriff Ward had withdrawn $7,000 for himself and a few close friends, even after authorities ordered him to see that no employee or depositor removed anything. The failure of Hudson & Company created deep bitterness. Support for the strike began to fade as businessmen watched the money supply dry up. Frustrated, they met with union leaders and persuaded them to temporarily accept a lower wage. Approaching the superintendents as intermediaries, they hoped for a return to a $4 day after the resumption of full-scale production and the installation of large-capacity pumps to fight water at lower levels. Somebody came up with the idea of a $3.50 compromise. Superintendents, sensing victory, brushed that suggestion aside, telling everyone they must accept $3 regardless of future production figures. Strikers refused and the standoff continued.

With owners conspiring against them, miners dreamed up some retaliation of their own. These sadly futile maneuvers included a boycott against N. K. Fairbank & Company of Chicago. Fairbank, a self-made millionaire lard and cooking oil refiner, was the major stockholder in the Grand Central Company. The plan had little chance of success since Tombstone remained an isolated center of conflict, whereas Fairbank, also a director of the sprawling Maxwell Land Grant in New Mexico, enjoyed a nationwide reputation and vast sales base.[21] Besides, his sense of civic generosity did not extend much beyond Chicago's city limits.

Ruthlessly exploiting a situation they themselves created, owners understood the value of pressure and simply awaited the inevitable. Walking along deserted streets during that second month, one felt the business depression brought on by the absence of miners' wages. Public opinion, fickle during the best of times, turned sour. Even the *Epitaph*, so vocal in its support at the outset, began advocating that the miners give up for the promised $3. No longer did anyone talk of a $3.50 compromise.

But the strike had become a symbol. At no other place in the West had miners so successfully defied their employers. Having sacrificed too

---

[20]Ibid.

[21]*The Biographical Dictionary and Portrait Gallery of Representative Men of Chicago*, 740-743; and Flinn, ed., *Hand-Book of Chicago Biography*, 142-143.

much already, the more militant simply refused to accept defeat. Others, humbled by exhausted savings and hungry families, surrendered. The superintendent of the Head Center, part of the Grand Central Company, hired fifty of these desperate men at $3 and reopened the shafts in late July. During an unruly meeting at Schieffelin Hall, Stevens admitted the union was losing control. Talk turned ugly, punctuated by threats and recrimination. Sheriff Ward, still defensive about the bank crisis, posted deputies at the Head Center to protect company property and its employees. Officers did what they could to assure the men's safety on city streets.

Feelings hardened as superintendents announced no one would be rehired unless they disavowed the union. Grand Central officials advertised throughout the country for nonunion workers. Men drifted in from other western camps and from mining regions in the East. By early August the Grand Central reopened. It seemed as if Tombstone's experiment with labor unions had run its course. One union man got into a nasty row with a city officer and was pistol-whipped for his trouble. Another member exchanged shots with one of the imported miners. Frustration ran so high that police locked up the nonunion man for his own protection.

As the night shift left the Grand Central at three o'clock on the morning of August 8, guards noticed a crowd of several dozen making its way up the road from Fifth and Toughnut. Fearing trouble, those at the mine prepared to fight. One of the crowd stepped forward and demanded to know from the shift foreman if he intended to settle for $3. The speaker was taken hostage, then escaped in the confusion of gunfire. There were no casualties, but this attack—it never was determined who fired first—broke remaining public support for the union's cause. Tombstone had seen enough gunplay when the Earps were in town.

Sheriff Ward, using the incident at the Grand Central as his excuse, hastily requested federal troops. On August 12 two fully equipped companies from Fort Huachuca set up camp on Contention Hill. With soldiers in place to assure the peace—and in effect protect mine owners and their property—further union resistance seemed pointless. Within two weeks even the most steadfast supporters realized they had lost the fight. On August 25, 1884, the Tombstone Miners' Union officially disbanded. The defeated men could either accept $3 or try and find work elsewhere; no easy task in western mining camps prejudiced against union membership. Even discounting lost profits the owners saw the breaking of the $4 day as a triumph. Wages quickly fell in other Arizona mining camps.

★ ★ ★

Senseless violence, labor strife, boycotts and greed could not break Tombstone. Not even the volatile precious metal market alone could bring the district down. In the end, strangely ironic for desert country, water doomed bullion production. Falling silver prices only aggravated the decline. It all came about rather slowly. Yet compared with this tragedy the Earps were remembered only as an embarrassment.

Most mining engineers from the beginning had predicted the possibility of serious water strikes somewhere below the thousand-foot mark. No one was prepared for what actually happened. In late March 1881 miners hit water at 520 feet in the main shaft of the Sulphuret, north of the Contention and Flora Morrison and directly west of the Head Center and Tranquility. One intrigued reporter, scribbling notes as he stood alongside the hoisting works, watched "the 50-gallon bucket ascending and descending with great regularity, and a small creek was flowing into the gulch."[22]

Experts, for the most part, reacted well to the news: "they were unanimous in regarding it as a great benefit to the camp in many ways should it prove to be permanent."[23] Some owners believed they had found an excuse for moving milling operations on site, thus eliminating fees demanded by freighters hauling ore to the San Pedro. Townspeople hoped cheap water could be used to dampen streets (forcing people into bathtubs on a regular basis was still an insurmountable problem; cheap water had little to do with that).

No one wanted to consider the dangers. Ore strikes were still being made, and within months the Girard Gold and Silver Mining Company of Philadelphia built a twenty-stamp mill using water from the West

[22] *Tombstone Epitaph*, March 25, 1881.

[23] Ibid.

Side Mine. Besides, everybody argued, 1882 proved the most productive year in the history of the Tombstone Mining District. But in March workers again struck water, this time in a new shaft at the Grand Central: "it agrees with the point where water was encountered in both Head Center and Sulphuret."[24] The decision on pumping equipment was delayed by Superintendent E. B. Gage going east for the funeral of his mother. Even so, experts predicted it would take at least four months to install the necessary machinery. With the future of the district tied to production below six hundred feet, repeated water strikes began causing concern.

Gage ordered pumps, but as Tough Nut Superintendent William F. Staunton recalled some years later:

> The Grand Central Company installed a line of direct-acting steam pumps capable of raising 500,000 gallons in 24 hours; but, to the surprise of all, the withdrawal of this amount of water produced no appreciable effect. The Contention Company then put in a plant of 12-inch Cornish pumps at an expense of about $150,000, and capable of raising 1,000,000 gallons in 24 hours, and again the attempt to sink was made, but it soon became evident that the combined capacity of the pumps was inadequate. The Grand Central then put in a line of 14-inch Cornish pumps of 1,500,000 gallons capacity, and at a cost of in the neighborhood of $200,000, and together the two Cornish plants gained steadily on the water and sinking below began.

Staunton astutely outlined other dangers: "much valuable time had been lost, and from a lack of appreciation of the seriousness of the problem the rate of dividends had gone on undiminished, without retaining an adequate reserve for contingencies. Furthermore, there was a lack of harmony among those concerned which prevented the attainment of the best results."[25]

At first the pumps seemed to work. Then the miners' strike closed the major producers for four months and water levels rose. In early 1885 the Grand Central and Contention combined their pumping operations. To reduce expenses the Tombstone Mill and Mining Company bought the Girard Mill and began closing its facilities at Millville. By 1886 only the Grand Central still worked ore on the San Pedro. That year its hoisting

---

[24]Ibid., March 14, 1882.

[25]Blake, *Tombstone and Its Mines,* 17.

Superintendent E. B. Gage installed massive Cornish pumps to fight the serious threat to deep mining posed by repeated water strikes. The challenge involved in hauling all these weighty sections from railroad shipping points is obvious by comparing the size of the apparatus finally assembled with the young woman seated in the foreground. *Courtesy, Arizona State Library.*

works and pump house burned to the ground, leaving the Contention to fight the water alone.

Then, as Staunton explained, "There is no doubt that the Contention pumps could have held the water in check . . . until other machinery could have been put on the Grand Central, but differences arose between the companies and pending the settlement of these the pumps were stopped and the shafts allowed to fill." Meanwhile, "through the carelessness of a watchman, the Contention plant took fire, and its complete destruction postponed indefinitely the working of the mines below the water."[26]

The Contention closed in late 1886, leaving only the Grand Central and Tombstone Mill and Mining Company—of the old giants—still working ore. A year later the Grand Central suspended operations,

[26]Ibid., 18.

although their mill continued to work tailings for another two years. By the end of the decade most of the district's production came from the Bunker Hill and Tranquility and from leases on the Old Guard and Vizina properties. Even the Tombstone Mill and Mining Company suspended the last of its marginal operations by 1896. Clearly Tombstone was on the skids. By the turn of the century the town's population had fallen to 646; a sad reversal for a community that once saw itself as the next Virginia City.

Added to the costly water problem, declining silver prices crippled Tombstone. During the district's early years the price per troy ounce averaged between $1.15 and $1.20. Between 1881 and 1886 this figure dropped to a low of sixty-three cents and before 1900 bottomed out at fifty-two. By then all mining activity was small-scale. Only Artemus L. Grow, superintendent of the Tranquility, "kept that company employing a force of men when everything else in this district was closed, thus veritably keeping this community from total industrial death."[27] Still, production continued to fall. From 1887 yearly totals registered well under the million-dollar mark. By the mid-1890s they had slipped below a $300,000 gross for the entire district.

Amidst the decline, and at the suggestion of New York publishers Harper & Brothers, Owen Wister visited with hopes of gathering material for a book—never written—on the Earp-Clanton feud. He traveled by stage from Benson through Fairbank, but was forced to admit on this arrival, "Tombstone is quite the most depressing town I have ever seen. 'The glory is departed' is written on every street and building." Wister, however, made a good impression on the remaining locals, who hoped he would come again. "But," the future author of *The Virginian* confided in a letter to his mother, "I shall never do that."[28]

Soon afterward E. B. Gage, together with Frank M. Murphy and others, slowly began consolidating Tombstone's major holdings in one final effort to revive the district's glory days. By 1900 the dormant Tombstone Mill and Mining Company and the Contention joined with the Grand Central in a complete reorganization called the Tombstone Consolidated Mines Company Ltd. Gage, as the scheme's chief architect,

[27] *Tombstone in History, Romance and Wealth,* 38.

[28] Owen Wister to Sarah Butler Wister, June 25, 1894 (two letters, same date), Wister, ed., *Owen Wister Out West,* 212.

Part of a huge boiler brought in by the Tombstone Consolidated Mines Co., Ltd. to try and revitalize the District at the turn of the century, seen here passing the San Jose House on the northwest corner of Fifth and Fremont.
Courtesy, Arizona State Museum, Tombstone.

became president with William Staunton serving as general manager. Holdings included twenty-six claims of the old Grand Central Company; eighteen from the Tombstone Mill and Mining Company; four from the Contention Consolidated; an equal number from the Head Center and Tranquility; together with assorted smaller mines, mills, mill sites, buildings, and heavy machinery.

Gage linked the new venture with the Development Company of America, a Delaware corporation with offices in New York City. In Arizona this company controlled the Congress Mine and Poland Mining Company in Yavapai County and would soon take over the Silverbell copper properties. It also operated such diverse enterprises as the gas works at Coffeyville, Kansas, and the Chiricahua Timber & Land Company in Mexico. Frank Murphy, an old Gage colleague, was helped by the fact that his brother occupied the governor's chair. Frank and E. B. had developed the Santa Fe, Prescott and Phoenix Railroad in 1893. Together they now planned to use their political connections and reputations as business and mining men, along with their positions with the Prescott National Bank, to finance the Tombstone Consolidated by interesting investors in, and funneling money through, the Development Company of America.

At Tombstone Gage and other engineers decided to sink a four-compartment pump shaft down one thousand feet and prospect deep within the district's principal mines. By August 1902 they reached the water table. That year the price of silver dropped briefly to forty-seven cents. In November workmen installed a large direct-acting triple-expansion steam pump ordered from Milwaukee. The Tombstone company used four two-hundred-horsepower Morrison corrugated internal furnace boilers, each ten feet in diameter, fifteen feet long, and fueled with crude oil. As in the old days, all this massive equipment was hauled in by freight wagons and assembled on site. They succeeded in pulling up well over 2 million gallons every twenty-four hours.

Optimism traveled beyond the pumping stations. Officials of the El Paso & Southwestern Railroad decided to lay track down from Fairbank. Townspeople excitedly followed daily progress reports. Gold was even discovered by grading crews in one cut near Watervale. A few claims were filed but nothing came of it. Then, on the afternoon of March 25, 1903, Tombstone finally saw its dream fulfilled as workers drove the last spike, followed by the first locomotive reaching the Toughnut Street depot.

Before long the railroad brought in a group of visiting congressmen. Then, as now, few of them could pass up an all-expense-paid trip under the guise of official business. Even the remoteness of Tombstone could not discourage them. Ferried about in the rather flimsy automobiles of the day, these men toured the town and mines, even going below ground to inspect the newly opened pumping stations.

By 1905 Gage and the others incorporated the Tombstone & Southern Railroad Company, linking their mining property to the track of the El Paso & Southwestern. One spur joined the company's warehouse and pumping stations with the old Girard mill site, while a second ran to the Tranquility shaft.[29]

As Tombstone welcomed the railroad, miners continued to recondition shafts in the Contention, Grand Central, Lucky Cuss and other once-famous properties. Some ore was shipped to El Paso for processing. Pumps now in place at the seven-hundred-foot level brought up 2.3

---

[29]For a detailed look at Tombstone and its railroads, see Myrick, *Railroads of Arizona,* 1:443-459.

million gallons each day. Despite the elaborate setup the water fell agonizingly slow. At the end of 1905 workers installed pumps at the eight-hundred-foot level and opened a 125-ton cyanide mill to handle low-grade ore stockpiled nearby. The next year shafts reached a thousand feet. The mill, now using improved Wilfley concentrating tables, increased its capacity. Smaller operators still shipped to Texas. In 1908 the Tombstone mines produced 51,266 tons of ore.

That bright outlook soon gave way to reality. As the Arizona Bureau of Mines later described:

> The year 1909 proved to be disastrous. On June 1, due to a defect in the fuel supply for the boilers, the 1,000-foot-level pumps were submerged. Despite efforts to retrieve them by the aid of eight sinkers, the water raised to the 900-foot level. Due to overload, the six boilers of the power plant went out of commission simultaneously. Large sinkers were installed, but the heat from the exhaust steam proved to be excessive, and in 1910 a 4,000-cubic-foot compressor was installed to run the sinkers by air. After installing new boilers, the pumps were recovered, and drifting on the 1,000-foot level was resumed by the end of the year.[30]

Since costs eventually "proved too great for the Development Company of America, pumping was discontinued on January 19, 1911. The pumps on the 1,000-, 800-, 700-, and 600-foot levels were left on the stations, where they remain to this day. . . . After unsuccessful attempts to refinance . . . the mines were turned over to lessees who reworked the old dumps and upper-level stopes."[31] Between October 1908 and September 1909 the pumping system had raised an average 5 million gallons a day and reached 6.8 million a year later; but nature won the battle. Deep mining ended, although small-scale operations continued for many years. The Phelps Dodge Corporation, a major creditor, took over and placed the property under its subsidiary, the Bunker Hill Mining Company, and continued marginal operations until April 1918. Others toyed with the area well into the 1930s, but clearly the bonanza days were over. The glory of Tombstone's mines now belonged to history.

★ ★ ★

With the mines shutting down, many old timers returned to California. Watching it all fall apart, early resident Carlisle S. Abbott described

[30]Butler, Wilson, and Rasor, *Geology and Ore Deposits of the Tombstone District*, 47. [31]Ibid.

the process: "what had been the thriving city of Tombstone took on the resemblance of an army breaking camp, and everybody seemed anxious to get away."[32] Abbott had seen it all, first as an overland pioneer, then coming as he did from Salinas to establish a stock and dairy ranch on the San Pedro. He suffered Cowboy raids in the early 1880s but endured. In 1885 one of his daughters married Dr. Nelson S. Giberson, the same man who coldly announced Billy Clanton had no chance to live after his encounter with the Earps and Doc Holliday. Tombstone's decline forced even Abbott's retreat.

Many who did not flee for more distant places moved on to Bisbee or Pearce, a mining camp founded in the 1890s with the discovery of what became the Commonwealth Mine on the east side of the Dragoons. Some of the old town went with them. Eager expatriates dismantled wood-frame buildings, loaded a procession of wagons, and freighted them to their new home. Tombstone acquired an abandoned look that persisted for years. Fire later destroyed most of these relics at Pearce.

But the removal of buildings only changed the look of Tombstone. The departure of its pioneer citizens changed its character. Taking with them either a sense of bitterness or fond memories of their youthful days in one of the West's great mining camps, they went on about their lives with the same determination that had brought them into the San Pedro Valley in the first place.

[32]Abbott, *Recollections of a California Pioneer*, 234. For more on Abbott's life, see *History and Biographical Record of Monterey and San Benito Counties*, 523-527.

Looking west along Allen Street from the Oriental saloon, long after the wild days had ended. Comparing this view with those showing the same scene years before, it is obvious that the glory is gone.
Courtesy, Arizona State Library.

Epilogue

To the End of the Trail

Edward L. Schieffelin, the man who started it all, returned often after selling out. He was in town for a few days at the time of the big gunfight on Fremont Street. Financially comfortable but not content, he and his younger brother Effingham visited Juneau, Alaska, in March 1882. Effingham then recruited Charles Farciot, Jack Young, and Charles Sauerbrey at Tombstone to help Ed organize an expedition to examine the mining possibilities of the Far North. They returned empty-handed after a colorful excursion up the Yukon River. "It is the worst country for mosquitos that I ever saw," Ed observed. "So, after spending two summers and a winter in the country we came back satisfied that we wanted no more of Alaska."[1]

Schieffelin prospected on and off for the rest of his life, dying alone on May 14, 1897, in a miner's cabin twenty miles east of Canyonville, Oregon. Following detailed written instructions, his brother Charles arranged for the body's return to Tombstone. Ed wanted to be buried on the site of one of his early camps, not entombed in some formal graveyard. A stone monument, representing a giant claim marker, stands in silent tribute to this resolute man.

Ed outlived his brother Albert, who died of consumption at age thirty-six on October 13, 1885, at their mother's home in East Los Angeles. The *Los Angeles Times* noted, "A Prospector Gone." Their father, Clinton Schieffelin, had killed himself with a revolver the year before. Ed's mother, true to her Irish vitality, outlived all three. She died in 1916 at the age of ninety-two.

☆ ☆ ☆

[1]Edward Schieffelin, "Trip to Alaska, 1882-1883," typescript [circa 1885], 8, 10, Bancroft Library, University of California, Berkeley.

The grave of Edward Lawrence Schieffelin, viewed from a respectful distance—thus avoiding the graffiti and discarded beer cans. It would have comforted the tough old prospector to know that his final resting place proved as strangely enigmatic as did his colorful but lonely life.

Richard Gird left Tombstone relatively wealthy in 1881. He maintained interests there, primarily in the Huachuca Water Company and with some miscellaneous mining claims, but directed most of his energies elsewhere. Returning to California, he purchased the sprawling Rancho del Santa Ana del Chino. He soon increased his holdings by several thousand acres and brought in blooded stock. Not satisfied with life as a country squire, Gird became involved in other enterprises. He developed sugar beets as a southern California industry and founded the Agricultural Experimental Station under university auspices. Other moneys he invested in Mexican mines and railroads. Unfortunately many of these ventures turned sour, so that after his death on May 29, 1910, his estate showed a balance of less than $1,500.[2] Gird's wife, Nellie, who as a bride moved into the adobe house at Millville, died in Los Angeles in 1921. His brother William K. Gird had passed away the year before.

☆ ☆ ☆

[2]In the Matter of the Estate of Richard Gird, Deceased; First and Final Account, Report and Petition for Distribution; Superior Court, County of Los Angeles, No. 18040; filed, August 10, 1912.

Josiah Howe White, the man who bought the Contention from the Schieffelins and Gird for $10,000, prospered from his Tombstone adventures. Despite his engineering background, White purchased 1,500 acres in Sonoma County, California, and became a gentleman farmer. He planted vineyards, raised premium cattle (from a small herd he bought in New York), bred horses, set up dairy operations, and became president of the Sonoma and Marin District Agricultural Society. With his wife, Annie Daniels—they had married in 1879—he helped raise four sons and a daughter. White died at Alameda, California, on September 23, 1897. His friend and Tombstone investment mentor, capitalist Walter E. Dean, died on July 13, 1925, at the age of eighty-six.

☆ ☆ ☆

John S. Vosburg also went to California. He owned a ranch near San Gabriel, but by the late 1880s resided in Los Angeles with his sons, a brother, and sister-in-law. After a divorce, he finally married a woman twenty-one years his junior and lived quietly off investments. He was clever to the end, sheltering most of his estate so that the amount available for probate was only $11,280. Vosburg had already given his three children and other relatives what he intended. Earlier he transferred $374,000 in bonds and some real estate to his new wife, winning a ruling from the inheritance tax appraiser that he did so without anticipating his own death, thus saving it from any rapacious raid by the legal profession. John Vosburg died on January 9, 1931, at age ninety.

☆ ☆ ☆

Vosburg's Arizona mining partner Anson P. K. Safford also left Tombstone early but headed in the opposite direction. Selling his interest in the bank he had started with Charles Hudson, Safford went to Florida. There he and several friends bought vast tracts of land and became real estate developers. They founded Tarpon Springs, where Safford died in 1891.

☆ ☆ ☆

Philip Corbin and his energetic siblings profited from their investments in the Tombstone Mill and Mining Company and from other

Arizona properties. Not content with this, or his sprawling hardware empire, Corbin also became president of, among other enterprises, the Corbin Screw Corporation, the New Britain Machine Company, the Calumet Building Company, and in 1903, the Corbin Motor Vehicle Company. Together with all this, he was vice president of one bank and director of another, as well as holding the same position with a Hartford insurance company. He then became involved in politics and served in a number of local and state offices. Philip Corbin died in 1910, presumably not from boredom.

★ ★ ★

As for major figures in the Townsite Company, Joseph C. Palmer died in 1882. The *Epitaph* printed a special dispatch on May 5. The more laudatory passages must have struck some citizens as odd; especially those still fighting to clean up the mess in local real estate titles. James S. Clark struggled with those issues until he left for Washington, where he petitioned Congress—unsuccessfully as it turned out—for redress over his and Palmer's cotton scheme during the Civil War. Clark died there on October 3, 1889. Two years earlier, back in Tombstone, someone stuffed the family stove with explosives. Mrs. Clark was seriously injured when she lit a fire to prepare dinner and blew the house to bits. Sheriff John Slaughter offered a $250 reward, but no bombing suspect was ever apprehended.

★ ★ ★

Mike Gray got himself elected to the legislature from Benson in 1887. Six years later he repeated the feat from Rucker. Near that old military site he tried unsuccessfully to develop mining property. By 1899 he was back in the legislature, this time from Pearce. He got reelected in 1901. On the 1900 census (son John was the enumerator) Gray is listed as a capitalist, living with John and his family. Decades later the Nature Conservancy bought, but only briefly held, Gray's former New Mexico ranch. They seemed proud to be associated with that old Texas and California pioneer, apparently unaware of his maneuvering as part of Clark, Gray & Company. The price involved a huge sum, which would have pleased the old man. His only regret would have been not living long enough to get his hands on some of the money. Time finally caught up with the ex-townsite manipulator at the St. Francis Hospital in San Francisco. Michael Gray died there on September 8, 1905.

★ ★ ★

And Alder Randall? He left for California in the early 1880s and died a decade later, leaving his wife, Mary, to support their five surviving children as a copyist for the Los Angeles County Assessor's Office—helped by a small inheritance they received from their grandmother's San Diego estate.[3] Mary died in 1929. Randall's namesake son died in 1984 at the age of ninety-three.

★ ★ ★

By 1884 one of Clark, Gray & Company's harshest critics, diarist George W. Parsons, went into business with the ubiquitous Maurice E. Clark. From an office on Fremont Street near Sixth the two dealt in "Mines, Money and Real Estate." A year later Maurice was in Calico, California, as a deputy U.S. mineral surveyor, before moving to Los Angeles as co-proprietor of a hotel called the Belmont. Dissatisfied, he traveled north to Oakland where he advertised himself as a mining engineer.

George Parsons returned to California in 1887. He kept busy as a charter member of the Los Angeles Chamber of Commerce, serving for many years as chairman of its Mines, Mining, and Transportation Committee. Parsons also became a director of the local Y.M.C.A., treasurer of the Los Angeles Episcopal Diocese, a member of the Free Harbor League, and for a time sat as president of the old Los Angeles Mining and Stock Exchange. At the turn of the century he traveled to Alaska and once again ran into Wyatt Earp and John Clum. George Whitwell Parsons died in Los Angeles on January 5, 1933.

★ ★ ★

Parsons's friend Dr. Goodfellow stayed in Tombstone for a number of years. Although an outspoken supporter of the Earp brothers' efforts

[3]1900 United States Census, Los Angeles, California, E.D. 10, Sheet 7-B, Lines 73-78; and In the Matter of the Estate and Guardianship of Marion Todd Randall, Arthur Randall, Alder Randall, Edward David Randall, No. 3760, April 30, 1900, Superior Court, Dept. 2, County of Los Angeles.

against the Cowboys, he maintained a warm personal friendship with Johnny Behan. They owned several race horses together. Goodfellow left for Tucson in 1891. Two years later he was appointed the territory's first quarantine officer. By the late 1890s he returned to California, remarried, and began practicing medicine in San Francisco and Oakland. George Goodfellow eventually moved to Los Angeles, where he died on December 7, 1910.

★ ★ ★

Soon after the death of his wife, pioneer Tombstone editor Artemus E. Fay left Dos Cabezas for northern Arizona. He founded the *Arizona Champion* at Peach Springs, on the southern edge of the Hualapai Indian Reservation in Mohave County, then moved the paper to Flagstaff. Fay's son Dayton, a small-time gambler, was arrested for killing a man in a brothel at Kingman in 1888. Artemus moved to Fresno, California. He died there on February 10, 1906, and was buried under the auspices of the Masonic Lodge.

★ ★ ★

John P. Clum could not get Tombstone out of his system, even after becoming an inspector with the Post Office Department at Washington. He remarried in early 1883. Two years later he returned to Tombstone, again as postmaster. Facing local Democratic opposition, he resigned in August 1885 but stayed on as city clerk until late the following year. Moving to San Bernardino, California, Clum worked as an editor and earned extra money as a part-time insurance dealer and real estate agent. He joined the *San Francisco Examiner*'s staff before going to Alaska as chief inspector for the Post Office Department. While at Nome he served as a special informant to President Theodore Roosevelt during a particularly interesting case of jury tampering. By 1915 Clum ran a date farm near Indio, California. In later years he wrote and lectured extensively on his western career, often overstating his role as Indian agent at San Carlos and glossing over his days in Tombstone. As an old man he came back for a visit and was treated as a hero from the past—treated far better than he had been in 1882. John Clum died in Los Angeles on May 2, 1932, somewhat short of his eighty-first birthday.

★ ★ ★

Wyatt Earp never answered to the law for the men slain following Morgan's murder. Instead he wandered the West as a gambler and saloon operator. Avoiding extradition to Arizona, he left his Gunnison, Colorado, sanctuary for Silverton and other camps, but soon traveled to San Francisco by way of Salt Lake City. Wyatt returned briefly to Dodge City in 1883 for its famous saloon war. He, James, and Warren then showed up at Eagle City, Idaho, where they ran the White Elephant Saloon. Wyatt also dabbled in real estate and mining claims around that place and Murrayville, getting involved in lawsuits over alleged claim jumping and property forfeitures for delinquent taxes. He even served briefly as a deputy sheriff.

By the late 1880s Wyatt was back in southern California. He owned a number of profitable saloons and gambling joints in San Diego but also scooped up extra cash from several rigged boxing matches at nearby Tijuana. Describing himself as a capitalist, he raced horses at San Diego's Mission Beach. Then, after a year in Los Angeles, he returned to San Francisco in the1890s; patronizing gambling dens and running horses on local tracks. Earp took time to visit Chicago, where he suffered a mugging at the Columbian Exposition, before searching for easier marks at gaming tables in various western towns. Back in San Francisco he endured an unwanted burst of publicity following his highly controversial decision as referee of the Sharkey-Fitzsimmons heavyweight bout in December 1896.

Wyatt traveled to Yuma, Arizona, for a brief stay, then went to Alaska. Drawn by the excitement of this newest gold rush, he opened two saloons at Nome, the Dexter and the Second Class, and owned a share in a beer distributorship. Wyatt showed up next in Nevada, seduced by the Goldfield and Tonopah boom camps. At Tonopah he gambled and ran a small saloon called the Northern. He also prospected, worked for nearly four months hauling ore for a major company, and served briefly as a deputy U.S. marshal, mainly as a process server, before going to Los Angeles. There he lived on and off for the rest of his life, using that city as home base for trips throughout the West, unsuccessfully searching, as always, for fortune. With his San Diego and Alaska funds exhausted, Wyatt Earp found himself facing hard times and old age. He held his life together gambling and working marginal mining claims in the Mohave Desert northwest of Parker, Arizona.

Hoping to make some money, Wyatt began dictating an autobiography to his close friend John H. Flood Jr. But since the story contained more melodrama than substance no publishers were interested. During those sessions of recollection, however, Flood learned much about his friend's life on the frontier—mostly facts Wyatt would never allow in the manuscript. In the end John Flood understood Wyatt Earp better than anyone alive; certainly more so than Sadie, a woman too shallow and self-absorbed to ever decipher her husband's convoluted twists of character. As an old man Wyatt traveled back to Tombstone with Mr. Flood. Although no one recognized him, he saw many people he remembered from fifty years before. Walking the now-deserted streets, Earp quietly talked of the wild days, pointing out landmarks and locations of his various adventures.

Disappointed with the reaction to his own manuscript, Wyatt, despite repeated claims of disinterest, willing spoke and corresponded with professional writers Walter Noble Burns and Stuart N. Lake. Burns's *Tombstone* (1927) painted Wyatt as a fearless hero. Dying in Los Angeles of chronic cystitis on January 13, 1929, Earp missed the results of Lake's writing, the highly successful *Wyatt Earp, Frontier Marshal* (1931). That book convinced countless readers of Wyatt Earp's place in taming the lawless West. Only later did researchers, first intrigued by *Frontier Marshal,* begin uncovering truths far more challenging, by contrast, than any cleverly invented legend.

★ ★ ★

The legend made no mention of Wyatt's wife in Tombstone. How could it? He had deserted her for Miss Marcus. After supporting herself as a prostitute, Mattie Earp committed suicide in 1888 at Pinal, Arizona, with an overdose of alcohol and laudanum.[4] Unable to accept the implications of Wyatt's desertion, hardened admirers justify the maneuver by claiming—based on the drugs implicated in her suicide—that she was an

[4]Coroner's Inquest on the Body of Mattie Earp, Deceased, July 4, 1888, Clerk of the Superior Court, Pinal County, Arizona.

addict while in Tombstone. Actually she used no opiates there. Mattie's biggest mistake was falling in love with Wyatt Earp. It ruined her life.

★ ★ ★

After Wyatt's death in 1929 his wife, Sadie, proved more disagreeable than ever, sponging off relatives and Earp's old friends and acquaintances. To the quiet relief of those she constantly hit up for money, Josephine Sarah Marcus Earp died in Los Angeles on December 19, 1944, leaving as her estate one trunk, a radio, and five small boxes of personal effects. Total value—$175.[5]

★ ★ ★

Virgil Earp, later joined by Wyatt and Warren, stayed in San Francisco for several months after leaving Tombstone. In August 1882 he was arrested for dealing faro. After that momentary embarrassment, he left for Colton before visiting the desert mining camp of Calico. In San Bernardino during the summer of 1884 he was elected a delegate to the Republican county convention. Soon he began drifting to various sites in Nevada, Utah, Idaho, Colorado, and Texas. Back in Colton by 1886, Earp ran a detective agency and a burlesque theater, and was elected constable and finally city marshal the following year. He was reelected but resigned in 1889. Moving for a time to Los Angeles, Virgil soon began outfitting for a prospecting trip into California's Mohave Desert. At Vanderbilt he owned the only two-story building in town, a gambling joint known as Earp's Hall. He also ran for constable but was defeated.

Virgil spent time at San Luis Obispo before joining Wyatt at Cripple Creek, Colorado, in the spring of 1895. By the fall he was back in Arizona. Virgil linked up with his brother again at Yuma in April 1897, then returned to Prescott. There, in early 1899, he was visited by a daughter from a brief first marriage. He had been unaware of her existence, and this was the only time he saw her. By late 1900, after announcing and then withdrawing as a candidate for Yavapai County sheriff (an office once held by Johnny Behan), he moved to Sawtelle, California, but soon came back to Arizona.

[5]In the Matter of the Estate of Josephine Earp, Petition for Probate of Will, No. 240608, January 19, 1945, Superior Court, County of Los Angeles.

The excitement of Nevada's mining resurgence at the turn of the century tempted the old Tombstone officer. Virgil traveled to Goldfield and became a deputy sheriff of Esmeralda County in January 1905, hired chiefly to keep the peace at the National Club. He died ten months later of pneumonia. Allie shipped his body to the Riverview Cemetery at Portland, Oregon, to be nearer his daughter.

★ ★ ★

Allie Earp outlived her husband by forty-two years. Deeply offended by books that ignored Virgil while casting Wyatt as the great hero of the frontier, she dictated her own reminiscences of the family's days in Tombstone to writer Frank Waters. The resulting book, finally published in 1960 as *The Earp Brothers of Tombstone,* still upsets devoted Wyatt Earp enthusiasts, who have invented all sorts of stories to try and disavow its simple truths. Allie would have loved the controversy but did not live to see it. She died in Los Angeles on November 14, 1947. Even the *New York Times* noted her passing.

★ ★ ★

Crippled by a shoulder wound during the Civil War, James Earp stood on the sidelines during much of his family's troubles in Tombstone. Returning to California after Morgan's death, he grabbed odd jobs, gambled, drove a hack, and ran a boarding house at Colton. Over the years he traveled to Utah, Idaho, Colorado, Texas, and other points on the gambling circuit. By the late 1880s he was back in California, running a saloon at San Bernardino.

Toward the end of his life he endured "Paralysis due to extreme old age and high blood pressure, having lost control of the arm and shoulder. . . . He requires the constant attendance of aid."[6] James Earp died at Los Angeles of cerebral apoplexy on January 25, 1926, at the age of eighty-four. He had outlived his wife—who died in 1887—and his two stepchildren. Hattie, the teenager Virgil's wife suspected of having a crush on one of the McLaurys, had married Tombstone Foundry co-proprietor Thaddeus S. Harris, a man twenty years her senior. Because of her continued interest in other men, however, they eventually divorced.

[6]Declaration for Pension (Supplemental), James C. Earp, June 9, 1924, National Archives and Records Service.

★ ★ ★

Warren, youngest of the brothers (born Baxter Warren Earp at Pella, Iowa, in 1855), played only a minor role in Tombstone. In July 1900 he was shot to death by a man named John Boyett at the Headquarter Saloon in Willcox, Arizona. Virgil, although still crippled from the December 1881 attack at Tombstone, avenged his brother's murder, killing the man he felt even more responsible than the actual triggerman. The former deputy U.S. marshal calmly noted in his diary: "Warren may rest in peace. His assassin is just resting permanently."[7]

★ ★ ★

Morgan's wife, Louisa, left Tombstone before her husband's murder. At the First Baptist Church in Los Angeles on December 31, 1885, she married Gustave H. Peters, a rather disagreeable fellow who worked as an agent for the L.A. Soda Water Works in Pasadena. Louisa died at the Los Angeles County Hospital on June 12, 1894. She had been living in Long Beach with her husband, then employed as a wiper for the Southern Pacific Railroad. Her cause of death was given as nephritis, an inflammation of the kidneys. Louisa Earp Peters had not yet reached her fortieth birthday.

★ ★ ★

After finally escaping Wyatt's influence, Doc Holliday carried his Tombstone notoriety with him, getting into trouble at Leadville, Colorado, in 1884 by wounding William J. Allen, a bartender at the Monarch Saloon, over a $5 debt.[8] Fired as a faro dealer at the Monarch, Holliday found himself broke and unable to pay what he owed. As it turned out Allen was friendly with a group of Leadville sports that included Johnny Tyler, whom Doc had argued with so violently at the Oriental four years before. Now, encouraged by Tyler, Allen threatened to beat Holliday to a pulp. Doc had good reason to worry. Physically weakened—a condition not helped by the town's 10,200-foot altitude—he only weighed 122 pounds, 50 less than his surly adversary.

[7]Gilchriese, "The Odyssey of Virgil Earp," and numerous conversations between John Gilchriese and the author.

[8]The People of the State of Colorado vs. John Holliday, alias Doc Holliday, August 1884-March 1885, No. 258, Records of the Criminal Court, Lake County, Colorado; and *Leadville Daily Herald*, August 20, 26, 1884.

Finding Holliday standing behind the cigar case at Hyman's Saloon, Allen charged through the door with fight in his eye. Fearing a beating, Doc pulled his revolver and fired. He missed, sending the bullet through the door's upper glass panel, lodging in the frame. Billy Allen, his courage quickly evaporating, turned to run but in the confusion stumbled and fell. Holliday leaned over the cigar case and fired again, striking the back of Allen's right arm, missing the bone some inches below the shoulder. A bystander disarmed the gunfighting dentist before he could fire a third time. Arrested and jailed, Doc survived this brush with the law. Authorities ordered him out of Leadville. Continually pursued by hard times (including incarceration at Denver on a vagrancy charge in 1886), he drank and gambled until tuberculosis finally overtook him. Dr. John Henry Holliday died at a Glenwood Springs sanitarium on November 8, 1887, at the age of thirty-six.

☆ ☆ ☆

Kate Holliday reverted to her maiden name when marrying fifty-two-year-old George M. Cummings at Aspen, Colorado, on March 2, 1890. They eventually moved to Cochise County and lived in Bisbee for two years. Cummings worked as a blacksmith while Kate earned money as a cook. Tiring of his alcoholic excesses, she left her husband late in the decade and found employment at Wells Fargo agent James Rath's small hotel at Cochise. During this period she visited Tombstone many times. Afterward, until 1930 Kate worked ostensibly as a housekeeper for the irascible Englishman John J. Howard at Dos Cabezas.[9] He had been in the area since the late 1870s and owned several nearly worthless mining claims, which Kate inherited. After his death she settled his meager estate, paying for Howard's funeral and distributing token cash payments to each of his two daughters.[10] Kate then applied to the Arizona Pioneers' Home at Prescott.

While there she became incensed, as had Virgil's wife, reading accounts glorifying Wyatt Earp's role in Tombstone. Kate dictated at least four versions of her own recollections, as well as an intriguing sketch of John Ringo, claiming, "Every time I think of [him], my eyes

---

[9]1900 United States Census, Cochise County, Arizona, E.D. 6, Sheet 4, Lines 1-4; and 1920 United States Census, Dos Cabezas Precinct, Cochise County, Arizona, E.D. 8, Sheet 5, Lines 13-14.

[10]de la Garza, *The Story of Dos Cabezas,* 41-44.

will fill with tears, he was a loyal friend to my husband, Doc Holliday, and to myself." She also described how Ringo warned her to leave town during the Spicer hearing: "'They are watching to get [Doc] in your apartment and they may get you too.' I told him I wanted to go back to Globe but did not think I had enough money." Ringo gave her "two twenty dollar pieces and one ten dollar gold. He said, 'This will take you back to Globe.'"[11]

Mary Katherine died of arteriosclerosis on November 2, 1940, less than a month before reaching ninety. She is buried in the Pioneers' Home Cemetery alongside other Tombstone veterans, including stagecoach operator J. D. Kinnear, who died in 1916.

☆ ☆ ☆

John Yoast found Ringo's body on the afternoon of July 14, 1882. He was slumped against a tree in a clump of oaks twenty yards off the road and a quarter mile from B. F. Smith's house near the mouth of Morse's Canyon in the Chiricahuas. John Ringo had been shot through the right temple, apparently at close range. A portion of his upper skull on the left side was missing from the force of the bullet passing through his brain. His hand clutched a .45 Colt revolver. A model 1876 Winchester rifle was propped up next to the body. He had been dead about twenty-four hours.[12]

The description of the wound strongly suggested suicide. Ringo had been on a protracted drunk, first in Tombstone and then at Galeyville. Over the years, with legend forever dissatisfied with one of its heroes killing himself, a number of candidates from Buckskin Frank Leslie to Johnny-Behind-the-Deuce have been suggested as Ringo's killer. Of them all the most intriguing is Wyatt Earp's claim of having done the deed himself. But John D. Gilchriese, the foremost authority on Earp's life, remains unconvinced.[13]

☆ ☆ ☆

Suspected of rustling in 1887, Ike Clanton was shot and killed along the Blue River in eastern Arizona by Detective Jonas V. Brighton.

---

[11]Mrs. Mary K. Cummings, "Character Picture of the Late John Riggold [*sic*]."

[12]Inquest on the Body of John Ringo, No. 170; filed November 3, 1882, District Court of the First Judicial District, County of Cochise.

[13]"Did Earp Kill Ringo?"; and numerous conversations between John Gilchriese and the author.

★ ★ ★

Following his 1880 defeat for reelection as Pima County sheriff, Charles A. Shibell engaged in private business, taking over the Palace Hotel at Tucson (later renamed the Occidental). After a couple years as a deputy sheriff in the late 1880s, Shibell was elected Pima County recorder. He held that office for ten consecutive terms—even with Republican support—until his death on October 21, 1908.

★ ★ ★

Robert H. Paul would be elected twice more as Pima County sheriff, but not without controversy—including, oddly enough, a charge of ballot tampering during his last campaign. Finally forced from office, he found work as a Southern Pacific Railroad detective before President Harrison appointed him Arizona's United States marshal. Paul owned a small share of the *Arizona Republican* at Phoenix, but fell on hard times toward the end of his life, finishing his career as Pima County undersheriff, an appointment received with the help of his former rival Charlie Shibell. Bob Paul died in Tucson on March 26, 1901.

★ ★ ★

John H. Behan never recovered the office of Cochise County sheriff. Instead he became superintendent of the Yuma Territorial Prison. President Cleveland then appointed him a special Treasury Department agent at El Paso, Texas. He even planned on running for sheriff. During the Spanish-American War Behan worked as a quartermaster in Cuba. Later he witnessed the Boxer Rebellion in China. At the time of his death at Tucson on June 7, 1912, Johnny Behan enjoyed strong personal friendships, even among some old Tombstone adversaries.

Behan's young son Albert, who suffered slightly from deafness, had been mothered during visits to Tombstone by his father's paramour, Josephine Sarah Marcus. Sadie and Albert remained friends even after she took up with Wyatt Earp. Albert often visited the Earps at their home in Los Angeles. Holding no grudge because of his earlier difficulties with the father, Wyatt developed a fondness for young Behan. Still, Sadie could not forget the dashing Cochise County sheriff, and she never let Wyatt forget it either. Even after Johnny Behan's death theirs remained a bizarre triangle.

★ ★ ★

David Neagle, Behan's deputy and Tombstone's city marshal, returned to California and became involved in a number of controversial episodes. These included the shooting of David S. Terry in 1889, while serving as a hastily appointed deputy U.S. marshal guarding Supreme Court Associate Justice Stephen J. Field. The circumstances surrounding this killing led to an important Supreme Court decision examining the doctrine of implied powers.[14] Later, Neagle was embarrassed by an encounter with Wyatt Earp in San Francisco, before becoming bodyguard for railroad mogul Henry Huntington. As such he became embroiled in an ugly dispute with local editor James H. Barry, prompting the remark, "He will find it convenient to avoid Barry hereafter, just as he finds it convenient to avoid the Earp boys when they visit San Francisco."[15] Chagrined, the old Tombstone officer later worked as an investigator for the celebrated criminal lawyer Earl Rogers during San Francisco's notorious graft trials. David Butler Neagle died in Oakland, California, at age seventy-eight on November 28, 1925.

★ ★ ★

Leslie F. Blackburn also ended up in Oakland as a self-styled horseman, capitalist, and mining speculator. In the early 1890s he even served as an Alameda County deputy sheriff. Blackburn died in Oakland on January 22, 1913.

★ ★ ★

Wells Spicer, after years chasing wealth and influence and finding neither, committed suicide in 1887 near Quijotoa, eighty miles west of Tucson. For a time Tombstone pioneer John B. Allen ran a hotel there. Allen died of cancer on June 13, 1899. Spicer's long-suffering wife, Abbie, died in Chicago in 1924.

★ ★ ★

[14]Thomas Cunningham, Sheriff of the County of San Joaquin, *Appt.* v. *David Neagle, The Decisions of the Supreme Court of the United States at October Term, 1889.* For some background and details of this bizarre episode, see Gould, *A Cast of Hawks.*

[15]*San Francisco Chronicle,* August 6, 1896. Later Barry happily published a book in which its author described a group of unsavory characters, including Neagle, "the gun-fighter, who numbered among his accomplishments the slaying of Judge Terry." Hichborn, "*The System,*" 280.

Of Wyatt Earp's lawyers, Thomas Fitch spent part of his fee from the Spicer hearing building himself a comfortable home in Tucson. He spent little time, however, congratulating himself. Later Fitch got involved in such diverse activities as writing novels, serving as attorney for Queen Liliuokalani of Hawaii, becoming a civic booster for the new town of Santa Monica, and working as an editorial writer for the *Los Angeles Times.* He died at the Masonic home at Decoto, California, on November 12, 1923, ten weeks short of eighty-six.

Colonel William Herring was appointed attorney general of the territory even before leaving Tombstone in 1897. He moved his law practice to Tucson, helped draft the first state constitution, and was named chancellor of the University of Arizona's board of regents. Herring died in 1912. One of his daughters, Sarah, married Thomas Sorin, an original co-proprietor of the *Epitaph.* After several years teaching and serving as school principal at Tombstone, she followed her father's example and became a lawyer, the first woman admitted to the Arizona bar. As an expert on mining law, she helped her husband—who had abandoned journalism—develop properties in the Swisshelm Mountains and elsewhere.

☆ ☆ ☆

Wyatt Earp's friend Albert Bilicke stayed in the hotel business after returning to California in 1885. At first he ran the Ross House at Modesto, and six years later became proprietor of the Pacific Ocean House in Santa Cruz. He moved to Los Angeles in 1893 and together with his father managed the Hollenbeck, a popular rendezvous for the old Arizona crowd. After the turn of the century he organized the Bilicke-Rowan Fireproof Building Company and built the Alexandria Hotel, a residence for many early Los Angeles celebrities. Albert died a millionaire on May 7, 1915, just short of his fifty-fourth birthday. He was traveling to Europe aboard the *Lusitania* when that unfortunate vessel was torpedoed by the German submarine U-20 off the southern coast of Ireland. His wife, Gladys, luckily survived the sinking.[16] Albert's father, Carl Gustav Frederick Bilicke, had died in Los Angeles in 1896.

[16]Estate of Albert C. Bilicke, Superior Court, County of Los Angeles, No. 29418, Dept. 2, Final Account, Report and Petition for Distribution, January 27, 1922; *Press Reference Library, Notables of the Southwest,* 59; and McGroarty, *Los Angeles,* 2:104-105.

★ ★ ★

In 1884 Nellie Cashman joined Milton Joyce, Tombstone attorney Marcus Aurelius Smith (later a United States senator from Arizona), and ten others searching for gold in Mexico. Arrested at Guaymas, they escaped only through the surreptitious assistance of the U.S. consul. Nellie went on to New Mexico, where she leased the Stone House Hotel at Kingston. Next she traveled to several Colorado mining camps, including Cripple Creek, then on to the Yukon in 1898 via the treacherous Chilkoot Pass. She ran a grocery store in the Donovan Hotel at Dawson City before finally moving to Fairbanks, Alaska. Nellie staked out some claims for herself under the Midnight Sun Mining Company, her last located at Coldfoot, the northernmost mining camp in Alaska during those wild days. She visited Tombstone only once, hoping to see an old boyfriend, then lived out the rest of her life in the Far North among the rough mining crowd she so loved. Hospitalized with double pneumonia at Victoria, British Columbia, Nellie Cashman died on January 4, 1925.

★ ★ ★

After the Guaymas adventure with Nellie Cashman, Milt Joyce returned to San Francisco. There he leased the Baldwin Hotel bar and billiard room in partnership with James W. Orndorff, a mining expert and sporting man from Virginia City who had operated a keno game with policeman George Magee at Danner and Owens's saloon at Tombstone. Joyce died in November 1889, but Orndorff proved more resilient, continuing to invest in mines, run saloons, billiard parlors, and gamble until his death in 1923, seven weeks short of eighty-nine.

★ ★ ★

Mining man James Vizina, whose building at Fifth and Allen housed the Oriental Saloon, returned to San Francisco by 1884. He had lived in the Bay City before going to Arizona, working as a steward for R. R. Swain & Company's bakery and restaurant. After clerking for A. J. Mitchell at Tucson's Palace Hotel he went to Tombstone. Facing hard times back in California, he managed the billiard parlor at the Baldwin Hotel for Milt Joyce and Jimmy Orndorff. Vizina survived the great San Francisco earthquake and fire, but died a year later on April 16, 1907.

★ ★ ★

Although scouting for the army again with General George Crook, Frank Leslie's luck finally ran out. Divorced from the former Mrs. Killeen—she died in 1947—he took up with Mollie Bradshaw. At his Magnolia Ranch in the Sulphur Springs Valley, Leslie shot and killed her during a drunken rage. Frank spent seven years in the Yuma Territorial Prison before winning parole. Trying to rebuild his life, he managed a grocery store in San Francisco, claimed mining interests while living across the bay at Berkeley, but mostly supplemented his meager income with small-stake card games. As an old man he hired on as a servant to Mrs. Clara Younkins, recently back from Alaska with her two small children. Later he lived off the generosity of James Orndorff, then running a billiard parlor in Sausalito. Frank Leslie tried to recoup his fortunes but ultimately died in penniless obscurity. George M. Perine, his defender during the battle with Mike Killeen on the balcony of the Cosmopolitan, became, of all things, a successful banker and insurance executive. He died of a cerebral hemorrhage at the age of eighty-one.

★ ★ ★

Others from the early days stayed on. C. S. Fly served a term as Cochise County sheriff in the mid-1890s. Earlier he had traveled with the army taking photographs during the final stages of the Apache Wars. Although seldom given credit, many of the most famous images of Geronimo and his followers are examples of Fly's work.[17] One picture of Bisbee pioneer George Warren, in full prospector's garb, was used as the model on Arizona's state seal. With the mines closing and the population drifting away, Camillus moved to Bisbee. He died there on October 12, 1901. Mollie Fly, who kept the Tombstone gallery open, brought his body back for burial in the city cemetery. She stayed until fire destroyed the old Fremont Street boarding house and gallery in 1912, then returned to California. After some years in Los Angeles, she died at Santa Ana in 1922 but is buried in Berkeley.

★ ★ ★

William A. Harwood, one of the original Signal crowd and Tomb-

[17]Fly, *Geronimo.*

stone's first mayor, never did leave. He served on the city council then turned to mining toward the end of the century. Harwood died of cancer on March 5, 1913, at age sixty-five and is also buried in the City Cemetery. He at least outlived the O.K. Corral's proprietor. John Montgomery died on May 26, 1909.

★ ★ ★

James Reilly stayed in Tombstone, practicing law and arguing with anyone willing to exchange verbal barbs with the one-time justice of the peace. In 1893 he married a Mexican woman from California thirty-seven years his junior. They had no children, but the experience seemed to mellow the old warrior. Reilly died a wealthy man at Long Beach, California, on June 8, 1909.

★ ★ ★

The Grand Central Mining Company's chief stockholder, Nathaniel K. Fairbank, died in Chicago on March 27, 1903. Its superintendent, E. B. (Eliphalet Butler) Gage, who helped break the miners' union in 1884 and tried desperately to consolidate and reopen the Tombstone mines at the turn of the century, went on to San Francisco as president of the International Amet Gas Power Company. He died in that city on March 12, 1913.

★ ★ ★

With the failure of the Tombstone Consolidated the town again fell victim to lazy decay. Reports from the old camp no longer sparkled with the excitement of discovery or vendetta. Tired citizens seemed more interested in local baseball scores. Tombstone hung on anyway, mainly from its position as county seat. But even that changed after voters agreed in 1929 to move those functions to Bisbee.

During the Great Depression many could not afford property taxes. Since buildings without roofs were assessed the same as vacant lots, roofs were torn off. The harsh climate and passing years did the rest. Thus many more historic landmarks were lost, including the original two-story *Epitaph*, the Mining Exchange, and the Gird Block. Hard times even encroached upon the old burial ground, now glorified as Boot Hill. Interments had long since been shifted to the city cemetery, west of town off Allen Street. The abandoned site became a convenient place to dump trash, before finding new life as a fanciful tourist shrine.

Charleston, Millville, and Contention City fared even worse. After the mills closed it was all over. Time and vandals have nearly obliterated those locations. The Department of the Interior recently took over the waterway as a nature preserve. One hopes it will be successful in preventing further destruction along the San Pedro.

After the fall Tombstone tried becoming, of all things, a health resort. The plan did not go well; its name proved too serious an impediment. In the end what saved the town was a handful of books, several popular motion pictures, piles of pulp magazine articles, and the nightly parade of television westerns. Seduced by six-shooter heroes and make-believe situations, audiences accepted the excitement and dismissed questions of accuracy. Tourism still keeps Tombstone alive, even after the state destroyed much of its ambiance by ordering Fremont Street gutted into a curbed four-lane highway. Gone now are the shallow earthen embankments and fragments of old wooden sidewalks. Only afterward did the federal government place the town on its list of National Historical Sites.

But something more important has been overlooked in the mad rush to deify men of violence. What is left behind is a mere shadow of the past, a faded image of a once vibrant place. Clothed in caricature and fantasy, the story has no room for real people, those forgotten dreamers who made Tombstone what it was. Tricked into accepting "The Wild West" as invented by the entertainment industry, devotees easily succumbed to the legend's intoxication. Even those described as serious historians often refuse to take Tombstone seriously. Wrapping themselves in a comfortable cloak of academic snobbery, they either ignore the place or see only a childish playground for the hopelessly gullible.

Few of today's visitors come for the actual history of the place, whether staring at the sights with mock solemnity or having a casual drink or two at the rebuilt Wehrfritz Building (known now as the Crystal Palace). Ed Schieffelin's dogged determination is of little interest to them, nor do they wonder about the nameless few lucky enough to have pocketed the profits, or even notice the land still scarred from the 1880s. Playing out an already stylized drama, they substitute Wyatt

Two young pedestrians offer the only signs of life. This once bustling thoroughfare quietly mirrors the shabbiness and sad reality of Tombstone's decline. Except for Schieffelin Hall, hardly any of the buildings seen here along Fremont Street are still standing.
Courtesy, Arizona State Library.

Earp and Doc Holliday as larger-than-life heroes strolling center stage. The real story is pushed aside.

But the legend still lures the faithful. Fooled by decades of false images, tourists descend upon the O.K. Corral as if on pilgrimage. Others wander up and down Allen Street in an endless promenade, listening to one crazy story after another, all the while stuffing bags with souvenirs more reflective of Hollywood than the wild days. Only then, deceived but satisfied, do they jump into their cars for the journey out. Pondering their brief excursion—some sporting new boots, hats, and stern looks—they somehow seem convinced of reliving part of America's frontier heritage, without ever understanding what that heritage was. Sizing up this phenomenon from beyond the grave, the old timers must be smiling, their ghosts huddled together calculating yet another way to make money off that place called Tombstone.

Bibliography

Primary Sources

Arizona Historical Society, Tucson: Biographical files, manuscript collections, and miscellaneous records relating to Pima and Cochise counties.

Bancroft Library, University of California, Berkeley, Manuscript Collections, Arizona interviews: J. E. Durkee, Virgil W. Earp, Wyatt S. Earp, Artemus E. Fay, S. L. Hart, James P. McAllister, W. K. Meade, J. J. Patton, Charles A. Shibell, William H. Stilwell, J. V. Vickers, and Paul B. Warnekros.

California State Library, Sacramento: Biographical files, early city directories, and incorporation papers involving Arizona companies: Contention Consolidated Mining Company; Contention Hill Mining and Milling Company; Grand Central South Mining Company; Head Center and Tranquility Mining Company; San Pedro Mining Company; Tombstone Consolidated Gold and Silver Mining Company; Tombstone Gold and Silver Mining Company; Tombstone Water, Mill and Lumber Company; and Western Mining Company.

Campana contra los Trjanos o Cow-boys, Biblioteca y Museo de Sonora Archivo Historico, Hermosillo, Mexico.

Cochise County, Arizona (Bisbee): Board of Supervisors, County Clerk, Clerk of the Superior Court, Recorder's Office.

Cochise County Coroner's Inquests (Clerk of the Superior Court, Bisbee, Arizona): Wm. Clanton, F. & T. McLowery [*sic*], Document No. 48, (1881); Morgan Earp, Document No. 68, (1882); Warren Earp, Document No. 434 (1900); John Ringo, Document No. 170 (1882).

Cox Library, Tucson: Genealogical collections; territorial, state, and federal census records; county histories; and city directories.

Mrs. Mary K. [Holliday] Cummings, "Character Picture of the Late John Ringgold [*sic*]," typescript, John D. Gilchriese Collection.

Memoirs of Mary Katherine [Holliday] Cummings, typescript, John D. Gilchriese Collection.

Dedication of Tombstone as a National Historical Site, Address of Arizona Governor Paul Fannin, September 30, 1962, copy in the author's collection.

Wyatt Earp's Autobiography (four unpublished versions), John D. Gilchriese Collection.

Wyatt Earp's Correspondence with Walter Noble Burns, Special Collections, University of Arizona Library.

Wyatt Earp's Correspondence with Stuart N. Lake, The Huntington Library, San Marino, California.

James Y. Eccleston, unpublished manuscript, copy in the author's collection.

Estate Proceedings (Superior Court, County of Los Angeles): Albert Bilicke, No. 29418; J. E. Durkee, No. 1662; Josephine Earp, No. 240608; Richard Gird, No. 18040; James Reilly, No. 17027; J. V. Vickers, No. 22592; John S. Vosburg, No. 116992; and Estate and Guardianship of Marion Todd Randall, Arthur Randall, Alder Randall, Edward David Randall, No. 3760.

W. B. Garner, as Administrator of the Estate of Thomas J. Bidwell, Deceased, vs. Richard Gird, *Transcript on Appeal,* In the Supreme Court of the State of California, April 24, 1885.

Great Registers: Alameda County, California, 1894, 1900; Cochise County, Arizona, 1881, 1882, 1884, 1886, 1892, and 1894; and Pima County, Arizona, 1876-1881.

Kansas State Census, 1875, Dodge City, Ford County.

Kelleher, M., M. R. Peel, and Frank S. Ingoldsby, *Map of the Tombstone Mining District, Cochise Co. Arizona Ter.* San Francisco: Lith. H. S. Crocker & Co., 1881.

William R. McLaury Letters, 1880–1884, New-York Historical Society and Arizona Historical Society.

Minute Books A-E, Common Council, Village of Tombstone, September 10, 1880–August 20, 1887.

Pima County, Arizona (Tucson): Board of Supervisors, County Clerk, Clerk of the Superior Court, Recorder's Office.

Pinal County Coroner's Inquest (Clerk of the Superior Court, Florence, Arizona): Mattie Earp, filed July 21, 1888.

Schieffelin, Edward L., "Trip to Alaska, 1882–1883," typescript, Bancroft Library, University of California, Berkeley.

_____. "History of the Discovery of Tombstone, Arizona, As Told by the Discoverer," typescript, Bancroft Library, University of California, Berkeley.

Territory of Arizona vs. Morgan Earp, et al, Defendants, Document No. 94, In Justice Court, Township No. 1, Cochise County, A. T. (Hayhurst typescript, WPA), Courtesy Robert N. Mullin.

Tombstone City Ordinance No. 24, Huachuca Water Company Franchise, September 9, 1881. Original in the author's collection.Tombstone City Records: Assessment Rolls, 1881–1911; City License Book, 1881–1888; City Treasurer Ledger, 1881–1887; Day Book, 1883–1892; Register of Warrants, 1880–1883, Courtesy Everett L. Brownsey.

Tombstone Townsite Title Abstract. Bisbee, Ariz: Pioneer Abstract Corporation, 1946.

University of Arizona, Special Collections, Tucson: Biographical files, manuscript collections, and miscellaneous records relating to Pima and Cochise counties.

Primary Sources, U.S. Government

Thomas Cunningham, Sheriff of the County of San Joaquin vs. David Neagle, Decisions of the Supreme Court of the United States, October Term, 1889.

Letters from Collectors, Custom House Nominations, General Records of the Treasury Department, Record Group 56, National Archives and Records Service.

General Records of the Department of Justice, Record Group 60, National Archives and Records Service.

Gosper, John J. "Report of the Acting Governor of Arizona to the Secretary of the Interior." Washington: Government Printing Office, 1881.

House of Representatives, 47th Cong., 1st Sess., Ex. Doc. No. 58, "Lawlessness in Parts of Arizona; Message from the President of the United States," 1882.

_____, 47th Cong., 1st Sess., Ex. Doc. No. 188, "Lawlessness in Arizona; Message from the President of the United States," 1882.

_____, 47th Cong., 2nd Sess., Report No. 2031, "James S. Clark & Co.," 1883.

_____, 51st Cong., 1st Sess., Report No. 2380, "Estate of James S. Clark," 1890.

_____, 52nd Cong. 1st Sess., Report No. 377, "Estate of James S. Clark, Deceased," 1892.

Military Records and/or Pension Files: James C. Earp, Virgil W. Earp, John J. Gosper, Henry G. Howe, Patrick Kelly, Thomas Moses, Andrew S. Neff, Martin Ringo, Joseph H. Tuttle, National Archives and Records Service.

Old Townsite Files, Tombstone, A. T., Records of Former General Land Office, Record Group 49, National Archives and Records Service.

Plat of the Tombstone Townsite Claim, March 31, 1880, Federal Records Center, Laguna Niguel, California.

Post Office Department, Records of Appointment of Postmasters, Vols. 48, 58, Cochise County, Arizona; Vol. 48, Pima County, Arizona, National Archives and Records Service.

_____, Appointment Office, Application of Richard Gird, October 25, 1878, National Archives and Records Service.

Post Returns, Fort Bowie and Camp Huachuca, Arizona Territory; Returns from U.S. Military Posts, National Archives and Records Service.

Records of the Establishment of True Meridians, Field Notes of Tombstone Mining District, 1880–1905, Records of the Bureau of Land Management, Surveyor General of Arizona, Administrative Records, Record Group 49, Federal Records Center, Laguna Niguel, California.

Records of the Office of the Secretary of the Interior, Record Group 48, National Archives and Records Service.

Survey Field Notes of the 1st Northern Extension of the Mountain Maid Mining Claim (Tombstone Mining District), General Land Office, No. 6716, Mineral Certificate No. 78, Lot No. 62, November 10–16, 1880.

United States Census, 1850: Tioga County, Pennsylvania; 1860: Jackson County, Oregon, Dallas County, Texas; 1870: Santa Barbara County, California; Rock Island County, Illinois; 1880: Arizona Territory, Alameda County, Napa County, San Bernardino County, Santa Clara County, California, Buchanan County, Iowa, Storey County, Nevada, Tarrant County, Texas; 1900: Cochise County Arizona, Alameda County, Los Angeles County, San Bernardino County, San Francisco, California, Tarrant County, Texas; 1910: Cochise County, Arizona, Alameda County, Los Angeles County, California; 1920: Cochise County, Arizona, Alameda County, Los Angeles County, Marin County, San Francisco, California.

United States Customs Service, Special Agents Reports, Record Group 36, National Archives and Records Service.

U.S. Senate, 89th Cong., 1st Sess., Doc. No. 13, *Federal Census—Territory of New Mexico and Territory of Arizona, Excerpts . . . 1860, 1864, 1870.* Washington: United States Government Printing Office, 1965.

Newspapers

Coeur d'Alene (Idaho) *Weekly Eagle*
Cripple Creek (Colo.) *Morning Journal*
Denver Republican
Dodge City (Kan.) *Ford County Globe*
Florence *Arizona Citizen*
Gunnison (Colo.) *Daily News-Democrat*
Las Vegas (N. Mex.) *Daily Optic*
Leadville (Colo.) *Daily Herald*
Los Angeles Times
Napa, (Calif.), *Daily Register*
New York Times
Oxnard, (Calif.), *Daily Courier*
Phoenix Gazette
Phoenix Herald
Phoenix *Arizona Republican*
Pueblo, (Colo.), *Daily Chieftain*
San Diego Union
San Francisco Chronicle
San Francisco *Daily California Chronicle*
San Francisco Examiner
San Francisco *Mining and Scientific Press*
Santa Rosa (Calif.) *Press Democrat*
Tombstone Commercial Advertiser

Tombstone Epitaph
Tombstone Nugget
Tucson *Arizona Citizen*
Tucson *Arizona Quarterly Illustrated*
Tucson *Arizona Star*
Virginia City (Nev.) *Daily Territorial Enterprise*

ARTICLES

Anderson, Mike. "Posses and Politics in Pima County: The Administration of Sheriff Charles Shibell." *The Journal of Arizona History* 27, 3 (Autumn 1986): 253-282.

Chamberlain, D. S. "Tombstone in 1879: The Lighter Side." *The Journal of Arizona History* 13, 4 (Winter 1972): 229-234.

Clum, John P. "Nellie Cashman." *Arizona Historical Review* 3, 4 (January 1931): 9-34.

____. "It All Happened in Tombstone." *Arizona Historical Review* 2, 3 (October 1929): 46-72.

"Did Earp Kill Ringo?" Parts 1, 2. *Brewery Gulch Gazette* 42, No. 15 (July 29, 1971); 42, No 16 (August 5, 1971): 1, 3 and 8.

"Ed Schieffelin: The Founding of Tombstone, Arizona & the Story of the Red Blanket Mine." *The Table Rock Sentinel, Newsletter of the Southern Oregon Historical Society* (January 1983): 13-18.

Fulton, Richard W., "Millville-Charleston, Cochise County, 1878-1889." *The Journal of Arizona History* 7, 1 (Spring 1966): 9-22.

Fulton, Richard W., and Conrad J. Bahre. "Charleston, Arizona: A Documentary Reconstruction." *Arizona and the West* 9, 1 (Spring 1967): 41-64.

Gilchriese, John D. "The Life and Times of Wyatt Earp." *Cochise Quarterly* 1, 1 (March 1971): 3-6.

____. "The Odyssey of Virgil Earp." *Tombstone Epitaph*, National Edition (Fall 1968).

Gird, Richard. "True Story of the Discovery of Tombstone." *Out West* 27, 1 (July 1907): 39-50.

Goodfellow, George E., M.D. "Cases of Gunshot Wound of the Abdomen Treated by Operation." *The Southern California Practitioner* 4, 5 (May 1889): 209-217.

Hattich, William. "Highlights on Arizona's First Printing Press." *Arizona Historical Review* 3, 3 (October 1930): 67-72.

Parker, Marjorie Clum. "John P. Clum: The Inside Story of an Inimitable Westerner." *The American West* 9, 1 (January 1972): 32-37.

Walker, Henry P. "Arizona Land Fraud: Model 1880, The Tombstone Townsite Company." *Arizona and the West* 29, 1 (Spring 1979): 5-36.

Walter, Paul A. F. "Necrology-John P. Clum." *New Mexico Historical Review* 7, 3 (July 1932): 292-296.

Books and Pamphlets

Abbott, Carlisle S. *Recollections of a California Pioneer.* New York: The Neale Publishing Company, 1917.

Acts and Resolutions, Eleventh Legislative Assembly. Prescott, Arizona, 1881.

Adams, Ward R. *History of Arizona.* 4 vols. Phoenix: Record Publishing Co., 1930.

Ahnert, Gerald T. *Retracing the Butterfield Overland Trail Through Arizona, A Guide to the Route of 1857–1861.* Los Angeles: Westernlore Press, 1973.

Altshuler, Constance Wynn. *Cavalry Yellow & Infantry Blue: Army Officers in Arizona Between 1851 and 1886.* Tucson: The Arizona Historical Society, 1991.

_____, ed. *Latest From Arizona! The Hesperian Letters, 1859-1861.* Tucson: Arizona Pioneers' Historical Society, 1969.

_____. *Starting With Defiance: Nineteenth Century Arizona Military Posts.* Tucson: The Arizona Historical Society, 1983.

Ashburn, Frank D. *Peabody of Groton.* New York: Coward McCann, Inc., 1944.

Ball, Larry D. *Desert Lawmen: The High Sheriffs of New Mexico and Arizona, 1846-1912.* Albuquerque: University of New Mexico Press, 1992.

_____. *The United States Marshals of New Mexico and Arizona Territories, 1846-1912.* Albuquerque: University of New Mexico Press, 1978.

Bancroft, Hubert Howe. *Chronicles of the Builders of the Commonwealth: Historical Character Study.* Vols. 3 and 4. San Francisco: The History Company, Publishers, 1892.

______. *History of Arizona and New Mexico, 1530-1888.* San Francisco: The History Company, Publishers, 1889.

______. *History of California, 1848-1859.* Vol 6. San Francisco: The History Company, Publishers, 1888.

______. *History of California, 1860-1890.* Vol. 7. San Francisco: The History Company, Publishers, 1890.

Barrett, Walter. *The Old Merchants of New York City.* New York: Carlelton, Publishers, 1870. Vol. 5. Reprinted, Greenwood Press, Publishers, New York, 1968.

Barry, T. A., and B. A. Patten. *Men and Memories of San Francisco, in the "Spring of '50."* San Francisco: A. L. Bancroft & Company, 1873.

Barter, G. W., comp. *Directory of the City of Tucson for the Year 1881.* San Francisco: H. S. Crocker & Co., Printers, 1881.

Bartholomew, Ed. *Wyatt Earp, 1879 to 1882: The Man & The Myth.* Toyahvale, Tex: Frontier Book Company, 1964.

Bates, Samuel P. *Our County and Its People: A Historical and Memorial Record of Crawford County, Pennsylvania.* N.p.: W. A. Fergusson & Company, Publishers, 1899.

The Bay of San Francisco: The Metropolis of the Pacific Coast and its Suburban Cities. Vol I. Chicago: The Lewis Publishing Company, 1892.

Benham's New Haven City Directory. New Haven, Conn: Price, Lee & Co., 1877-1881.

Bieber, Ralph P., ed. *Exploring Southwestern Trails, 1846-1854.* Glendale, Calif: The Arthur H. Clark Company, 1938.

The Biographical Dictionary and Portrait Gallery of Representative Men of Chicago. Chicago: American Biographical Publishing Company, 1892.

Biographical Directory of the American Congress, 1774-1927. Washington: Government Printing Office, 1928.

Bishop, William Henry. *Old Mexico and Her Lost Provinces: A Journey in Mexico, Southern California, and Arizona By Way of Cuba.* New York: Harper & Brothers, 1883.

Blake, William P. *Tombstone and its Mines: A Report upon the Past and Present Condition of the Mines of Tombstone, Cochise County, Arizona.* New York: The Cheltenham Press, 1902.

Blanchard, Geo. A., and Edward P. Weeks. *The Law of Mines, Minerals, and Mining Water Rights.* San Francisco: Sumner Whitney & Co., 1877.

Bourke, John G. *On the Border With Crook.* New York: Charles Schribner's Sons, 1891.

Brandes, Ray. *Frontier Military Posts of Arizona.* Globe, Ariz: Dale Stuart King, Publisher, 1960.

Breakenridge, William M. *Helldorado: Bringing the Law to the Mesquite.* Boston: Houghton Mifflin Company, 1928.

Briggs, L. Vernon. *Arizona and New Mexico 1882, California 1886, Mexico 1891.* Boston: privately printed, 1932.

The Brooklyn City and Business Directory. New York: Lain & Company, 1874-1875.

Brown, George Rothwell, ed. *Reminiscences of Senator William M. Stewart of Nevada.* New York: The Neale Publishing Company, 1908.

Brown, Ronald C. *Hard-Rock Miners: The Intermountain West, 1860-1920.* College Station: Texas A&M University Press, 1979.

The Bunker Hill Mining Company of Tombstone. Phoenix: Press of the J. H. McNeil Co. [1902].

Burnham, Major Frederick Russell. *Scouting on Two Continents.* Garden City, N.Y.: Doubleday, Page & Company, 1926.

Burns, Walter Noble. *Tombstone: An Iliad of the Southwest.* Garden City, N.Y: Doubleday, Page & Company, 1927.

Butler, B. S., E. D. Wilson, and C. A. Rasor. *Geology and Ore Deposits of the Tombstone District, Arizona.* University of Arizona Bulletin, Arizona Bureau of Mines, Vol. 9, No. 1 (January 1, 1938).

Caldwell, Hon. J. R. *A History of Tama County Iowa.* Chicago: The Lewis Publishing Company, 1910.

Carr, John. *Pioneer Days in California.* Eureka, Calif: Times Publishing Company, 1891.

Carter, Capt. Robert G. *The Old Sergeant's Story.* New York: Frederick H. Hitchcock, Publisher, 1926.

Chisholm, Joe. *Brewery Gulch: Frontier Days of Old Arizona—Last Outpost of the Great Southwest.* San Antonio, Tex: The Naylor Company, 1949.

The City and County of San Diego. San Diego: Leberthon & Taylor, 1888.

Clemens, Samuel. *Mark Twain's Autobiography.* Vol. 1. New York: Harper & Brothers, Publishers, 1924.

Clum, Woodworth. *Apache Agent: The Story of John P. Clum.* Boston: Houghton Mifflin Company, 1936.

Colahan, W. J., and Julian Pomery. *The San Jose City Directory and Business Guide of Santa Clara Co.* San Francisco: Excelsior Press, Bacon & Company Printers, 1870.

Colville, Samuel. *Colville's Marysville Directory.* San Francisco: Monson & Valentine, 1855.

Commemorative Biographical Record of Hartford County, Connecticut. Chicago: J. H. Beers & Co., 1901.

Conkling, Roscoe P., and Margaret B. Conkling, *The Butterfield Overland Mail, 1857-1869.* Vol. 2. Glendale, Calif: The Arthur H. Clark Company, 1947.

Cooke, P. St. Geo. *The Conquest of New Mexico and California; An Historical and Personal Narrative.* New York: G. P. Putnam's Sons, 1878.

Cross, Ira B. *Financing an Empire: History of Banking in California.* Vol. 1. Chicago: The S. J. Clarke Publishing Co., 1927.

Davis, Ellis A., ed. *Davis' Commercial Encyclopedia of the Pacific Southwest.* Oakland, Calif: Ellis A. Davis, 1915.

de la Garza, Phyllis. *The Story of Dos Cabezas.* Tucson: Westernlore Press, 1995.

DeLong, Sidney R. *The History of Arizona.* San Francisco: The Whitaker & Ray Company, 1905.

Dempsey, David, and Raymond P. Baldwin. *The Triumphs and Trials of Lotta Crabtree.* New York: William Morrow & Company, Inc., 1968.

Dillon, Richard. *Wells Fargo Detective: The Biography of James B. Hume.* New York: Coward-McCann, Inc., 1969.

Dunn, Jacob Piatt. *Indiana and Indianans: A History of Aboriginal and Territorial Indiana and the Century of Statehood.* Vol I. Chicago: The American Historical Society, 1919.

Dunning, Charles H. *Rock to Riches.* Phoenix: Southwest Publishing Company, Inc., 1959.

Elsing, Morris J., and Robert E. S. Heineman. *Arizona Metal Production.* University of Arizona Bulletin, Arizona Bureau of Mines, Vol. 7, No. 2 (February 15, 1936).

Erwin, Allen A. *The Southwest of John H. Slaughter, 1841-1922.* Glendale, California: The Arthur H. Clark Company, 1965.

Farish, Thomas Edwin. *History of Arizona.* 8 Vols. Phoenix: n.p., 1915-1916.

Flinn, John J., ed. *The Hand-Book of Chicago Biography.* Chicago: The Standard Guide Company, 1893.

Fly, Mrs. M. E. *Geronimo: The Apache Chief.* Tombstone, Arizona: n.p., [1906].

Foy, Eddie and Alvin F. Harlow. *Clowning Through Life.* New York: E. P. Dutton & Company, 1928.

Ganzhorn, Jack. *I've Killed Men.* London: Robert Hale Limited, [1940].

Genealogical and Family History of the State of Connecticut. New York: Lewis Historical Publishing Company, 1911.

Gopsill's Philadelphia City Directory. Philadelphia: James Gopsill, 1878-1881.

Gould, Milton S. *A Cast of Hawks: A Rowdy Tale of Greed, Violence, Scandal, and Corruption in the Early Days of San Francisco.* La Jolla, Calif: Copley Books, 1985.

Guinn, J. M. *Historical and Biographical Record of Los Angeles and Vicinity.* Chicago: Chapman Publishing Company, 1901.

Hale & Emory's Marysville City Directory. Marysville, Calif.: Marysville Herald Office, 1853.

Hamilton, Patrick. *The Resources of Arizona.* Prescott, Ariz.: n. p., 1881.

Hammond, John Hays. *The Autobiography of John Hays Hammond.* Vol. 1. New York. Farrar & Rinehart, Inc., 1935.

Hardin, George A., ed. *History of Herkimer County, New York.* Syracuse, N.Y.: D. Mason & Co., Publishers, 1893.

Hardy, Leslie C., comp. *Charter and Ordinances of the City of Tucson.* Tucson: The Star Job Rooms, 1910.

Harlan, Edgar Rubey. *A Narrative History of The People of Iowa.* Vol. 4. Chicago: The American Historical Society, Inc., 1931.

Hichborn, Franklin. *"The System" As Uncovered by The San Francisco Graft Prosecution.* San Francisco: Press of The James H. Barry Company, 1915.

Hinton, Richard J. *The Hand-Book of Arizona: Its Resources, History, Towns, Mines, Ruins and Scenery.* San Francisco: Payot, Upham & Co., 1878.

History of Arizona Territory. San Francisco: Wallace W. Elliott & Co., Publishers, 1884.

History and Biographical Record of Monterey and San Benito Counties, California. Los Angeles: Historic Record Co., 1910.

History of Callaway County, Missouri. St. Louis: National Historical Company, 1884.

History of Cedar County, Iowa. Chicago: Western Historical Company, 1878.

History of Clay County, Missouri. Topeka, Kans: Historical Publishing Company, 1920.

History of Contra Costa County, California. San Francisco: W. A. Slocum & Co., Publishers, 1882.

History of Crawford County, Pennsylvania. Chicago: Warner, Beers & Co., 1885.

History of Greene and Jersey Counties, Illinois. Springfield, Ill: Continental Historical Co., 1885.

History of Nevada. Oakland, Calif: Thompson & West, 1881.

History of Santa Barbara County, California. Oakland, Calif: Thompson & West, 1883.

History of Solano and Napa Counties, California. Los Angeles: Historic Record Company, 1912.

History of Sonoma County, California. Los Angeles: Historic Record Company, 1911.

History of Tama County, Iowa. Springfield, Ill: Union Publishing Company, 1883.

History of Tioga County, Pennsylvania. N.p.: R. C. Brown & Co., 1897.

History of Venango County, Pennsylvania: Its Past and Present. Chicago: Brown, Runk & Co., Publishers, 1890.

History of Yuba County, California. Oakland, Calif: Thompson & West, 1879.

Hittell, John S. *The Commerce and Industries of the Pacific Coast of North America.* San Francisco: A. L. Bancroft & Co., Publishers, 1882.

Hodge, Hiram C. *Arizona As It Is; Or, The Coming Country.* New York: Hurd and Houghton, 1877.

Hoffman, Ogden. *Reports of Land Cases Determined in the United States District Court for the Northern District of California.* San Francisco: Numa Hubert, Publisher, 1862.

Hoyt, John P., comp. *The Compiled Laws of the Territory of Arizona, 1864-1877.* Detroit: Richmond Backus & Co., Printers, 1877.

Hutchinson, W. H. *Oil, Land and Politics: The California Career of Thomas Robert Bard.* Vol. I. Norman: University of Oklahoma Press, 1965.

Illustrated History of Plumas, Lassen & Sierra Counties, with California From 1513 to 1850. San Francisco: Fariss & Smith, 1882.

An Illustrated History of Sonoma County, California. Chicago: The Lewis Publishing Company, 1889.

An Illustrated History of Southern California. Chicago: The Lewis Publishing Company, 1890.

An Illustrated History of the State of Oregon. Chicago: The Lewis Publishing Company, 1893.

Ingersoll's Century Annals of San Bernardino County, 1769 to 1904. Los Angeles: L. A. Ingersoll, 1904.

Ingram, J. S. *The Centennial Exposition, Described and Illustrated.* Philadelphia: Hubbard Bros., 1876.

Irvine, Leigh H. *History of Humboldt County, California.* Los Angeles: Historic Record Company, 1915.

Jahns, Pat. *The Frontier World of Doc Holliday: Faro Dealer From Dallas to Deadwood.* New York: Hastings House, Publishers, 1957.

Journals of the Eleventh Legislative Assembly. Prescott, Arizona, 1881.

Keith, Stanton B. *Index of Mining Properties in Cochise County, Arizona.* The Arizona Bureau of Mines, Bulletin 187. Tucson: The University of Arizona, 1973.

Kelly, Geo. H. *Legislative History: Arizona, 1864-1912.* Phoenix: The Manufacturing Stationers, Inc., 1926.

King, Frank M. *Wranglin' the Past.* Pasadena, Calif: Trail's End Publishing Co., Inc., 1946.

Kintop, Jeffery M., and Guy Louis Rocha. *The Earps' Last Frontier: Wyatt and Virgil Earp in the Nevada Mining Camps, 1902-1905.* Reno, Nev: Great Basin Press, 1989.

Lake, Carolyn, ed. *Under Cover for Wells Fargo: The Unvarnished Recollections of Fred Dodge.* Boston: Houghton Mifflin Company, 1969.

Lake, Stuart N. *Wyatt Earp, Frontier Marshal.* Boston: Houghton Mifflin Company, 1931.

Ledbetter, Suzann. *Nellie Cashman: Prospector and Trailblazer.* El Paso: Texas Western Press, 1993.

Leonard, John W., ed. *Men of America.* New York: L. R. Hammersley & Company, 1908.

Leshy, John D. *The Mining Law: A Study In Perpetual Motion.* Washington, D.C.: Resources for the Future, Inc., 1987.

Lingenfelter, Richard E. *The Hardrock Miners: A History of the Mining Labor Movement in the American West, 1863-1893.* Berkeley: University of California Press, 1974.

Lockwood, Frank C. *Life in Old Tucson, 1854-1864.* Los Angeles: Ward Ritchie Press, 1943.

_____. *Pioneer Days In Arizona: From the Spanish Occupation to Statehood.* New York: The Macmillan Company, 1932.

_____. *Pioneer Portraits: Selected Vignettes.* Tucson: The University of Arizona Press, 1968.

Los Angeles City Directory. Los Angeles: Los Angeles Directory Company, 1887-1942.

Lummis, Charles F. *General Crook and the Apache Wars.* Flagstaff, Ariz: Northland Press, 1966.

Lutrell, Estelle. *Newspapers and Periodicals of Arizona, 1859-1911.* Tucson: University of Arizona General Bulletin No. 15, Vol. 20, No. 3 (July 1949).

McClintock, James H. *Arizona: Prehistoric—Aboriginal—Pioneer—Modern.* 3 vols. Chicago: The S. J. Clarke Publishing Co., 1916.

____. *Mormon Settlement in Arizona.* Phoenix: The Manufacturing Stationers, Inc., 1921.

McGroarty, Steven. *Los Angeles: From the Mountains to the Sea.* Vols. 2 and 3. Chicago: The American Historical Society, 1921.

McIntire, Jim. *Early Days in Texas; A Trip to Hell and Heaven.* Kansas City, Mo: McIntire Publishing Company, 1902.

Marcosson, Isaac F. *Anaconda.* New York: Dodd, Mead & Company, 1957.

Martin, Douglas D. *Tombstone's Epitaph.* Albuquerque: The University of New Mexico Press, 1951.

____. *Silver, Sex, and Six Guns: Tombstone Saga of the Life of Buckskin Frank Leslie.* Tombstone, Ariz: Tombstone Epitaph, 1962.

Memorial and Biographical History of the Counties of Fresno, Tulare, and Kern, California. Chicago: The Lewis Publishing Company, 1892.

A Memorial and Biographical History of the Counties of Santa Barbara, San Luis Obispo and Ventura, California. Chicago: The Lewis Publishing Company, 1891.

Miles, General Nelson A. *Personal Recollections and Observations.* Chicago: The Werner Company, 1896.

Miller, Charles Wallace Jr. *Stake Your Claim! The Tale of America's Enduring Mining Laws.* Tucson: Westernlore Press, 1991.

Miller, Joseph. *Arizona: The Last Frontier.* New York: Hastings House, Publishers, 1956.

Moody, Eric N. *Western Carpetbagger: The Extraordinary Memoirs of "Senator" Thomas Fitch.* Reno: University of Nevada Press, 1978.

Murphy, James M. *Laws, Courts, and Lawyers: Through the Years in Arizona.* Tucson: The University of Arizona Press, 1970.

Myers, John Myers. *The Last Chance: Tombstone's Early Years.* New York: E. P. Dutton & Co., Inc., 1950.

Myrick, David F. *Railroads of Arizona.* Vol. 1, *The Southern Roads.* Berkeley, Calif: Howell-North Books, 1975.

North, Diane M. T. *Samuel Peter Heintzelman and the Sonora Exploring and Mining Company.* Tucson: The University of Arizona Press, 1980.

Nott, Charles C., and Archibald Hopkins, rep. *Cases Decided in the Court of Claims at the December Term, 1875, and the Decisions of the Supreme Court in the Appealed Cases.* Vol. 11. Washington: Government Printing Office, 1877.

Officer, James E. *Hispanic Arizona, 1536-1856.* Tucson: The University of Arizona Press, 1987.

Older, Mr. and Mrs. Fremont. *George Hearst: California Pioneer.* Los Angeles: Westernlore Press, 1966.

Ormsby, Waterman L. *The Butterfield Overland Mail: Only Through Passenger on the First Westbound Stage.* San Marino, Calif: The Huntington Library, 1942.

Orton, Richard H., comp. *Records of California Men in the War of the Rebellion, 1861-1867.* Sacramento: State Printing Office, 1890.

Otero, Miguel Antonio. *My Life on the Frontier, 1864-1882.* New York: The Press of the Pioneers, 1935.

Otto, William T., rep. *United States Reports: Cases Argued and Adjudged in The Supreme Court of the United States, October Term, 1878.* Vol. 9. New York: The Banks Law Publishing Co., 1902.

Parsons, George W. *The Private Journal of George Whitwell Parsons.* Phoenix: Arizona Statewide Archival and Records Project, September 1939.

Past and Present of Rock Island, Ill. Chicago: H. F. Kett & Co., 1877.

Pendleton, Albert S. Jr., and Susan McKey Thomas. *In Search of the Hollidays: The Story of Doc Holliday and His Holliday and McKey Families.* Valdosta, Ga: Little River Press, 1973.

Phelps, Alonzo. *Contemporary Biography of California's Representative Men.* 2 vols. San Francisco: A. L. Bancroft and Company, 1881 and 1882.

Polk's Medical Register and Directory of the United States and Canada. Detroit: R. L. Polk & Co., Publishers, 1902.

Portrait and Biographical Record of Arizona. Chicago: Chapman Publishing Co., 1901.

Press Reference Library: Notables of the Southwest. Los Angeles: The Los Angeles Examiner, 1912.

Proceedings of the M. W. Grand Lodge of Free and Accepted Masons of the Territory of Arizona. 4 vols. San Francisco: Frank Eastman & Co., 1882-1904.

Pumpelly, Raphael. *Across America and Asia.* New York: Leypoldt & Holt, 1870.

Quebbeman, Frances E. *Medicine in Territorial Arizona.* Phoenix: Arizona Historical Foundation, 1966.

Quinn, John Philip. *Gambling and Gambling Devices.* Canton, Ohio: J. P. Quinn Co., 1912.

Rasmussen, Louis J. *San Francisco Ship Passenger Lists, Volume IV, June 17, 1852 To January 6, 1853.* Colma, Calif: San Francisco Historic Records, 1970.

Reeve, Frank D., ed. *Albert Franklin Banta: Arizona Pioneer.* Albuquerque: Historical Society of New Mexico, Vol. 14 (September, 1953).

Rickards, Colin. *Buckskin Frank Leslie: Gunman of Tombstone.* El Paso: Texas Western Press, 1964.

Ringo, Mary. *The Journal of Mrs. Mary Ringo: A Diary of Her Trip Across the Great Plains in 1864, With a Forward and Conclusion by Her Daughter Mattie Bell Cushing.* Santa Ana, Calif: privately printed, 1956.

Robinson, Will H. *The Story of Arizona.* Phoenix: The Berryhill Company, 1919.

Rockfellow, John A. *Log of an Arizona Trail Blazer.* Tucson: Acme Printing Co., 1933.

Rowe, John. *The Hard-Rock Men: Cornish Immigrants and the North American Mining Frontier.* New York: Barnes & Noble Books, 1974.

Sacks, B., M.D. *Arizona's Angry Man: United States Marshal Milton B. Duffield.* Tempe, Ariz: Arizona Historical Foundation, 1970.

San Francisco City Directories. Various publishers, 1852-1930.

Scharf, J. Thomas, and Thompson Westcott. *History of Philadelphia, 1609-1884.* Philadelphia: L. H. Everts & Co., 1884.

Schellie, Don. *The Tucson Citizen: A Century of Arizona Journalism.* Tucson: Tucson Daily Citizen, 1970.

Session Laws of the Eleventh Legislative Assembly of the Territory of Arizona. Prescott, Ariz, 1881.

Shuck, Oscar T. *Bench and Bar in California.* San Francisco: The Occident Printing House, 1888.

Sloan, Edward L., comp. *Gazeteer [sic] of Utah, and Salt Lake City Directory.* Salt Lake City: Salt Lake City Herald Publishing Company, 1874.

Sloan, Richard E. *Memories of an Arizona Judge.* Stanford, Calif: Stanford University Press, 1932.

Smith, Cornelius C. Jr. *Fort Huachuca.* Fort Huachuca, Ariz: n.p., 1978.

_____. *William Sanders Oury: History-Maker of the Southwest.* Tucson: The University of Arizona Press, 1967.

Sonnichsen, C. L. *Billy King's Tombstone: The Private Life of an Arizona Boom Town.* Caldwell, Idaho: The Caxton Printers, Ltd., 1942.

Spaulding, William A. *History of Los Angeles, City and County, California.* Vols. 2 and 3. Los Angeles: J. R. Finnell & Sons Publishing Company, 1931.

Spence, Clark C. *Mining Engineers & the American West: The Lace-Boot Brigade, 1849-1890.* New Haven, Conn: Yale University Press, 1965.

Stewart, Watt. *Henry Meiggs, Yankee Pizarro.* Durham, N.C.: Duke University Press, 1946.

Stout, Donald Frank, and Dorothy Jean Miller Stout. *Cedar Land: A History of Living, 1836-1980.* Vol 1. Tipton, Iowa: Cedar Lake Librarie [*sic*], 1981.

Theobald, John, and Lillian Theobold. *Arizona Territory: Post Offices & Postmasters.* Phoenix: The Arizona Historical Foundation, 1961.

______. *Wells Fargo in Arizona Territory.* Tempe: The Arizona Historical Foundation, 1978.

The Tombstone Consolidated Mines Company, Limited. Annual Report, September 30, 1910.

Tombstone in History, Romance and Wealth. Tombstone, Ariz: Daily Prospector, 1903.

Tombstone Mill and Mining Company, Report to Stockholders, July 15, 1884.

Trow's New York City Directory. New York: The Trow City Directory Company, 1878-1881.

Trumbull, J. Hammond, ed. *The Memorial History of Hartford County, Connecticut, 1633-1884.* Boston: Edward L. Osgood, Publisher, 1886.

Tucson and Tombstone General and Business Directory, for 1883 and 1884. Tucson: Cobler & Co., 1883.

Twain, Mark. *Roughing It.* Hartford, Conn: American Publishing Company, 1872.

Upshur, George Lyttleton. *As I Recall Them: Memories of Crowded Years.* New York: Wilson-Erickson, Inc., 1936.

Vandor, Paul E. *History of Fresno County, California.* Los Angeles: Historic Record Company, 1919.

Wagoner, Jay J. *Arizona Territory, 1863-1912: A Political History.* Tucson: The University of Arizona Press, 1970.

_____. *History of the Cattle Industry in Southern Arizona, 1540-1940.* University of Arizona Social Science Bulletin No. 20, Vol. 23, No. 2 (April 1952).

Waters, Frank. *The Colorado.* New York: Rinehart & Company, Inc., 1946.

_____. *The Earp Brothers of Tombstone: The Story of Mrs. Virgil Earp.* New York: Clarkson N. Potter, Inc., 1960.

Weeks, Edward P. *A Commentary on the Mining Legislation of Congress.* San Francisco: Sumner Whitney & Co., 1880.

Westphall, Victor. *Thomas Benton Catron.* Tucson: The University of Arizona Press, 1973.

Wilkins, Thurman. *Clarence King: A Biography.* New York: The Macmillan Company, 1958.

Willson, Clair Eugene. *Mimes and Miners: A Historical Study of the Theater in Tombstone.* University of Arizona Fine Arts Bulletin No. 1, Vol. 6, No. 7 (October 1, 1935).

Wilson, James Grant, and John Fiske, ed. *Appleton's Cyclopaedia of American Biography.* Vol. 1. New York: D. Appleton and Company, 1888.

Wiltsee, Ernest A. *Gold Rush Steamers of the Pacific.* Lawrence, Mass: Quarterman Publications, Inc., 1976.

Wister, Fanny Kemble, ed. *Owen Wister Out West: His Journals and Letters.* Chicago: The University of Chicago Press, 1958.

Woodson, W. H. *History of Clay County, Missouri.* Topeka, Kans: Historical Publishing Company, 1920.

Woodward, Arthur, ed. *Man of the West: Reminiscences of George Washington Oaks, 1840-1917.* Tucson: Arizona Pioneers' Historical Society, 1956.

Wyllys, Rufus K. *Arizona: The History of a Frontier State.* Phoenix: Hobson & Herr, 1950.

Wyman, Mark. *Hard Rock Epic: Western Miners and the Industrial Revolution, 1860-1910.* Berkeley: University of California Press, 1979.

Young, Otis E. Jr. *Black Powder and Hand Steel: Miners and Machines on the Old Western Frontier.* Norman: University of Oklahoma Press, 1976.

_____. *Western Mining.* Norman: University of Oklahoma Press, 1970.

INDEX

TOMBST

Aztec Boarding House
Addie Bourland
Assay Office
Carpentry Shop
Library
Epitaph
Mining Exchange
West End Corral
Second Street
Hoisting Works and Ore Dump Mountain Maid Mining Claim
Tombstone Livery and Feed Stable
L. Graf Blacksmith
Blacksmith
Arizona Corral
Lodgings
Chinese
Third Street
W.A. Harwood
C.S. Fly's Boarding House
Photo Gallery
Union Meat & Poultry Mkt.
Papago Cash Store
County Recorder (City Hall after 1882)
Nugget
Lodgings
Sheds
Driveway
Sheds
O.K. Corral
Mattie Webb's Starlight Saloon
Blacksmith
Saloon
Office
Harness
Saloon
Chinese
Chinese
Smith's Wagon Shop
Charcoal Yard
Dwellings
P.W. Smith's Corral
Lodgings
Restaurant
Carpenter
Dooling's Saloon
Dexter Livery, Feed & Sale Stable
Storage
Storage
Chinese
Blacksmith
Site of Tombstone First Sch
Courthouse and Jail (after 1882)
Hoisting
Vizina Cons
Mining C